COLLEGE ALGEBRA, VOLUME 2

$$\{r(\cos\theta + i\sin\theta)\}^n = r^n\{\cos(n\theta) + i\sin(n\theta)\}$$

$$ax^2 + bx + c = 0$$

$$x_{1,2} = \frac{-b \pm \sqrt{b^2 - 4ac}}{2a}$$

$$\begin{cases} \dfrac{x^5 + y^5}{x^3 + y^3} = \dfrac{33}{9} \\ x^2 + xy + y^2 = 7 \end{cases}$$

1) An excellent supplementary textbook for all Mathematics, Physics and Engineering students

2) 310 solved examples and 1050 problems for solution

3) Odd numbered problems are provided with answers

4) Hints or detailed outlines are given for the more involved problems

Demetrios P. Kanoussis

ABOUT THE AUTHOR

Dr. Kanoussis is a professional Electrical Engineer and Mathematician. He received his Ph.D degree in Engineering and Master's degree in Mathematics from Tennessee Technological University, U.S.A, and his Bachelor's degree in Electrical Engineering from National Technical University of Athens (N.T.U.A), Greece.

As a professional Electrical Engineer, Dr. Kanoussis has been actively involved in the design and in the implementation of various projects, mainly in the area of Integrated Control Systems.

Regarding his teaching experience, Dr. Kanoussis has a long teaching experience in areas of Applied Mathematics and Electrical Engineering.

His original scientific research and contribution, in Mathematics and Electrical Engineering, is published in various, high impact international journals.

In addition to his professional activities, teaching and research, Demetrios P. Kanoussis is the author of several textbooks in Electrical Engineering and Applied Mathematics.

A complete list of Dr. Kanoussis textbooks in Mathematics and Engineering can be found in the Author's page at Amazon Author Central (https://www.amazon.com/Demetrios-P.-Kanoussis/e/B071GZ215Z).

COLLEGE ALGEBRA, Vol. 2

All rights reserved. No part of this publication may be reproduced, distributed or transmitted in any form or by any means, electronic or mechanical, without the prior written permission of the author, except in the case of brief quotations and certain other noncommercial uses permitted by copyright law.

Inquires should be addressed directly to the author:

Demetrios P. Kanoussis

dkanoussis@gmail.com

This book is licensed for your personal use only. This book is not be resold or given away to other people. If you would like to share this book with another person, please purchase an additional copy for each recipient.

Thank you for respecting the work of the author.

PREFACE

Algebra, traditionally, deals with equations, systems of equations, inequalities, polynomials, etc, and develops methods and techniques which serve as an introduction to higher Mathematics.

This book was written to provide an essential help to all university students, in the areas of Mathematics, Physics and Engineering. A knowledge of introductory College Algebra is desirable, and can be found in my book, "**College Algebra, Vol. 1**". This first volume, is devoted to set theory, set of real numbers, algebraic operations, ratios and proportions, inequalities, absolute values, identities, factorization and permanent inequalities. The current volume, "**College Algebra, Vol. 2**" is, by far, more advanced, and covers several topics on higher degree equations and inequalities, systems of equations (linear and non linear), polynomials, complex numbers, progressions, logarithmic and exponential equations, etc.

The book contains 19 chapters, as shown analytically in the table of contents. **Chapter 1** is devoted to mappings and functions, Cartesian coordinates and graphs of functions. **Chapter 2** treats first degree equations in one unknown, factored equations and equations involving absolute values. **Chapter 3** covers first degree inequalities in one unknown and inequalities with absolute values. **Chapter 4** concentrates on systems of linear equations, ($2 \times 2, 3 \times 3$, etc). Useful and powerful methods and techniques are developed, (method of substitution, Cramer's rule, Gauss's elimination method, the generalized method of substitution, etc), for the solution of linear systems and various special types of linear systems are considered. Graphical solution of linear systems and linear inequalities are studied in **chapter 5**, while rational equations and rational inequalities are considered in **chapter 6**. Irrational equations are studied in **chapter 7**. The theory of complex numbers and related properties are developed in **chapter 8**. Quadratic equations are studied in considerable depth and details in **chapter 9**, while the theory of quadratic trinomial is developed in **chapter 10**. **Chapter 11** is devoted to equations and inequalities transformable to quadratic equations and inequalities, (for example, biquadratic equations, reciprocal equations, binomial and trinomial equations, etc). Non linear algebraic systems are considered in **chapter 12**. Polynomials in one variable and related theorems are studied in **chapter 13**, while **chapter 14** is devoted to the general properties of polynomial equations, (theorem of conjugate roots, theorem of rational roots, theorem of irrational roots, Vieta's theorem, etc). Polynomials in several variables and related theorems are studied in **chapter 15**. Arithmetic, harmonic and geometric progressions and various applications are introduced in **chapter 16**. Logarithms, logarithmic equations and exponential equations are developed in **chapter 17**. **Chapter 18** is devoted to the theory of conditional maxima and minima of functions of several variables. Finally, in **chapter 19**, we study some special topics, related to the application of complex numbers in polynomials and trigonometry. The famous, Cote's theorem, is proved easily, with the aid of complex numbers. At the end of the book, there is a list of 256 supplementary problems, covering all topics developed in the book.

The book contains, in total, 310 solved examples and 1050 problems for solution. The examples and the problems have been selected to help students develop a solid background in Algebra, broaden their knowledge and sharpen their analytical skills, and finally, prepare them to pursue successfully more advanced studies in Mathematics and Engineering.

Hints or detailed instructions are given for the more involved problems, while answers to odd-numbered problems are provided, so that the students can check their progress and understating of the material studied.

TABLE OF CONTENTS

CHAPTER 1:..**11**

Mappings and functions: The set of real numbers, functions and function notation, real functions of one real variable, functions one-to-one and inverse functions, real functions of several real variables, composite functions, Cartesian coordinates, graphs of functions, linear equations (first degree equations in two variables), solved examples, problems.

CHAPTER 2:..**50**

First degree equations in one unknown: Fundamental concepts and definitions, equivalent equations, first degree equations, factored equations, equations with absolute values, word problems, solved examples, problems.

CHAPTER 3:..**70**

First degree inequalities in one unknown: Intervals of real numbers, inequalities in one unknown, solving first degree inequalities in one unknown, the sign of the product $(x - a)(x - b) \cdots (x - p)$, inequalities with absolute values, solved examples, problems.

CHAPTER 4:..**88**

Systems of linear equations: Equations with many unknowns, linear equations in two unknowns, systems of equations and equivalent systems, solving a linear system of two equations with two unknowns, linear 2×2 systems, the general case, investigation, solving linear 2×2 systems using determinants (Cramer's rule), linear systems with more than two unknowns, third order determinants, evaluation and properties, solving linear 3×3 systems using determinants (Cramer's rule), consistent equations, linear and homogeneous systems, special types of linear systems, word problems, solved examples, problems.

CHAPTER 5:..**152**

Graphical solution of linear systems and linear inequalities: Graphical solution of a linear system of two equations in two unknowns, graphical solution of a linear inequality of the form $ax + by + c > 0$, equation of the circle, solved examples, problems.

CHAPTER 6: 160

Rational equations and inequalities: Rational algebraic fractions, operations with rational algebraic fractions, rational equations of the form $(P/Q) = 0$, rational inequalities of the form $(P/Q) > 0$ or $(P/Q) < 0$, solved examples, problems.

CHAPTER 7: 175

Irrational equations: Basic definitions and operations with radicals (a brief summary), rationalization of the denominator, double radicals, binomials of the form $a + b\sqrt{c}$, irrational equations (equations with radicals), equations of the form $\sqrt[3]{P(x)} + \sqrt[3]{Q(x)} + \sqrt[3]{R(x)} = 0$, solved examples, problems.

CHAPTER 8: 202

Complex numbers: Introduction, the fundamental laws in the algebra of complex numbers, the imaginary unit $i = (0,1) = \sqrt{-1}$, complex conjugate numbers, the absolute value (modulus) of a complex number, the trigonometric or polar form of a complex number, De Moivre's theorem, roots of complex numbers, the roots of unity, the complex plane (or the Argand's diagram), solved examples, problems.

CHAPTER 9: 247

Quadratic equations: The solution of a quadratic equation, symmetric expressions of the roots r_1 and r_2, conditions between the coefficients of a quadratic equation when the roots satisfy a given relation, solved examples, problems.

CHAPTER 10: 264

The quadratic trinomial: General concepts and definitions, the sign of a quadratic trinomial as x runs over the set of real numbers, quadratic inequalities, position of a real number relative to the roots of a trinomial, position of two real numbers relative to the roots of a trinomial, maximum or minimum value of a trinomial, when two trinomials have common roots, inequalities with radicals (irrational inequalities), solved examples, problems.

CHAPTER 11: **289**

Equations and inequalities transformable to quadratics: Biquadratic equations, biquadratic inequalities, reciprocal equations, binomial and trinomial equations, the general method of substitution, solved examples, problems.

CHAPTER 12: **308**

Non linear algebraic systems: One of the equations is linear in one of the unknowns, homogeneous systems, symmetric systems, some general techniques and miscellaneous examples, solved examples, problems.

CHAPTER 13: **324**

Polynomials in one variable: Introduction, the division of polynomials, the identity of algorithmic division, the remainder and the factor theorems, polynomials identically equal, solved examples, problems.

CHAPTER 14: **346**

General properties of polynomial equations: Some general theorems on polynomial equations (the conjugate roots theorem, the rational roots theorem, the irrational roots theorem, etc), relation between the coefficients of a polynomial and its roots (Vieta's theorem), some general remarks in solving polynomial equations, algebraic numbers, solved examples, problems.

CHAPTER 15: **368**

Polynomials in several variables: Introduction, the division of polynomials in several variables, symmetric polynomials, homogeneous polynomials and related theorems, solved examples, problems.

CHAPTER 16: 381

Progressions: Arithmetic progressions, harmonic progressions, geometric progressions, the problem of interpolation, mixed progressions, the three means (arithmetic, harmonic, geometric) and Cauchy's inequality, solved examples, problems.

CHAPTER 17: 406

Logarithms, logarithmic and exponential equations: Powers with irrational exponents, the logarithm of a positive number, properties of logarithms, logarithmic and exponential equations and systems, solved examples, problems.

CHAPTER 18: 422

Conditional maxima and minima: Introduction, some general theorems on conditional maxima and minima, solved examples, problems.

CHAPTER 19: 434

Some special topics on complex numbers, polynomials and trigonometry: The Σ and Π notation, evaluation of trigonometric sums of a special type, factorization of the polynomials $x^n - 1$ and $x^n + 1$, factorization of the polynomial $x^{2n} - 2a^n \cos(n\theta) x^n + a^{2n}$, Cote's theorem, solved examples, problems.

SUPLEMENTARY PROBLEMS: 453

A collection of 256 problems is provided, covering all topics studied in the book.

FUNDAMENTAL ALGEBRAIC IDENTITIES

1) $(x \pm y)^2 = x^2 \pm 2xy + y^2$

2) $x^2 - y^2 = (x - y)(x + y)$

3) $(x + y + z)^2 = x^2 + y^2 + z^2 + 2xy + 2yz + 2zx$

4) $(x \pm y)^3 = x^3 \pm 3x^2 y + 3xy^2 \pm y^3$

5) $x^3 - y^3 = (x - y)(x^2 + xy + y^2)$

6) $x^3 + y^3 = (x + y)(x^2 - xy + y^2)$

7) $(x + y + z)^3$
$$= x^3 + y^3 + z^3 + 3x^2(y + z) + 3y^2(z + x) + 3z^2(x + y) + 6xyz$$

8) $x^2 + y^2 + z^2 - xy - yz - zx = \dfrac{1}{2}\{(x - y)^2 + (y - z)^2 + (z - x)^2\}$

9) $x^n - a^n = (x - a)(x^{n-1} + x^{n-2}a + x^{n-3}a^2 \ldots + x^2 a^{n-3} + x a^{n-2} + a^{n-1})$, $\quad n$ natural number

10) $x^n + a^n = (x + a)(x^{n-1} - x^{n-2}a + x^{n-3}a^2 - \cdots + x^2 a^{n-3} - x a^{n-2} + a^{n-1})$, $\quad n$ **odd** natural number

11) Cauchy's identity: $x^3 + y^3 + z^3 - 3xyz$
$$= (x + y + z)(x^2 + y^2 + z^2 - xy - yz - zx)$$

12) Lagrange's identity: $(a^2 + b^2 + c^2)(x^2 + y^2 + z^2) - (ax + by + cz)^2$
$$= \begin{vmatrix} a & b \\ x & y \end{vmatrix}^2 + \begin{vmatrix} a & c \\ x & z \end{vmatrix}^2 + \begin{vmatrix} b & c \\ y & z \end{vmatrix}^2$$

CHAPTER 1: MAPPINGS AND FUNCTIONS

1-1) The set of real numbers

Most of our work in Algebra deals with the set of **real** numbers $\mathbb{R}$, which is considered to be the universal set U, (see my book "**College Algebra, Vol. 1**"). Special subsets of the set of real numbers are:

1) The set of **natural** numbers $\mathbb{N} = \{1,2,3,\cdots\}$ or in the equivalent notation, $\mathbb{N} = \{x/x \text{ is a natural number}\}$,

2) The set of **integer** numbers $\mathbb{Z} = \{\cdots,-3,-2,-1,0,1,2,3,\cdots\}$ or in the equivalent notation, $\mathbb{Z} = \{x/x \text{ is an integer number}\}$,

3) The set of **rational** numbers $\mathbb{Q} = \{x/x \text{ is a rational numbe}\}$.

Recall that **rational** number is a number which can be expressed as a **fraction of two integers**. For instance, the numbers $3/8, 13/9, -5/7$ etc. are rational numbers.

There are numbers that **cannot be expressed as fractions of integers**, and these numbers are called **irrational numbers**. For example, the numbers $\sqrt{2}, \sqrt{3}, \sqrt{5}$ are all irrational. The number π which is defined as the ratio of the circumference L of a circle to its diameter D, i.e. $\pi = L/D$, is an irrational number, (actually π is more than just an irrational number, it is a **transcendental number**). We shall give the precise definition of transcendental numbers in Chapter 14, section 14-4.

The discovery of irrational numbers goes back to the ancient Greeks, who proved that the number $\sqrt{2}$ is irrational. This of course means that **the number $\sqrt{2}$ cannot be expressed as the ratio of two positive integers**. The discovery of the irrationality of $\sqrt{2}$ had a great impact on the Mathematical and Philosophical thought of ancient Greeks. We just mention that the term irrational means insane, i.e. for the ancient Greeks the number $\sqrt{2}$ was an insane (crazy) number! (For a proof that $\sqrt{2}$ is irrational, see Example 1-1-3).

The set of the rational numbers and the set of the irrational numbers together comprise the set of **real** numbers $\mathbb{R}$. This is shown graphically in Figure 1-1.

The set of real numbers $\mathbb{R}$ is **closed under the operations of addition, subtraction, multiplication and division**. This simply means that if we add, subtract, multiply or divide any two real numbers, the outcome will be a real number as well. (Division is allowed provided that the denominator is not zero. **Division by zero is not allowed**).

However, **the set of real numbers is not closed under the operation of extracting square roots**. There exist real numbers whose square root is not real. For example $\sqrt{-4}, \sqrt{-7}$ and in general **the square root of any negative number is not real**. Indeed, if $\sqrt{-4}$ were a real number x, then we should have $x^2 = -4$ and this is **impossible**, since the square of any real number is a non-negative number, $(x^2 \geq 0)$. For this reason, mathematicians were eventually forced to extend the set $\mathbb{R}$ to a more general set, known as the set of **Complex numbers** $\mathbb{C}$. Obviously the set of real numbers is a subset of the set of complex numbers ($\mathbb{R} \subset \mathbb{C}$). Complex numbers will be studied in Chapter 8.

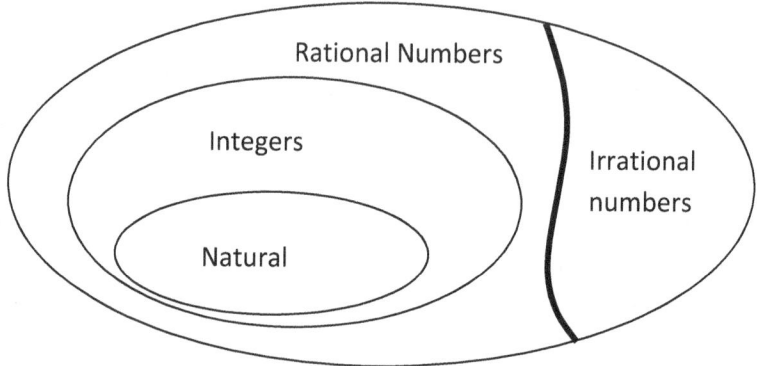

Fig. 1-1: The set of Real Numbers.

Note that $\mathbb{Z} \subset \mathbb{Q}$ since any integer number a can be written as $a/1$, which by definition is a rational number. For instance, $5 = 5/1, -7 = -7/1$, etc.

The decimal form of any rational number will either terminate or will form an infinite periodic (repeating) pattern. For example:

$$\frac{1}{4} = 0.25, \quad \frac{2}{5} = 0.4, \quad \frac{45}{15000} = 0.003$$

while

$$\frac{1}{9} = 0.1111\ldots, \quad \frac{2}{7} = 0.2857128571\ldots$$

For brevity we write $1/9 = 0.\dot{1}$, where the dot on top of 1, means that the digit 1 is **repeated indefinitely**, and similarly we may write $2/7 = 0.\dot{2}85\dot{7}\dot{1}$, meaning that the digits "28571" are repeated periodically forever. The converse is also true. **Every rational number with a repeating pattern represents a rational number.** Indeed, let for instance $x = 0.\dot{3}\dot{7} = 0.3737\ldots$, then $100x = 37.3737\ldots$, and subtracting term wise we get, (check it),

$$99x = 37 \Longrightarrow x = \frac{37}{99}$$

and this shows clearly that the number $0.\dot{3}\dot{7}$ is the rational number $37/99$.

The irrational number $\sqrt{2}$ has also a decimal representation, $\sqrt{2} = 1.41421356\ldots$, but **the decimal digits do not form a periodic pattern**.

Example 1-1-1: Which rational number is the number $x = 3.\dot{2}7\dot{5}$ equal to?

Solution

$$x = 3.\dot{2}7\dot{5} = 3.275275275\ldots \Longrightarrow 1000x = 3275.275275\ldots \Longrightarrow$$

$$1000x - x = 3272 \Longrightarrow 999x = 3272 \Longrightarrow x = 3272/999$$

Example 1-1-2: If $y \in \mathbb{Z}_0^+ \stackrel{\text{def}}{=} \{0,1,2,3,\cdots\}$ and $A = \{a/a = \sqrt{y}\}$, is $A \subset \mathbb{Z}_0^+$?

Solution

The set $A = \{0, \sqrt{1}, \sqrt{2}, \sqrt{3}, \cdots\}$ and obviously A is **not** a subset of $\mathbb{Z}_0^+$, since for example $\sqrt{2}$ is not an integer, i.e. $\sqrt{2} \notin \mathbb{Z}_0^+$.

Example 1-1-3: Show that the real number $\sqrt{2}$ is irrational.

Solution

We prove the irrationality of $\sqrt{2}$ by **contradiction**, (a popular method of proving of the ancient Greeks). We assume that $\sqrt{2}$ is a rational number, i.e. $\sqrt{2}$ can be expressed as a fraction (k/λ), where k and λ are **positive integers**.

We further assume that the fraction (k/λ) is in its lowest terms, i.e. **the integers k and λ do not have any common divisor** (are relatively prime numbers). Then,

$$\sqrt{2} = \frac{k}{\lambda} \Rightarrow k = \sqrt{2}\,\lambda \Rightarrow k^2 = 2\lambda^2 \qquad (*)$$

Since the right-hand side member in (*) is even (multiple of 2), the number 2 should also divide the left-hand side member, i.e. 2 should divide k^2 and this implies that 2 should divide k as well, i.e. $k = 2\rho$, where ρ is another **positive integer**. Substituting this expression of k into (*) we get

$$(2\rho)^2 = 2\lambda^2 \Rightarrow 2\rho^2 = \lambda^2 \qquad (**)$$

and reasoning similarly, we conclude that 2 should divide λ.

In summary, assuming that $\sqrt{2} = k/\lambda$ where the positive integers are relatively prime (**do not have any divisors in common**), we end up with the necessary conclusion that 2 divides both, k and λ simultaneously, i.e. 2 is a common divisor of k and λ, which however contradicts our hypothesis (that k and λ do not have any divisors in common). To avoid this contradiction we are forced to accept that **our hypothesis is not correct**, i.e. $\sqrt{2}$ **cannot** be written as a fraction of two integers, in other words the number $\sqrt{2}$ is an irrational number, and this completes the proof.

PROBLEMS

1-1-1) Which rational number is the number $x = 0.\dot{9}$ equal to?

(**Ans:** $x = 1$).

1-1-2) If $a = 2 + \sqrt{3}$ find the numbers a^2 and a^3.

1-1-3) Show that $1/(\sqrt{2} - 1) = \sqrt{2} + 1$.

1-1-4) Which rational number is the number $x = 2.3\dot{5}\dot{7}$ equal to?

1-1-5) If $x = 2 + \sqrt{3}$ and $y = 1 + \sqrt{5}$ show that $x^2 + y^2 = a\sqrt{3} + b\sqrt{5}$, where a and b are rational numbers to be determined.

1-2) Functions and function notation

a) The symbols $\forall$ and $\exists$.

The symbol $\forall$ is read **"for every"** and is used to denote that a mathematical proposition is true for all values of the variables involved. For example, to denote that for every real value of x the quantity $x^2 + 3 > 0$, we may write,

$$\forall x \in \mathbb{R}: \quad x^2 + 3 > 0$$

Similarly, we may express the permanent inequality $x^2 + 1 \geq 2x$ as follows:

$$\forall x \in \mathbb{R}: \quad x^2 + 1 \geq 2x$$

The symbol $\exists$ is read **"there exists at least one ….such that"**. For example, the proposition "there exists at least one integer x such that $x^2 = 25$" may be written as

$$\exists x \in \mathbb{Z}: \quad x^2 = 25$$

(Obviously, there are two integers, the 5 and the -5, whose squares are equal to 25).

The symbol $\nexists$ is read **"there is no….such that"**. For example, to state that there is no real number x such that $x^2 + 1 = 0$, we may write,

$$\nexists x \in \mathbb{R}: \quad x^2 + 1 = 0$$

b) The sets $\mathbb{R}, \mathbb{R}^2, \mathbb{R}^3, ...$

If $\mathbb{R}$ is the set of real numbers, then by the symbol $\mathbb{R}^2$ (or $\mathbb{R} \times \mathbb{R}$) we shall denote the set of **ordered pairs** (x, y) of real numbers.

Similarly the symbol $\mathbb{R}^3$ (or $\mathbb{R} \times \mathbb{R} \times \mathbb{R}$) designates the set of **ordered triads** (x, y, z) of real numbers.

Thus, $(x, y) \in \mathbb{R}^2$ means that the pair x, y is a pair of real numbers, **x being the first number and y the second**. Similarly, $(x, y, z) \in \mathbb{R}^3$ means that the triad x, y, z is a triad of real numbers, **x being the first number, y the second and z the third**.

c) Functions: a brief introduction.

Let A and B be two sets of real numbers.

Definition 1: A function is a rule f which, at each element x in the set A, ($x \in A$), assigns exactly one element y in the set B, ($y \in B$). The element y is called the image of x (via the function f), while x is the preimage of y.

The set of all the preimages is called the **domain** of the function and the set of all images is called the **range** of the function.

Since we are free to choose the values of x from the set A, we call x the **independent variable**. The element y, which is associated with x via the rule f, is called the **dependent variable**.

The important thing to notice is that the function assigns **a unique (one and only one)** value of y to each element x chosen from the set A. To denote that y is the image of x under the function (rule) f, we write $y = f(x)$, or more precisely,

$$A \ni x \xrightarrow{f} y = f(x) \in B \qquad (1-2-1)$$

Let us consider the following, illustrative examples:

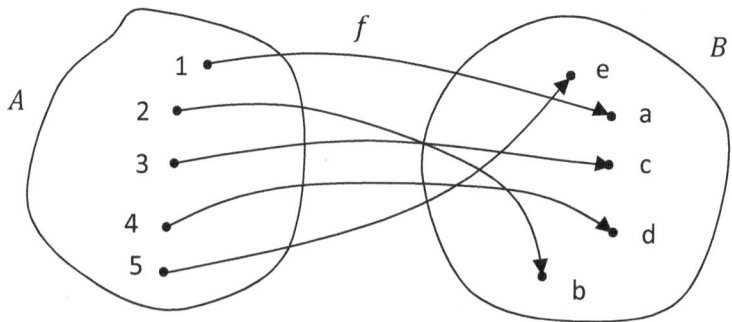

Fig. 1-2: A mapping (function) of the set A to the set B.

The set A is $A = \{1,2,3,4,5\}$ and the set B is $B = \{a, b, c, d, e\}$. Suppose now that there is a rule, which associates elements of A with elements of B in the following fashion:

$$1 \to a, \quad 2 \to b, \quad 3 \to c, \quad 4 \to d, \quad 5 \to e$$

This rule is **a function**, since to each element of the set A assigns **exactly one** element of the set B. The set A is the **domain** of the function and the set B is the **range** of the function.

To denote that a is the element of B corresponding to the element 1 of A, we may write $a = f(1)$, and similarly, $b = f(2), c = f(3), d = f(4)$ and $e = f(5)$. The numbers 1,2,3,4,5 constitute the set of **independent variables** (the domain of definition) while the elements a, b, c, d, e are the **depended variables** (the range of the function).

Note that the function **is completely determined** once we know the domain of definition, the range and the rule which assigns **a unique element** of the range to each element of the domain of definition.

This simple example shows that we may give an alternative, but equivalent definition of the function.

Definition 2: A function is a set of ordered pairs (x, y) such that to each value of the first element x there corresponds a unique (one and only one) value of the second element y, (i.e. there are no pairs with same first but different second elements).

In this example the function may be expressed as follows:

$$\{(1, a), (2, b), (3, c), (4, d), (5, e)\}$$

As a **second example** let us consider the set of ordered pairs

$$\{(1,10), (2,11), (1,12), (4,13), (5,14), (6,15), (7,16)\}$$

Does this set of ordered pairs define a function?

Since we have two pairs, the pairs $(1,10)$ and $(1,12)$ which have the same first element (1) but different second elements (10 and 12) the given set of ordered pairs **is not a function**.

The same conclusion follows from the first definition of the function, since in this example the rule assigns two values (**and not a unique one**) to the element 1, i.e. assigns the elements 10 and 12 to the same element 1.

On the contrary, the set of ordered pairs

$$\{(1,15),(2,15),(3,15),(4,15),(5,15),(6,15)\}$$

does define a function, since to each element of the domain of definition $A = \{1,2,3,4,5,6\}$ assigns a unique value of the set $B = \{15\}$.

As **a third example**, let us consider the function f which is defined by the formula $y = f(x) = x^5$, $x \in \mathbb{R}$. We may formally write,

$$f: x \in \mathbb{R} \xrightarrow{f} y = f(x) = x^5 \in \mathbb{R}$$

The domain of definition is the set of real numbers $\mathbb{R}$ and the range is also $\mathbb{R}$. The **rule** which allows us to determine a unique y (once x is given) is

$$f: rise\ the\ number\ x\ to\ the\ fifth\ power\ and\ the\ result\ will\ be\ y.$$

In this book, the domain of definition and the range will be sets of real numbers. However, note that in general, functions may be defined over other sets as well, for instance **sets of points**, as is shown in Fig, 1-3.

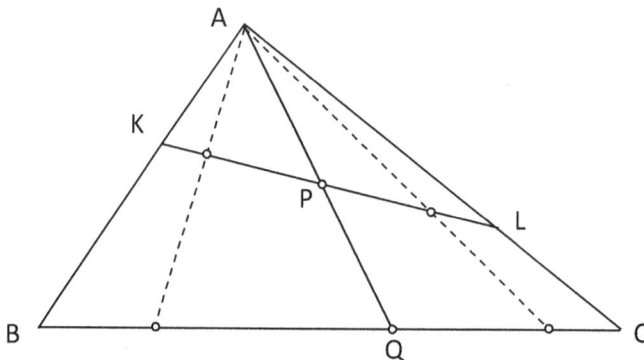

Fig. 1-3: A function with domain and range sets of points.

In Fig. 1-3, a ray AQ issued from the vertex A intersects KL at P and BC at Q, and it is obvious that (in this process) each point of the line segment KL is associated with one point of BC. This association defines a function, and we may formally write,

$$P \in KL \xrightarrow{f} Q = f(P) \in BC$$

The domain of this function is the set of points in KL and the range is the set of points in BC. We may say that this function **maps** the line segment KL onto BC. The term **mapping** is often used as a synonym for functions.

There are cases where the domain and the range of a function are implied indirectly from the form of the function, (see Examples 1-2-1 and 1-2-2).

Example 1-2-1: Find the domain and the range of the function $y = \sqrt{x-2}$.

Solution

Since the radicand must be **non-negative** (the square root of a negative number does **not** exist in the set of real numbers), we must have $x - 2 \geq 0$, i.e. $x \geq 2$. The domain of definition is $x \geq 2$ and the range is $y \geq 0$.

Example 1-2-2: Find the domain and the range of the function $y = 1/(x-3)$.

Solution

The function is defined for all real numbers x, except $x = 3$, at which the denominator vanishes. This means that the domain of the function is the set of all real numbers except the number 3, i.e. is the set $\mathbb{R} - \{3\}$. To find the range, we solve the given equation for x, i.e. express the x in terms of y as follows:

$$y = \frac{1}{x-3} \Leftrightarrow x - 3 = \frac{1}{y} \Leftrightarrow x = 3 + \frac{1}{y}$$

and this shows that the range is the set of all real numbers except the number 0, i.e. is the set $\mathbb{R} - \{0\}$.

1-3) Real functions of one real variable

A mapping of a subset of the set of real numbers $\mathbb{R}$ into $\mathbb{R}$ is called a real function of a real variable, (the word real here means that both the independent and the dependent variables take **real values**). In other words, if to each element x of a set A (which is the set of real numbers $\mathbb{R}$ or a subset of $\mathbb{R}$) there corresponds **a unique** element $y \in \mathbb{R}$, then this correspondence defines a function $f = f(x)$ which is called **a real function of a real variable**.

The following are examples of real functions of one real variable:

a) $y = 2x + 3$, domain $\mathbb{R}$, range $\mathbb{R}$. The rule which assigns a unique y to a given $x \in \mathbb{R}$ is, "multiply x by 2 and then to this product add the number 3".

b) $y = 1/(2x - 10)$, domain $\mathbb{R} - \{5\}$, range $\mathbb{R} - \{0\}$, (see Example 1-2-2).

c) $y = 2x + x^2 + 3x^3$, domain $\mathbb{R}$, range $\mathbb{R}$.

d) $y = x^2 + x^4$, domain $\mathbb{R}$, range $y \geq 0$, (due to the even powers).

e) $y = \sqrt{x + 3}$, domain $x \geq -3$, range $y \geq 0$.

f) $y = \sqrt[3]{x - 5}$, domain $\mathbb{R}$, range $\mathbb{R}$, (why?).

Elementary operations between real functions of a real variable

Let us consider two functions $x \xrightarrow{f} f(x)$ and $x \xrightarrow{g} g(x)$ having the same domain A, which could be the set of real numbers or part (a subset) of it. Then we may define the sum, the difference, the product and the quotient of these two functions as follows:

1) The sum of the functions $(f + g)$: $x \in A \xrightarrow{(f+g)} y = f(x) + g(x)$

2) The difference of the functions $(f - g)$: $x \in A \xrightarrow{(f-g)} y = f(x) - g(x)$

3) The product of the functions $(f \cdot g)$: $x \in A \xrightarrow{(f \cdot g)} y = f(x) \cdot g(x)$

4) The quotient of the functions (f/g): $x \in A \xrightarrow{(f/g)} y = f(x)/g(x)$

The quotient is defined for all values of $x \in A$ for which $g(x) \neq 0$.

We may multiply a function $y = f(x)$ by itself n times to get the function $\{f(x)\}^n$.

In particular the function $y = cx^n$ where c is a constant, n is a natural number and $x \in \mathbb{R}$ is called **a monomial in x of degree n**. Any constant c can be considered as a monomial of degree zero. For instance, $y = 3x^5$ is a monomial of degree 5, $y = (-7/5)x^8$ is a monomial of degree 8 and

$y = \sqrt{7}\,x^3$ is a monomial of degree 3. The function $y = x^{1/2} = \sqrt{x}$ is not a monomial, since the exponent of x (the number $1/2$) is not a natural number. For the same reason $y = 5x^{-3}$ is not a monomial. The function $y = 1$ (constant) is a monomial of degree zero.

A polynomial in x is an expression that is **the sum of monomials**. Any polynomial $P(x)$ can therefore be expressed as follows:

$$P(x) = a_n x^n + a_{n-1} x^{n-1} + \cdots + a_2 x^2 + a_1 x + a_0 \qquad (1-3-1)$$

where n is **a natural number** and $x \in \mathbb{R}$. Each of the monomials in (1-3-1) is called **a term** of the polynomial, and the real numbers $a_n, a_{n-1}, \cdots, a_2, a_1, a_0$ are the **coefficients** of the polynomial. The **degree** of $P(x)$ is n, the coefficient a_n is the **leading coefficient** and the coefficient a_0 is **the constant term**. As an example, the polynomial $P(x) = 5x^2 - 2x + 3$ is a second degree polynomial with leading coefficient 5 and constant term 3. The polynomial $P(x) = -\sqrt{3}\,x^7 + 15x^3 + 3$ is a seventh degree polynomial with leading coefficient the number $(-\sqrt{3})$ and constant term 3.

Example 1-3-1: If $P(x) = x^3 + 1$ and $Q(x) = 2x^2 - 5x + 3$ find: **a)** $P(x) + Q(x)$, **b)** $3P(x) - Q(x)$, **c)** $P(x) \cdot Q(x)$.

Solution

$$P(x) + Q(x) = (x^3 + 1) + (2x^2 - 5x + 3) = x^3 + 2x^2 - 5x + 4$$

$$3P(x) - Q(x) = 3(x^3 + 1) - (2x^2 - 5x + 3) = 3x^3 + 3 - 2x^2 + 5x - 3$$
$$= 3x^3 - 2x^2 + 5x$$

$$P(x) \cdot Q(x) = (x^3 + 1) \cdot (2x^2 - 5x + 3) \Rightarrow$$

$$P(x) \cdot Q(x) = 2x^5 - 5x^4 + 3x^3 + 2x^2 - 5x + 3$$

Example 1-3-2: If $P(x) = x^2 - 3x + 2$ find the numbers $P(1)$ and $P(-1)$.

Solution

$$P(1) = 1^2 - 3 \cdot 1 + 2 = 1 - 3 + 2 = 0$$

$$P(-1) = (-1)^2 - 3 \cdot (-1) + 2 = 1 + 3 + 2 = 6$$

Example 1-3-3: Consider the function $f(x) = x(x-3) + |2x-1|$, $x \in \mathbb{R}$. Find the numbers $f(0), f(1), f(-1)$ and $f(-3)$.

Solution

$$f(0) = 0 \cdot (0-3) + |2 \cdot 0 - 1| = 0 + |-1| = 0 + 1 = 1$$

$$f(1) = 1 \cdot (1-3) + |2 \cdot 1 - 1| = 1 \cdot (-2) + |1| = -2 + 1 = -1$$

$$f(-1) = (-1)(-1-3) + |2(-1) - 1| = 4 + |-3| = 4 + 3 = 7$$

$$f(-3) = (-3)(-3-3) + |2 \cdot (-3) - 1| = 18 + |-7| = 18 + 7 = 25$$

Example 1-3-4: If $g(x) = x^2$ find $(g(b) - g(a))/(b-a)$, assuming $a \neq b$.

Solution

$$\frac{g(b) - g(a)}{b-a} = \frac{b^2 - a^2}{b-a} = \frac{(b-a)(b+a)}{b-a} = b + a$$

Example 1-3-5: a) If $n \in \mathbb{N} = \{1,2,3,\cdots\}$ and $f(n) = \frac{1}{n(n+1)}$, $g(n) = \frac{1}{n}$, show that $f(n) = g(n) - g(n+1)$, and **b)** Show that $f(1) + f(2) + \cdots + f(n) = g(1) - g(n+1)$.

Solution

$$g(n) - g(n+1) = \frac{1}{n} - \frac{1}{n+1} = \frac{(n+1) - n}{n(n+1)} = \frac{1}{n(n+1)} = f(n) \quad (*)$$

and this completes the proof of part (a).

b) Application of formula (*) for $n = 1,2,3,\cdots,n$ yields,

$$f(1) = g(1) - g(2)$$
$$f(2) = g(2) - g(3)$$
$$\cdots \quad \cdots \quad \cdots$$
$$f(n-1) = g(n-1) - g(n)$$
$$f(n) = g(n) - g(n+1)$$

and adding term wise we get, (due to the pair wise cancelation),

$$f(1) + f(2) + \cdots + f(n) = g(1) - g(n+1) \quad (**)$$

Example 1-3-6: Using formula (**) obtained in Example 1-3-5, find the sum

$$S = \frac{1}{1 \cdot 2} + \frac{1}{2 \cdot 3} + \frac{1}{3 \cdot 4} + \cdots + \frac{1}{198 \cdot 199} + \frac{1}{199 \cdot 200}$$

Solution

Application of formula (**) in Example 1-3-5, with $f(n) = \frac{1}{n(n+1)}$, $g(n) = \frac{1}{n}$ and $n = 199$, yields $S = 1 - \frac{1}{200} = \frac{199}{200}$.

PROBLEMS

1-3-1) If $P(x) = x^3 - 3x^2 + 3x - 1$ find $P(1), P(-1)$ and $P(\sqrt{2})$.

(**Ans:** $P(1) = 0, P(-1) = -8, P(\sqrt{2}) = -7 + 5\sqrt{2}$).

1-3-2) If $P(x) = x^3 - 1$ and $Q(x) = x^2 + 5$, find $2P(x) - 3Q(x)$ and $P(x) - xQ(x)$.

1-3-3) If $P(x) = x^3$ find $(P(b) - P(a))/(b - a)$, (assume $a \neq b$).

(**Ans:** $b^2 + ab + a^2$).

1-3-4) Let f be the function

$$x \in \mathbb{R} \xrightarrow{f} y = f(x) = (x-1)(x+2)|2x+7| \in \mathbb{R}$$

Find the numbers $f(0), f(-2), f(-1), f(-1/2)$ and $f(-4)$.

1-3-5) If $S(r)$ is the area of a circle of radius r, express the area as a function of r. Repeat for the circumference $L(r)$ of the circle.

(**Ans:** $S(r) = \pi r^2$, $L(r) = 2\pi r$).

1-3-6) If $f(x) = x^2 - 3|x| + 2$ and $g(x) = x^2 - 2$ find the numbers $f(-2) + g(0), f(3) - 2g(5), f(0) \cdot g(1)$ and $f(-3)/g(-1)$.

1-3-7) If $f(x) = \frac{x+1}{x-1}$ and b is a real number, find the expression $\left(f(b) + f\left(\frac{1}{2}\right)\right) \Big/ \left(1 - f(b)f\left(\frac{1}{2}\right)\right)$, where $b \neq 1$ and $b \neq -\frac{1}{2}$.

(**Ans:** $(2-b)/(1+2b)$).

1-3-8) If $f(n) = 3^n$ and $g(n) = 3^n/2$, where n is a natural number, show that $g(n+1) - g(n) = f(n)$ and then use this result to evaluate the sum $S = 3 + 3^2 + 3^3 + \cdots + 3^{n-1} + 3^n$.

1-3-9) If $f(x) = x^2 + 1$ and $g(x) = x - 1$ find the functions $f(g(x))$ and $g(f(x))$.

(**Ans:** $f(g(x)) = x^2 - 2x + 2$, $g(f(x)) = x^2$).

1-4) Functions one-to-one and inverse functions

According to the definition of a function (mapping), it is possible two or more different elements from the domain of definition to be associated with the same element in the range, (the converse is not allowed, i.e. one element in the domain of definition to be associated with two different elements in the range).

If we demand that every element in the domain to be assigned to **a different** element in the range, then this function is called **one-to-one**.

If a and b be any two elements in the domain of a **one-to-one** function f, then,

$$\begin{cases} f(a) = f(b) \Leftrightarrow a = b \\ f(a) \neq f(b) \Leftrightarrow a \neq b \end{cases} \qquad (1-4-1)$$

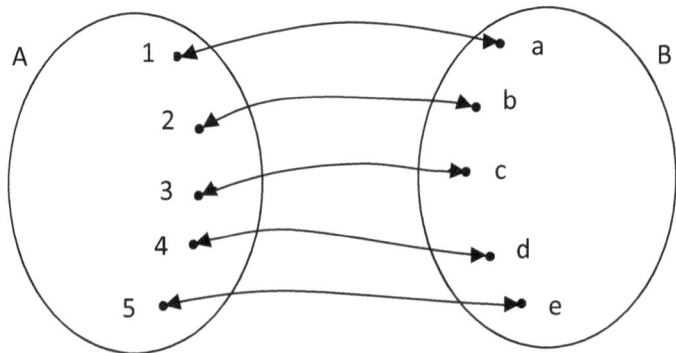

Fig. 1-4: Function one-to-one.

For the function shown in Fig. 1-4, the domain is $A = \{1,2,3,4,5\}$ and the range is $B = \{a, b, c, d, e\}$. The one-to-one function f is the set of ordered pairs

$$f: \quad \{(1, a), (2, b), (3, c), (4, d), (5, e)\}$$

Inverse functions: Assuming that the function f is a one-to-one function, each element in the range is associated to **a unique element of the domain,** for example the element $a \in B \to 1 \in A$, $b \in B \to 2 \in A$, etc. This association defines a new function, called **the inverse function of f**, and denoted by f^{-1}. The domain of f^{-1} is the range of f and the range of f^{-1} is the domain of f. In our case the inverse function f^{-1} is thus defined as follows:

$$f^{-1}: \quad \begin{array}{l} Domain = \{a, b, c, d, e\} \\ Range = \{1,2,3,4,5\} \\ \{(a, 1), (b, 2), (c, 3), (d, 4), (e, 5)\} \end{array} \qquad (1-4-2)$$

Remember that inverse functions are defined only when the function f is a one-to-one function.

From the very definition of the inverse functions, it follows that if $y = f(x)$ then $x = f^{-1}(y)$, i.e.

$$y = f(x) \Longleftrightarrow x = f^{-1}(y) \qquad (1-4-3)$$

From formula (1-4-3), if f^{-1} is the inverse function of f, the following identities hold true:

$$\begin{array}{l} y = f(f^{-1}(y)) \quad \forall y \in range\ of\ (f) \\ x = f^{-1}(f(x)) \quad \forall x \in domain\ of\ (f) \end{array} \qquad (1-4-4)$$

Since, traditionally, we denote the independent variable by x and the dependent variable by y, we usually write the second equation in (1-4-3) in the form $y = g(x)$, where g stands for the inverse function f^{-1}.

Example 1-4-1: If $y = f(x) = 2x + 3$, $x \in \mathbb{R}$, find the inverse function $y = g(x)$.

Solution

The function $y = f(x) = 2x + 3$ is a one-to-one function. To find its inverse we solve for x, i.e. find an expression of x in terms of y.

$$y = 2x + 3 \Leftrightarrow 2x = y - 3 \Leftrightarrow x = \frac{y-3}{2} \qquad (*)$$

In equation (*), y is the independent variable and x is the dependent one. If, as it is customary, we denote by x the independent variable and by y the dependent one, then equation (*) implies that $y = (x-3)/2$, and this is the inverse of the given function.

Example 1-4-2: Find the inverse of the function $y = x^2$, $x \in \mathbb{R}$.

Solution

The given function is **not a one-to-one function** since, for example, the two different elements 2 and (-2) in the domain are associated with the same element $y = 4$ in the range. **However, the function can be made to be one-to-one, if we restrict its domain of definition.** Indeed, if we allow x to take only **non-negative** values $(x \geq 0)$, then the function becomes a one-to-one function, and thus we may define its inverse. In this case, we have,

$$y = x^2, \qquad x \geq 0 \Leftrightarrow x = \sqrt{y}, \quad y \geq 0$$

This shows that $y = \sqrt{x}$, $x \geq 0$, $y \geq 0$ is the inverse of $y = x^2$.

PROBLEMS

1-4-1) Is the function $f: \{(1,5), (2,4), (3,5), (4,6), (5,7)\}$ one-to-one?

(**Ans:** No, because of the two pairs (1,5) and (3,5)).

1-4-2) Find the inverse of the function $y = x^3$, $x \in \mathbb{R}$).

(**Ans:** $y = \sqrt[3]{x}$).

1-4-3) Find the inverse of the function $y = x^2 - 1$, $x \geq 0$.

(**Ans:** $y = \sqrt{x+1}$).

1-4-4) If $f(x) = 5x + 7$, show that $g(x) = (x-7)/5$ is the inverse of $f(x)$ and verify that $f(g(x)) = x$ and $g(f(x)) = x$.

1-4-5) If $f(x) = x^7$, show that $g(x) = \sqrt[7]{x}$ is the inverse of $f(x)$ and verify that $f(g(x)) = x$ and $g(f(x)) = x$.

1-5) Real functions of several real variables

a) A real function of two real variables.

Let A be a set of ordered pairs which could be the set $\mathbb{R} \times \mathbb{R}$ or a subset of it. If to each ordered pair $(x,y) \in A$ there corresponds (by means of some rule f) a **unique** number $z \in \mathbb{R}$, then we say that we have a real function of two real variables. As usually, the domain of f is the set $A \subset \mathbb{R} \times \mathbb{R}$ and its range is $\mathbb{R}$, (or a subset of $\mathbb{R}$). In other words, **a real function of two real variables is a mapping of a subset of $\mathbb{R}^2$ into $\mathbb{R}$**. To denote that the rule f assigns the element (x,y) to the real number z, we write $z = f(x,y)$, i.e.

$$(x,y) \in \mathbb{R} \times \mathbb{R} \xrightarrow{f} z = f(x,y) \in \mathbb{R} \qquad (1-5-1)$$

The following are examples of real functions of two real variables:

1) $f: (x,y) \in \mathbb{R} \times \mathbb{R} \xrightarrow{f} z = x + y = f(x,y) \in \mathbb{R}$, (domain $\mathbb{R} \times \mathbb{R}$, range $\mathbb{R}$).

2) $f: (x,y) \in \mathbb{R} \times \mathbb{R} \xrightarrow{f} z = x^2 + y^2 = f(x,y) \in \mathbb{R}_0^+$ ($\mathbb{R}_0^+$ is the set of the **non-negative** real numbers, i.e. $z \geq 0$).

3) $f: (x,y) \in \mathbb{R} \times \mathbb{R} - \{0,0\} \xrightarrow{f} z = \frac{1}{x^2+y^2} = f(x,y) \in \mathbb{R}^+$, (set of positive real numbers).

4) $f: (x,y) \in \mathbb{R} \times \mathbb{R} \xrightarrow{f} z = \sqrt[3]{x^3 + y^5} = f(x,y) \in \mathbb{R}$.

5) $f: (x,y) \in \mathbb{R} \times \mathbb{R} - \{1,-3\} \xrightarrow{f} z = \frac{1}{(x-1)(y+3)} = f(x,y) \in \mathbb{R}$.

Let us consider the function $z = f(x,y) = x^2 + y^3 - 2x + 3y + 1$. The value of the function at $(x = -1, y = 2)$ is

$$z = f(-1,2) = (-1)^2 + 2^3 - 2 \cdot (-1) + 3 \cdot 2 + 1 = 18$$

The value of the same function at $(x = 2, y = -1)$ is

$$z = f(2,-1) = 2^2 + (-1)^3 - 2 \cdot 2 + 3 \cdot (-1) + 1 = -3$$

b) A real function of three real variables.

A mapping of a subset of $\mathbb{R} \times \mathbb{R} \times \mathbb{R}$ into $\mathbb{R}$ is a real function of three real variables. In other words, **a real function of three real variables is a rule which assigns a real number to each ordered triad taken from a subset of** $\mathbb{R} \times \mathbb{R} \times \mathbb{R}$. If we call f the rule which assigns **a unique** real number $w \in \mathbb{R}$ to each ordered triad $(x, y, z) \in \mathbb{R} \times \mathbb{R} \times \mathbb{R}$ (or a subset of $\mathbb{R} \times \mathbb{R} \times \mathbb{R}$), then we may formally write,

$$(x, y, z) \in \mathbb{R} \times \mathbb{R} \times \mathbb{R} \xrightarrow{f} w = f(x, y, z) \in \mathbb{R} \qquad (1-5-2)$$

The following are examples of real functions of three real variables:

1) $f: (x, y, z) \in \mathbb{R} \times \mathbb{R} \times \mathbb{R} \xrightarrow{f} w = x + y + z \in \mathbb{R}$.

2) $f: (x, y, z) \in \mathbb{R} \times \mathbb{R} \times \mathbb{R} \xrightarrow{f} w = x^2 + y^2 + z^2 \in \mathbb{R}_0^+$.

3) $f: (x, y, z) \in \mathbb{R} \times \mathbb{R} \times \mathbb{R} - \{0,0,0\} \xrightarrow{f} w = \frac{1}{x^2+y^2+z^2} \in \mathbb{R}^+$.

4) $f: (x, y, z) \in \mathbb{R} \times \mathbb{R} \times \mathbb{R} - \{1,2,3\} \xrightarrow{f} w = \frac{1}{(x-1)(y-2)(z-3)} \in \mathbb{R}$.

In every ordered triad of real numbers, the function f assigns one real number. For example, if $w = f(x, y, z) = 3x^2 + 2y - z + 1$, then $f(0,0,0) = 1$, $f(0,1,2) = 1$, $f(1,1,1) = 5$, $f(2,-1,1) = 10$, etc.

c) Operations between real functions of many real variables.

The **elementary operations** between real functions of many real variables (sum, difference, product, and quotient) are defined as for functions of one real variable, (see Section 1-3).

For example, if $f(x, y) = x + y$ and $g(x, y) = x - y$, then:

$$f(x,y) + g(x,y) = 2x$$
$$f(x,y) - g(x,y) = 2y$$
$$f(x,y) \cdot g(x,y) = x^2 - y^2$$
$$\frac{f(x,y)}{g(x,y)} = \frac{x+y}{x-y}$$

Any **algebraic expression** is nothing else but a real function of real variables. For example, the algebraic expressions $f(x,y) = \sqrt{x^2 + y^2}$, $g(x,y,z) = 1/(2x - 3y + 5z)$, $F(x) = x^2 + 5$, $G(x,y,z) = \sqrt[3]{x} + \sqrt[3]{y} + \sqrt[3]{z}$ are real functions of real variables.

In particular, the function $cx^n y^m$ where c is a **constant** and n, m are **natural numbers**, is called **a monomial in two variables, x and y**. The degree of the monomial in x and y is the sum of the exponents of x and y. For example, the degree of the monomial $3x^2 y^7$ is $9 = 2 + 7$, the degree of the monomial $\sqrt{5}\, x^3 y$ is $4 = 3 + 1$, etc. The sum of monomials is called a **polynomial in x and y**. The degree of the polynomial in x and y is the degree of the monomial that has the **highest degree**.

The degree of the polynomial $P(x,y) = 3x^2 y^3 + 2x^2 y^2 - 7x^2 y + xy^2 + 5$ is 5, while the degree of $Q(x,y) = x^3 y^4 + x^2 y^4 + xy^5 + 10$ is 7.

The function $cx^n y^m z^k$ where c is **a constant** and n, m, k are **natural numbers** is called **a monomial in three variables x, y and z**. The degree of the monomial is the sum of the exponents of x, y and z. For example, the degree of the monomial $2x^2 y^3 z^4$ is $9 = 2 + 3 + 4$, the degree of $3x^4 y^6 z$ is $11 = 4 + 6 + 1$, etc. The sum of monomials is called **a polynomial in x, y and z**. The degree of the polynomial is the degree of the monomial that has the **highest degree**, for example the degree of $P(x,y,z) = 2x^3 y^4 z^3 + 3x^3 y^2 z^3 + x^5 yz - 1$ is $10 = 3 + 4 + 3$.

d) In a similar fashion we may define a real function of four real variables, a real function of five real variables, etc.

For instance, $u(x, y, z, w) = 2x - 3y^2 + 5z - w^3 + 7$ defines a function, since to each $(x, y, z, w) \in \mathbb{R} \times \mathbb{R} \times \mathbb{R} \times \mathbb{R}$ assigns a unique number $u \in \mathbb{R}$. The arithmetic value of u at $(x = 1, y = 0, z = 3, w = 2)$ is
$$u(1,0,3,2) = 2 \cdot 1 - 3 \cdot 0^2 + 5 \cdot 3 - 2^3 + 7 = 16$$

Example 1-5-1: Which of the following expressions are polynomials?

a) $A(x) = 2x^{1/2} + 2 + 3x^2$, b) $B(x) = x^{-2} + 2x$, c) $C(x) = x^2 + 3$

Solution

The expressions $A(x)$ and $B(x)$ are **not** polynomials since the exponents $(1/2)$ in $A(x)$ and (-2) in $B(x)$ are not natural numbers. The expression $C(x)$ is a polynomial of second degree.

Example 1-5-2: If $A(x) = 2x^2 + 3x + 1$ and $B(x) = x^3 + 2$ find the polynomial $P(x) = 2A(x) - 3B(x)$, and then arrange $P(x)$ **a)** In ascending powers of x and **b)** In descending powers of x.

Solution

$$P(x) = 2A(x) - 3B(x) = 2(2x^2 + 3x + 1) - 3(x^3 + 2) \Rightarrow$$

$$P(x) = 4x^2 + 6x - 4 - 3x^3$$

a) Ascending powers of x: $P(x) = -4 + 6x + 4x^2 - 3x^3$

b) Descending powers of x: $P(x) = -3x^3 + 4x^2 + 6x - 4$

Example 1-5-3: If $A(x,y) = \sqrt{x^2 + y^2}$ and $B(x,y) = \sqrt[3]{(x+y)^2}$ find $F(x,y) = \bigl(A(x,y)\bigr)^2 - \bigl(B(x,y)\bigr)^3$.

Solution

$$F(x,y) = \bigl(A(x,y)\bigr)^2 - \bigl(B(x,y)\bigr)^3 = \left(\sqrt{x^2+y^2}\right)^2 - \left(\sqrt[3]{(x+y)^2}\right)^3$$
$$= x^2 + y^2 - (x+y)^2 = x^2 + y^2 - (x^2 + y^2 + 2xy) = -2xy$$

Example 1-5-4: If $f: (x,y) \in \mathbb{R} \times \mathbb{R} \xrightarrow{f} f(x,y) = 2x + y - 3$, find the numbers $f(0,0), f(3,-1), f(-2,3)$ and $f(1,5)$.

Solution

$$f(0,0) = 2 \cdot 0 + 0 - 3 = -3$$
$$f(3,-1) = 2 \cdot 3 - 1 - 3 = 2$$
$$f(-2,3) = 2 \cdot (-2) + 3 - 3 = -4$$
$$f(1,5) = 2 \cdot 1 + 5 - 3 = 4$$

Example 1-5-5: What is the domain of definition of the functions: **a)** $f(x,y) = \frac{1}{(x-3)(y+10)}$ and **b)** $(x,y,z) = \frac{1}{(x+1)(x^2+1)y^2(z+5)}$.

Solution

a) The domain of definition of $f(x,y)$ is $\mathbb{R} \times \mathbb{R} - \{3,-10\}$ (since at these values the denominator vanishes, and **division by zero is not allowed in Mathematics**).

b) Since $x^2 + 1 \neq 0, \forall x \in \mathbb{R}$, the domain of definition of $g(x,y,z)$ is $\mathbb{R} \times \mathbb{R} \times \mathbb{R} - \{-1, 0, -5\}$.

Example 1-5-6: Consider the function $f: (x,y) \in \mathbb{R} \times \mathbb{R} \xrightarrow{f} f(x,y) = |x - y|$. Find the domain and the range of $f(x,y)$ and show that

$$f(x,y) = f(y,x)$$
$$f(x,y) = 0 \Leftrightarrow x = y$$
$$f(x,y) \leq f(x,z) + f(z,y)$$

Solution

The domain of $f(x,y)$ is $\mathbb{R} \times \mathbb{R}$ and the range is $\mathbb{R}_0^+$, i.e. the set of non-negative real numbers, since the absolute value of any real number is either zero or positive.

$$f(x,y) = |x - y| = |y - x| = f(y,x)$$

$$f(x,y) = 0 \Leftrightarrow |x - y| = 0 \Leftrightarrow x - y = 0 \Leftrightarrow x = y$$

$$f(x,y) = |x - y| = |(x - z) + (z - y)| \leq |x - z| + |z - y| \Rightarrow f(x,y) \leq f(x,z) + f(z,y)$$

PROBLEMS

1-5-1) If $f: (x, y, z) \in \mathbb{R} \times \mathbb{R} \times \mathbb{R} \xrightarrow{f} f(x, y, z) = xyz + x^2y^3z^4 + 3$ find the numbers $f(0,0,0)$ and $f(-1,1,-1)$.

(Ans: $3, 5$).

1-5-2) What is the degree of the monomials $P(x, y) = 3x^2y^6$ and $Q(x, y, z) = 2x^3y^2z$.

1-5-3) Find the domain of the functions $f(x, y) = \dfrac{5}{(2x+3)(3y-1)}$ and $g(x, y, z) = \dfrac{3}{(x-7)(2y+5)(5z+2)}$.

(Ans: $\mathbb{R}^2 - \left\{-\dfrac{3}{2}, \dfrac{1}{3}\right\}$, $\mathbb{R}^3 - \left\{7, -\dfrac{5}{2}, -\dfrac{2}{5}\right\}$).

1-5-4) Determine the constant c so that the polynomial $F(x, y) = x^2y^3 - 2xy^2 + cxy + 1$ vanishes at $(x = -1, y = 2)$.

1-5-5) What is the domain of definition of the function $f(x, y) = \dfrac{x^3y^2+3}{|x+5|+(y-7)^4}$?

(Ans: $\mathbb{R}^2 - \{-5, 7\}$).

1-5-6) If n is a natural number, show the identity

$$n^3 = \left(\frac{n(n+1)}{2}\right)^2 - \left(\frac{n(n-1)}{2}\right)^2$$

and then show that $1^3 + 2^3 + 3^3 + \cdots + n^3 = \left(\dfrac{n(n+1)}{2}\right)^2$.

1-5-7) Simplify the function $A(a, b) = \dfrac{a^2-b^2+2bc-c^2}{a^2+b^2+c^2+2ab-2ac-2bc}$.

(Ans: $A(a, b) = (a - b + c)/(a + b - c)$).

1-5-8) If $a = x^3 + 3x^2y - y^3$, $b = -3xy(x + y)$ and $c = x^2 + xy + y^2$ show that $a^2 + ab + b^2 = c^3$.

1-5-9) If $x = \frac{a-b}{a+b}, y = \frac{b-c}{b+c}, z = \frac{c-a}{c+a}$, show that $(1+x)(1+y)(1+z) = (1-x)(1-y)(1-z)$.

1-5-10) Simplify the function $f(x,y,z) = \frac{1}{x(x-y)(x-z)} + \frac{1}{y(y-x)(y-z)} + \frac{1}{z(z-x)(z-y)}$.

(Ans: $f(x,y,z) = \frac{1}{xyz}$**)**.

1-5-11) If $\frac{1}{x+y+z} = \frac{1}{x} + \frac{1}{y} + \frac{1}{z}$ show that $\frac{1}{x^3+y^3+z^3} = \frac{1}{x^3} + \frac{1}{y^3} + \frac{1}{z^3}$.

1-6) Composite functions

Let us consider a function f which assigns a value y in a set Y to a value x in a set D, and another function g which assigns a value z in set R to y in Y. The net effect of this combination of f and g is **a new function** w which assigns $x \in D$ **directly** to the value $z \in R$.

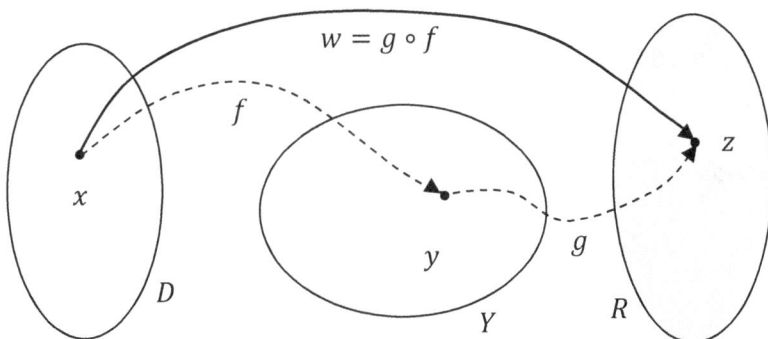

Fig. 1-5: Composition of two functions.

This new function is called the **composite function of f and g**, and is denoted by $g \circ f$. We may thus write:

$$\begin{cases} x \in D \xrightarrow{f} y = f(x) \in Y \\ y \in Y \xrightarrow{g} z = g(y) \in R \end{cases} \Rightarrow x \in D \xrightarrow{w = g \circ f} z \in R \qquad (1-6-1)$$

We may also write this composite function as

$$w(x) = (g \circ f)(x) \stackrel{\text{def}}{=} g(f(x)) \qquad (1-6-2)$$

(read g of $f(x)$).

Equation (1-6-2) shows how to construct the composite function $w = g \circ f$, once the expressions $f(x)$ and $g(x)$ are known. Notice that in general, $g \circ f \neq f \circ g$.

Example 1-6-1: Given $f(x) = 2x + 3$, $g(x) = 5x - 7$, find the following: $(g \circ f)(x) = g(f(x))$ and $(f \circ g)(x) = f(g(x))$.

Solution

$$g(f(x)) = 5 \cdot f(x) - 7 = 5 \cdot \underbrace{(2x+3)}_{f(x)} - 7 = 10x + 8 \qquad (*)$$

$$f(g(x)) = 2 \cdot g(x) + 3 = 2 \cdot \underbrace{(5x-7)}_{g(x)} + 3 = 10x - 11 \qquad (**)$$

From (*) and (**) we see that $g(f(x)) \neq f(g(x))$.

Example 1-6-2: Given $f(x) = \sqrt{2x+1}$, $g(x) = x + 3$, find $f(g(x))$ and $g(f(x))$.

Solution

$$f(g(x)) = \sqrt{2g(x)+1} = \sqrt{2(x+3)+1} = \sqrt{2x+7}$$

$$g(f(x)) = f(x) + 3 = \sqrt{2x+1} + 3$$

PROBLEMS

1-6-1) Given $f(x) = x^2$ and $g(x) = \sqrt{5x-7}$, find $f(g(x))$ and $g(f(x))$.

(Ans: $f(g(x)) = 5x - 7$, $g(f(x)) = \sqrt{5x^2 - 7}$).

1-6-2) Given $f(x) = x^2 + 5$ and $g(x) = \sqrt{x-1}$, find the domain and the range of $(f \circ g)$, the domain and the range of $(g \circ f)$ and the functions $f(g(x))$ and $g(f(x))$.

1-6-3) Given $f(x) = x^2$ and $g(x) = \sqrt{x-4}$, find the domain and the range of the functions $(f \circ g)$ and $(g \circ f)$ and the functions $f(g(x))$ and $g(f(x))$.

(Ans: $f(g(x)) = x - 4$, domain $\mathbb{R}$, range $\mathbb{R}$; $g(f(x)) = \sqrt{x^2 - 4}$, domain$(x^2 - 4) \geq 0$, range $\mathbb{R}_0^+$, (the set of non-negative real numbers)).

1-7) Cartesian coordinates

a) The line of real numbers: Let us consider a straight line, as shown in Fig. 1-6. We arbitrarily choose a point O on this line and call this point **the origin**. The direction to the right of the origin is defined to be the **positive direction**, while the direction to the left of the origin is defined to be the **negative direction**. Finally, we choose arbitrarily **a unit length** against which distances on the line are measured. The length of any line segment, lying on this line is measured (compared) with respect to the unit length.

Such a line for which we have defined,

1) The origin,

2) The positive direction, and

3) The unit length

is called **an axis, (the axis of real numbers)**.

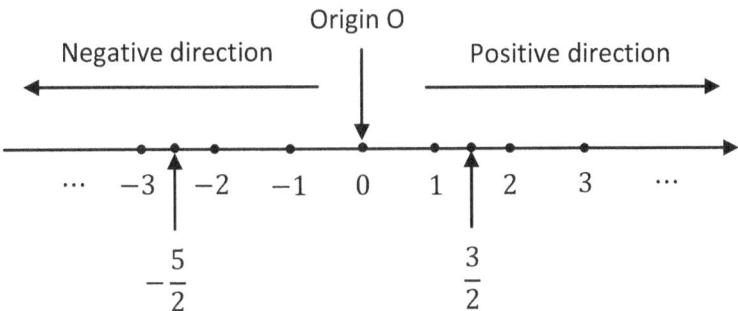

Fig. 1-6: The line of real numbers (the real axis).

The correspondence between the set of real numbers and the set of points on the real axis is established by means of the following axiom.

Axiom: There exists a one to one correspondence between the set of real numbers and the set of points on the real axis, such that:

1) Every real number x corresponds to a unique point on the real axis, and conversely,

2) Every point on the real axis corresponds to a unique real number.

To denote that the point M on the real axis corresponds to the real number x, we write $M(x)$. The number x associated with the point M, is called the **abscissa of M**. For example, $A(5)$ corresponds to the point A having abscissa 5, $B(-\sqrt{2})$ corresponds to the point B having abscissa $(-\sqrt{2})$, $C(\pi)$ corresponds to the point C having abscissa π, $O(0)$ corresponds to the origin O, etc.

b) The Cartesian plane (Cartesian coordinates): Let us consider two axes, Ox and Oy **perpendicular to each other**, as shown in Fig. 1-7.

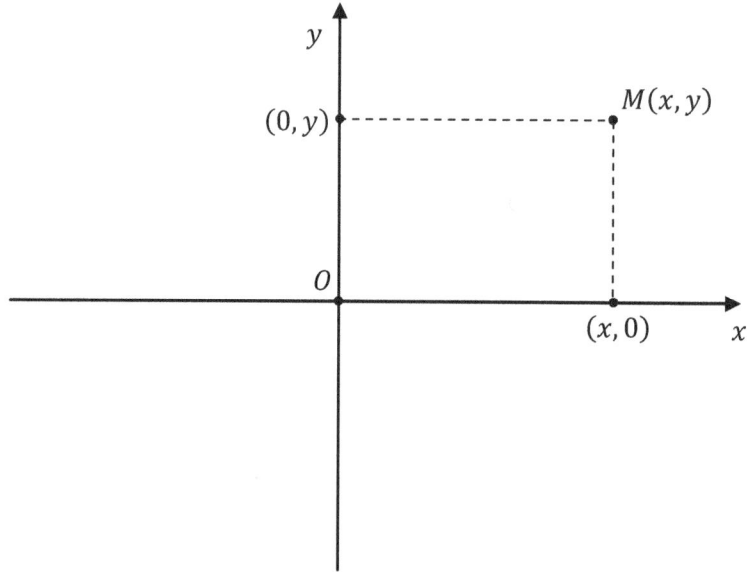

Fig. 1-7: The Cartesian plane, (Cartesian coordinates).

The point of intersection of the two axes is called **the origin**. Then, the two axes are scaled as real lines (as it was described in part (a)) by using the origin as the point 0 for each one of the two axes. The portion of Ox to the right of the origin is the **positive** direction and the portion to the left is the **negative** direction. The portion of Oy above the origin is the **positive** direction and the portion below the origin is the **negative** direction.

The system of the two axes just described is called a **coordinate system of axes (or coordinate lines)**. The axis Ox is referred to as the **horizontal axis or just the x – axis**, and the axis Oy is referred to as the **vertical axis or just the y – axis**.

Given an **ordered pair** of real numbers (x, y) we may associate **a unique point M on the xOy plane** in the following way: we locate x on the horizontal axis and draw a line perpendicular to this axis at the point x; then we locate y on the vertical axis and draw a line perpendicular to this axis at the point y. **The intersection of these two lines determines a point M on the plane, used to represent the ordered pair (x, y),** (see Fig. 1-7). To denote that the point M corresponds to the ordered pair (x, y) we write $M(x, y)$. The number x is called the **abscissa** of M while the number y is called the **ordinate** of M. Notice that the points $M(x, y)$ and $N(y, x)$ are different points.

The two axes in Fig. 1-7 divide the plane into four **disjoint regions**, called **quadrants**. The **first quadrant** includes all the points (x, y) such that $x > 0$ and $y > 0$; the **second quadrant** includes all the points (x, y) such that $x < 0$ and $y > 0$; the **third quadrant** includes all the points (x, y) such that $x < 0$ and $y < 0$ and the **fourth quadrant** includes all the points (x, y) such that $x > 0$ and $y < 0$.

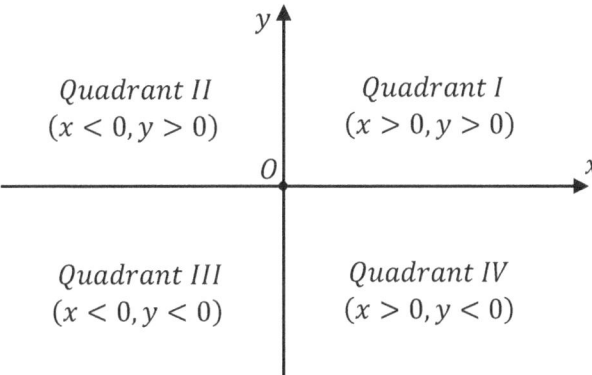

Fig. 1-8: The four quadrants in the plane.

Example 1-7-1: Graph the set of points, $A(3,2), B(-3,2), C(-3,-2)$ and $D(3,-2)$.

Solution

We first set up our axes, as shown in Fig. 1-9. The origin O corresponds to the pair $(0,0)$.

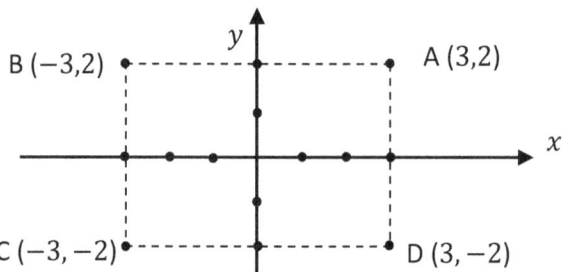

Fig. 1-9: Graph of points in the Cartesian plane.

Example 1-7-2: What are the set of points: $\{(x,0), x > 0\}, \{(x,0), x < 0\}, \{(0,y), y > 0\}, \{(0,y), y < 0\}$.

Solution

$\{(x,0), x > 0\}$, the positive x semi-axis,

$\{(x,0), x < 0\}$, the negative x semi-axis,

$\{(0,y), y > 0\}$, the positive y semi-axis,

$\{(0,y), y > 0\}$, the negative y semi-axis

PROBLEMS

1-7-1) Graph the set of points $A(1,4)$ and $B(4,1)$.

1-7-2) Determine a and b if $(2,a) = (b,-7)$.

1-7-3) Graph the set of points $A(4,3), B(-4,3), C(-4,-5)$ and $D(4,-5)$. What kind of figure is $ABCD$.

(**Ans:** Square).

1-7-4) Which of the following statements are true and which are false? Justify your answer: **a)** $(x, y) = (x, y)$, **b)** $\{x, y\} = \{y, x\}$, **c)** $(x, y) = (y, x)$, **d)** $(x, y) = (y, x)$ if and only if $x = y$.

1-7-5) Graph the set of points and indicate what quadrant, if any, contains each of the points: **a)** $(2, -3)$, **b)** $(1, 0)$, **c)** $(-7, -1)$, **d)** $(-5, 7)$, **e)** $(3, 4)$, **f)** $(-3, 0)$, **g)** $(4, -1)$, **h)** $O(0, 0)$.

(**Ans: a)** Quadrant IV, **c)** Quadrant III, **d)** Quadrant II, **e)** Quadrant I, **g)** Quadrant IV).

1-8) Graphs of functions

a) Let $y = f(x)$ be a real function of the real variable x. The real variable x, (the independent variable) takes values within some interval $[a, b]$, i.e. $a \leq x \leq b$.

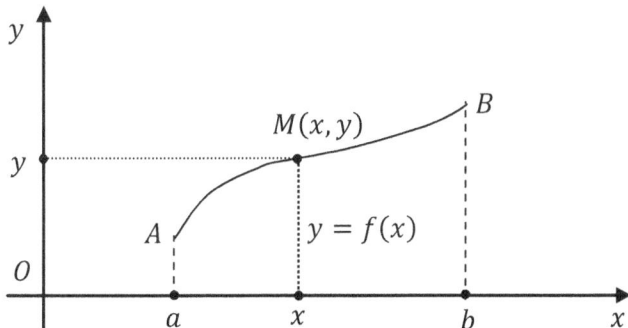

Fig. 1-10: The graph of a function.

Let us consider a point M whose abscissa is x, (a value between a and b) and whose ordinate is $y = f(x)$, i.e. the value of the function at the point x. We denote this point by writing $M = M(x, y) = M(x, f(x))$. As the independent variable x varies continuously within its domain of definition, the corresponding point M is moving on the plane xOy, tracing a curve AB, which is called **the graph of the function** $y = f(x)$. Every point on this curve, (**the graph of the function**) has abscissa equal to x and ordinate equal to the corresponding value of the function $y = f(x)$. In brief terms, **the graph of the function $y = f(x)$ is the set of points $M(x, f(x))$.**

The graph of a function provides **a geometrical picture** of the function, showing how the function behaves as the independent variable x varies continuously from a to b.

To construct, **approximately**, the graph of a function $y = f(x)$ we may divide the interval $[a, b]$ into a large number of sub intervals by means of the dividing points $\{a = x_0, x_1, x_2, \cdots, x_{n-2}, x_{n-1}, x_n = b\}$, where $\{a = x_0 < x_1 < x_2 < \cdots < x_{n-1} < x_n = b\}$, find the corresponding values of the function $\{y_0 = f(x_0), y_1 = f(x_1), y_2 = f(x_2), \cdots, y_{n-1} = f(x_{n-1}), y_n = f(x_n)\}$ and then graph the set of points:
$M_0(x_0, y_0), M_1(x_1, y_1), M_2(x_2, y_2), \cdots, M_{n-1}(x_{n-1}, y_{n-1}), M_n(x_n, y_n)$.

Evidently, as the number of dividing points increases, the approximate graph of the function becomes more and more accurate.

b) Increasing and decreasing functions

A function $y = f(x)$ defined in the interval $a \leq x \leq b$ is said to be **increasing** in $[a, b]$ if, as x increases, the corresponding values of the function $y = f(x)$ increase as well.

A function $y = f(x)$ defined in the interval $a \leq x \leq b$ is said to be **decreasing** in $[a, b]$ if, as x increases, the corresponding values of the function decrease.

A function $y = f(x)$ defined in the interval $a \leq x \leq b$ is said to be **constant** in $[a, b]$ if, the function takes on the same value, say c, for all values of $x \in [a, b]$.

In summary:

$$\begin{cases} y = f(x) \ \ increasing \ \ if \ f(x_1) < f(x_2) \ whenever \ x_1 < x_2 \\ y = f(x) \ \ decreasing \ \ if \ f(x_1) > f(x_2) \ whenever \ x_1 < x_2 \\ y = f(x) \ \ constant \ \ if \ f(x) = c \ \ \forall x \in [a, b] \end{cases} \quad (1-8-1)$$

The graph of an increasing function is a curve **ascending** as we move from the left towards the right; the graph of a decreasing function is a curve **descending** as we move from the left towards the right; the graph of a constant function is **a straight line parallel to the x axis**.

Finding the intervals over which a given function is either increasing or decreasing or remains constant, is termed "**the study of the function**".

Differential calculus, a branch of applied mathematics, develops powerful methods and techniques to investigate and study complicated functions.

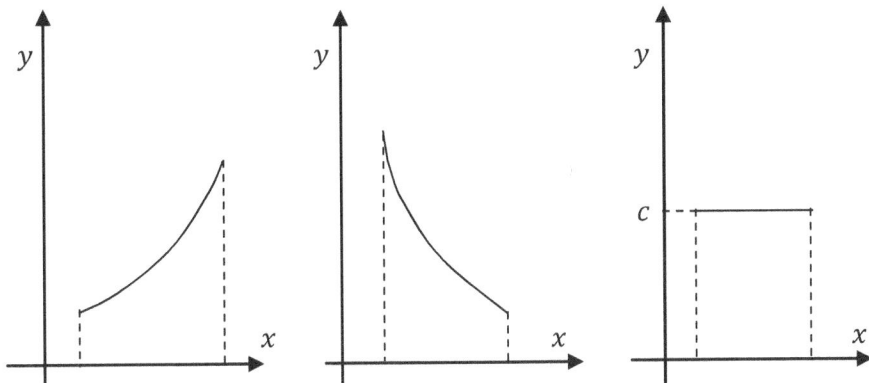

Fig. 1-11: Functions: increasing, decreasing, constant.

c) **Equation of a curve** (c)

Point satisfying an equation: We say that a point $M(\xi, \eta)$ satisfies the equation $f(x, y) = 0$ when the pair of coordinates (ξ, η) of the point M is a solution of the equation $f(x, y) = 0$, i.e. **when $f(\xi, \eta) = 0$, $((\xi, \eta)$ satisfies the equation $f(x, y) = 0$)**.

Equation of a curve: We say that a plane curve (c) (i.e. a curve lying on the xOy plane) has equation $f(x, y) = 0$, when all the points of the curve and only these points satisfy the equation $f(x, y) = 0$. This means that every point $M(\xi, \eta)$ of the curve satisfies $f(\xi, \eta) = 0$ and conversely, every point $M(\xi, \eta)$ satisfying the equation $f(x, y) = 0$ lies on the curve (c).

We say that **the curve (c) is represented by the equation** $f(x,y) = 0$. The branch of applied mathematics which studies geometrical curves by means of their corresponding equations is called **Analytic Geometry**. In Analytic Geometry we use algebraic methods and techniques to investigate geometrical properties of plane figures.

The simplest geometrical figure, **the straight line**, is studied in the next section.

Example 1-8-1: Which of the following points belong to the curve described by the equation $2x + 3y - 5 = 0$?

$$A(1,1), B\left(0, \frac{5}{3}\right), C\left(2, \frac{2}{3}\right), D\left(\frac{5}{2}, 0\right), E(10, -5)$$

Solution

A point belongs to a curve if and only if its coordinates satisfy the equation of the curve.

1) Point A: $2 \cdot 1 + 3 \cdot 2 - 5 = 0$, the point lies on the curve.

2) Point B: $2 \cdot 0 + 3 \cdot \frac{5}{3} - 5 = 0$, the point lies on the curve.

3) Point C: $2 \cdot 2 + 3 \cdot \frac{2}{3} - 5 = 1 \neq 0$, the point does not lie on the curve.

4) Point D: $2 \cdot \frac{5}{2} + 3 \cdot 0 - 5 = 0$, the point lies on the curve.

5) Point E: $2 \cdot 10 + 3 \cdot (-5) - 5 = 0$, the point lies on the curve.

Example 1-8-2: Show that the points $\{x = 2\cos t, y = 2\sin t\}$ where $0 \leq t < 2\pi$, lie on the curve $x^2 + y^2 - 4 = 0$, (this is the equation of **a circle** centered at the origin and having radius 2).

Solution

It suffices to show that the given set of points satisfy the equation of the curve.

$$x^2 + y^2 - 4 = (2\cos t)^2 + (2\sin t)^2 - 4 = 4\underbrace{\{(\cos t)^2 + (\sin t)^2\}}_{=1} - 4 = 0$$

PROBLEMS

1-8-1) Which of the following points belong to the curve $\sqrt[3]{x} + \sqrt[3]{y} = 3$?

$$A(1,8), B(0,27), C(27,-8), D(-27,-64), E(8,1)$$

(**Ans:** A, B, E).

1-8-2) Make a rough graph of the function $y = f(x)$ in the interval $[-4,7]$, given that: $f(-4) = 5, f(-3) = 4, f(-2) = 3, f(-1) = 2, f(0) = 1, f(1) = 0, f(2) = -2, f(3) = -3, f(4) = -3, f(5) = -2, f(6) = 0, f(7) = 3$.
Indicate the intervals where the function is increasing, decreasing or constant.

1-9) Linear equations (First degree equations in two variables)

Equations such as $2y = 5x - 7$ or $3x - 8y + 2 = 0$ are examples of first degree equations in two variables (unknowns) x and y. In general, any equation of the form $ax + by + c = 0$, where a, b, c are constant numbers and x and y are variables, is called **a first degree equation in two variables or a linear equation in two variables**. We assume that a and b are not zero simultaneously. **If at least one of the variables x or y is raised to a power other than one, then the equation is not linear.** For example, the equations $y - x^2 = 0$, $\sqrt{y} - 3x = 5$ are not linear; the first one because of the term x^2, the second one because of the term $\sqrt{y} = y^{\frac{1}{2}}$.

Any linear equation of the form $ax + by + c = 0$ can be cast in the equivalent form $y = kx + \lambda$. Indeed, from $ax + by + c = 0$ we have, $by = -ax - c$, and assuming that $b \neq 0$, $y = \underbrace{-\left(\frac{a}{b}\right)}_{k} x + \underbrace{\left(-\frac{c}{b}\right)}_{\lambda} = kx + \lambda$.

It can be shown that **the graph of any linear equation $ax + by + c = 0$ is a straight line**. Since a straight line **is completely determined by two distinct points**, it suffices to find two points of the line, connect them by a straight line, and then this line will be the graph of $ax + by + c = 0$, (see Example 1-9-1).

Let us consider two straight lines $(\varepsilon_1): y = k_1 x + \lambda_1$ and $(\varepsilon_2): y = k_2 x + \lambda_2$. The number k_1 is called **the slope** of the line (ε_1) and the number k_2 is **the slope** of the line (ε_2). It can be shown that:

1) If $k_1 = k_2$ then the two lines are **parallel**, $((\varepsilon_1) \parallel (\varepsilon_2))$,

2) If $k_1 k_2 = -1$, then the two lines are **perpendicular**, $((\varepsilon_1) \perp (\varepsilon_2))$.

For the **geometrical meaning of the slope**, see Example 1-9-2. For a proof of (1) and (2), see Example 1-9-3.

A straight line passing through the origin $O(0,0)$ has the general form $y = kx$, (since $x = 0, y = 0$ satisfies the equation of the straight line).

In general, to find the slope of the straight line $ax + by + c = 0$, **we solve this equation for y, and then the coefficient of x (i.e. the number which multiplies x), will be the slope of the line**. For example, to find the slope of $2x - 3y + 1 = 0$, we solve for y, i.e. $y = \frac{2}{3}x + \frac{1}{3}$ and the slope is $\frac{2}{3}$; to find the slope of $5x + 8y = 2$ we solve for y, i.e. $y = -\frac{5}{8}x + \frac{2}{8}$ and the slope is $\left(-\frac{5}{8}\right)$, etc. In general, the slope of $ax + by + c = 0$ is $\left(-\frac{a}{b}\right)$, assuming that $b \neq 0$. If $b = 0$, the equation of the straight line becomes $ax + c = 0$, and assuming that $a \neq 0$, this equation represents a straight line, perpendicular to the x axis at $x = -\frac{c}{a}$, or the same, a straight line parallel to the y axis at $x = -\frac{c}{a}$.

Example 1-9-1: Graph of the function $2x + 3y = 12$. Which ones of the points $A\left(1, \frac{10}{3}\right), B\left(2, \frac{7}{3}\right), C\left(-1, \frac{14}{3}\right), D\left(50, -\frac{88}{3}\right)$ belong to this line?

Solution

The easiest way to determine two points of the straight line is the following: We set $x = 0$ and then from the equation of the straight line we find $y = 4$. This means that the point $K(0, 4)$ belongs to the line. Then we set $y = 0$ and then from the equation of the line we find $x = 6$, and this shows that the point $L(6, 0)$ is another point of the line. The line determined by the segment KL is the sought for graph of the line $2x + 3y = 12$. **Notice that the point $K(0, 4)$ lies on the y axis, while the point $L(6, 0)$ lies on the x axis.**

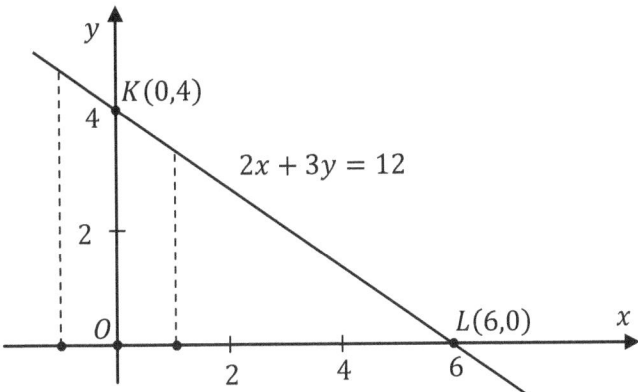

Fig. 1-12: Graph of the function $2x + 3y = 12$.

A point belongs to a curve, if the coordinates of the point satisfy the equation of the curve.

Point $A\left(1, \frac{10}{3}\right)$: $2 \cdot 1 + 3 \cdot \frac{10}{3} = 2 + 10 = 12$, the point A belongs to the line.

Point $B\left(2, \frac{7}{3}\right)$: $2 \cdot 2 + 3 \cdot \frac{7}{3} = 4 + 7 = 11 \neq 12$, the point B does not belong to the line.

Point $C\left(-1, \frac{14}{3}\right)$: $2 \cdot (-1) + 3 \cdot \frac{14}{3} = -2 + 14 = 12$, the point C belongs to the line.

Point $D\left(50, -\frac{88}{3}\right)$: $2 \cdot 50 + 3 \cdot \left(-\frac{88}{3}\right) = 100 - 88 = 12$, the point D belongs to the line.

Example 1-9-2: The geometric meaning of the slope.

In this Example and in the next Example, the reader is supposed to be familiar with some simple Trigonometry. Let us consider a straight line (ε) with equation $ax + by + c = 0$, as shown in Fig. 1-13.

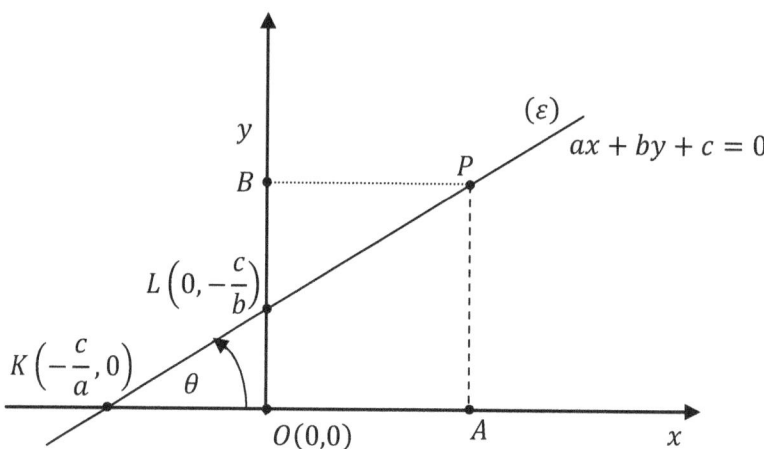

Fig. 1-13: The slope of a straight line.

To find the **x intercept** of the graph, i.e. the point where the line crosses the x axis, we set $y = 0$ in the equation of the line, and we get the corresponding value of $x = -c/a$; similarly, to find the **y intercept**, we set $x = 0$ in the equation of the line, and find $y = -c/b$. So, the points $K\left(-\frac{c}{a}, 0\right)$ and $L\left(0, -\frac{c}{b}\right)$ are **the x and y intercepts**, respectively, of the line.

Let us call θ the angle formed by the line (ε) and the positive x semi axis; From the right triangle KOL we have, (notice that the length $|OL| = -\frac{c}{b}$ and the length $|OK| = -\left(-\frac{c}{a}\right) = \frac{c}{a}$),

$$\tan\theta = \frac{|OL|}{|OK|} = \frac{-\frac{c}{b}}{-\frac{c}{a}} = -\frac{a}{b} \tag{$*$}$$

The slope of the straight line (ε) is the tangent of the angle θ, and this coincides with the coefficient of x, when the equation of the line is solve for y, $\left(y = -\frac{a}{b}x - \frac{c}{b}\right)$.

The slope of an **horizontal** line is 0, (since in this case $\theta = 0$ and $\tan 0 = 0$). The slope of a **vertical** line is undefined, (since in this case $\theta = 90°$ and the tangent of $90°$ is not defined).

The slope of an increasing linear function is a positive number, while the slope of a decreasing linear function is a negative number.

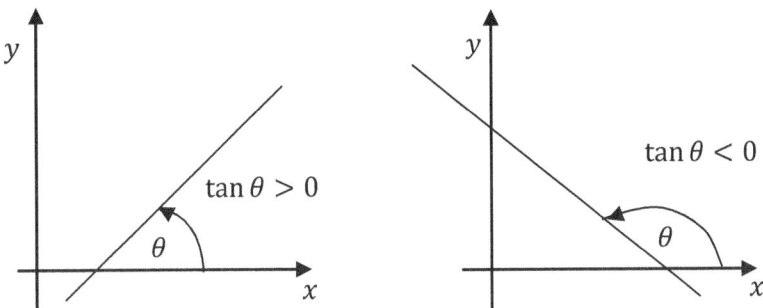

Fig. 1-14: Slope of increasing function-decreasing function.

In the case of an increasing function, $0 < \theta < 90°$, $\tan\theta > 0$ and therefore, the slope is positive; in the case of a decreasing function, $90° < \theta < 180°$, $\tan\theta < 0$ and the slope is negative.

If the equation of a straight line is $y = kx + \lambda$, then obviously, the slope of the line is k and the y intercept of the line is λ.

Example 1-9-3: Parallel and perpendicular straight lines.

Let $y = k_1 x + \lambda_1$ and $y = k_2 x + \lambda_2$ be the equations of two straight lines (ε_1) and (ε_2) respectively.

a) If the two lines are parallel, then they form **the same angle θ** with the positive x axis, and therefore, they must have the same slope, i.e. $k_1 = k_2 = \tan\theta$.

b) Suppose now that the two lines are perpendicular, i.e. suppose that $((\varepsilon_1) \perp (\varepsilon_2))$, as shown in Fig. 1-15, (we assume that the two lines are not parallel to the coordinate axes Ox and Oy, since in this case the slope of one of the two lines- the one parallel to the Oy axis-is not defined).

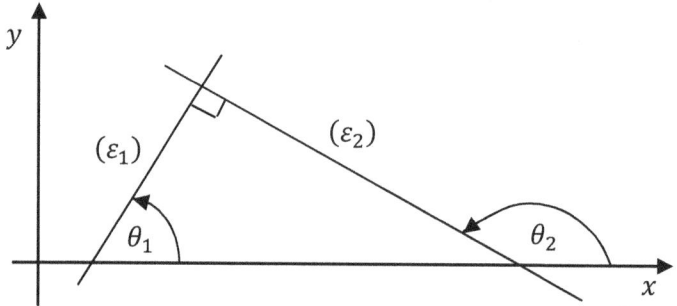

Fig. 1-15: Perpendicular straight lines.

From the right triangle in the figure, we have:

$$\theta_2 = 90° + \theta_1 \Rightarrow \theta_2 + (-\theta_1) = 90° \Rightarrow$$

$$\tan \theta_2 = \cot(-\theta_1) = -\cot \theta_1 = -\frac{1}{\tan \theta_1} \Rightarrow$$

$$(\tan \theta_1) \cdot (\tan \theta_2) = -1 \xrightarrow{(k_1 = \tan \theta_1)(k_2 = \tan \theta_2)} k_1 k_2 = -1$$

Example 1-9-4: Show that the straight lines $2x + 5y - 7 = 0$ and $5x - 2y + 3 = 0$ are perpendicular.

Solution

To slope of the first line is found if we solve for y, i.e. $y = -\frac{2}{5}x + \frac{7}{5}$. The slope is $k_1 = -\frac{2}{5}$. The slope of the second line is found similarly, i.e. $y = \frac{5}{2}x + \frac{3}{2}$, and the slope is $k_2 = \frac{5}{2}$. The product of the slopes is $k_1 k_2 = -1$, and this shows that the two lines are perpendicular.

PROBLEMS

1-9-1) Show that the straight lines $3x + 5y = 1$ and $6x + 10y + 12 = 0$ are parallel.

1-9-2) Show that the straight lines $x - y + 7 = 0$ and $x + y = 2$ are perpendicular.

1-9-3) Find the x and the y intercepts of the lines: $x + 2y = 0$, $4x - 3y + 5 = 0$, $3x - 7y + 12 = 0$.

(Ans: $\{O(0,0)\}, \left\{A\left(-\frac{5}{4},0\right), B\left(0,\frac{5}{3}\right)\right\}, \left\{A(-4,0), B\left(0,\frac{12}{7}\right)\right\}$).

1-9-4) For what values of m are the following two straight lines parallel?
$(\varepsilon_1): 2mx + 3y - 7 = 0$ and $(\varepsilon_2): mx - 2y + 5 = 0$.

(Ans: $m = \pm\sqrt{3}$).

1-9-5) For what values of m are the following straight lines perpendicular?
$(\varepsilon_1): 3mx + 3y - 7 = 0$ and $(\varepsilon_2): 9mx - y + 5 = 0$.

(Ans: $m = \pm 1/3$).

1-9-6) What are the integer and positive solutions of the equation $2x + 3y = 13$.

(Ans: $(x = 2, y = 3), (x = 5, y = 1)$).

CHAPTER 2: FIRST DEGREE EQUATIONS IN ONE UNKNOWN

2-1) Fundamental concepts and definitions

a) Expressions of the form

$$2x - 6 = 0, \qquad 3(x + 7) + 5 = 5x - 2, \qquad x^2 - 16 = 0, \qquad x^3 - 8 = 0$$

are examples of **equations in one unknown** x. The first and second equations are **first degree** equations, the third equation is a **second degree** equation (or **quadratic**) and the last one is a **third degree** equation (or **cubic equation**). **An equation states that two algebraic expressions are equal**. These expressions are known as the **left-hand side** and the **right-hand side** of the equation. In the second equation, the left-hand side is $3(x + 7) + 5$ and the right-hand side is $5x - 2$.

Solving an equation means to find the value(s) of x for which the equation holds true. A number that satisfies an equation is called **a solution** or **a root** of the equation. For example, $x = 1$ is a root of $2(x + 1) = 4$, since $2 \cdot (1 + 1) = 4$ is true; the numbers $x = 2$ and $x = 3$ are roots of the second degree equation $x^2 - 5x + 6 = 0$, since $2^2 - 5 \cdot 2 + 6 = 0$ and $3^2 - 5 \cdot 3 + 6 = 0$.

The set of all the solutions (roots) of an equation constitute the **solution set** of the equation. The solution set, for instance, of $x^2 - 5x + 6 = 0$ is $\{2,3\}$.

In general, if $F(x)$ and $G(x)$ are two functions with the same domain of definition, then we say that we have to solve the equation $\boldsymbol{F(x) = G(x)}$ when we seek to determine all the values of x for which the equation holds true.

If x takes a value, say $x = a$, then the **two numbers** $F(a)$ and $G(a)$, in general will not be equal, (will be different); the problem therefore is, for which a we have $F(a) = G(a)$?

The variable x is called "**the unknown of the equation**" $F(x) = G(x)$. When **both** functions $F(x)$ and $G(x)$ are polynomials, then the equation $F(x) = G(x)$ is called "**a polynomial equation**". The highest exponent of x appearing in the equation is called "**the degree of the equation**". For example, $x^5 + 2x^3 = x^2 - 5$ is a fifth degree equation, $x^3 + 3x + 7 = x^7 + 5x^4 - 3$ is a seventh degree equation, etc.

The equation $2^x + 1 = 5x + 3$ is **not** a polynomial equation, since the left-hand side is not a polynomial, (this equation is an "**exponential equation**", since the unknown x appears as an exponent).

b) Identities and impossible equations: An equation which is true for all values of the unknown x is called an **identity**. For example, $x + 1 = x + 1$, $(x - 1)^2 = x^2 - 2x + 1$ are identities, since they are satisfied for **all** values of x. On the other hand, we may come across to an equation, which has no solutions at all, i.e. there are no values of x which satisfy the equation. In such cases we say that the equation is **impossible**. For example, $x + 5 = x + 4$, $0 \cdot x = 5$, are impossible equations.

2-2) Equivalent equations

a) Two equations are called **equivalent**, when every solution of the first equation is a solution of the second and conversely, every solution of the second equation is a solution of the first. In other words, **two equations are equivalent when they have the same solution set**.

For example the equations $5x - 2 = 2x + 1$ (first equation) is equivalent to $3x = 3$ (second equation), since both are satisfied by $x = 1$, i.e. the solution set of both equations is $\{1\}$. Similarly the equations $x^2 - 5x + 6 = 0$ (first equation) and $2x^2 - 2x = x^2 + 3x - 6$ (second equation) are equivalent since they have the same solution set $\{2,3\}$ (check it).

Let us consider the equations $2x + 1 = 5$ and $x^2 - 5x + 6 = 0$. The solution set of the first one is $\{2\}$ and of the second one is $\{2,3\}$. These two equations do have one common root ($x = 2$) but they are **not equivalent** since the root $x = 3$ of the second is not a root of the first.

It is obvious that, when we try to solve an equation, it is desirable to replace the given equation by **an equivalent** one, which however is of a simpler form.

The following two theorems are used frequently, in order to replace an equation by an equivalent one.

Theorem 2-2-1: Let us consider an equation $F(x) = G(x)$. If we add or subtract the same function $A(x)$ to both sides of the equation we get an equation equivalent to the given. In symbols:

$$F(x) = G(x) \iff F(x) + A(x) = G(x) + A(x) \qquad (2-2-1)$$

(The symbol $\iff$ is read "**is equivalent to**").

Proof: Let r be a root (solution) of $F(x) = G(x)$. This means that
$$F(r) = G(r) \implies F(r) + A(r) = G(r) + A(r)$$
and this shows that r is also a root of $F(x) + A(x) = G(x) + A(x)$.

Conversely, assuming that a is a root of $F(x) + A(x) = G(x) + A(x)$, then
$$F(a) + A(a) = G(a) + A(a) \implies F(a) = G(a)$$
and this shows that a is a root of $F(x) = G(x)$.

We have thus shown that every solution of $F(x) = G(x)$ is a solution of $F(x) + A(x) = G(x) + A(x)$, and conversely, every root of the second is a solution of the first, i.e. these two equations are **equivalent**.

For example, the equations $\underbrace{x^3 + 3}_{F(x)} = \underbrace{2x^2 - 5}_{G(x)}$ and

$\underbrace{x^3 + 3}_{F(x)} + \underbrace{2x^4 - 5x + 3}_{A(x)} = \underbrace{2x^2 - 5}_{G(x)} + \underbrace{2x^4 - 5x + 3}_{A(x)}$ are equivalent.

Theorem 2-2-2: **Let us consider an equation $F(x) = G(x)$. If we multiply both sides of the equation by the same number $k \neq 0$, we get an equation equivalent to the given. In symbols:**

$$F(x) = G(x) \iff kF(x) = kG(x), \quad k \neq 0 \qquad (2-2-2)$$

Proof: If r is a root of $F(x) = G(x)$, then $F(r) = G(r)$ and therefore, for any $k \neq 0$, $kF(r) = kG(r)$, i.e. r is a root of $kF(x) = kG(x)$.

Conversely, assuming that a is a root of $kF(x) = kG(x), \; k \neq 0$, then $kF(a) = kG(a)$, which implies that $F(a) = G(a)$, and this means that a is a root of $F(x) = G(x)$.

We have thus shown that every solution of $F(x) = G(x)$ is a solution of $kF(x) = kG(x), k \neq 0$, and conversely, every root of $kF(x) = kG(x)$ is a solution of $F(x) = G(x)$, i.e. these two equations are **equivalent**.

For example, the equations $x^2 + 5 = x + 7$ is equivalent to the equation $4 \cdot (x^2 + 5) = 4 \cdot (x + 7)$.

b) Transfer of a term from one side to the other: According to Th. (2-2-1) the equation $f(x) = g(x) + w(x)$ is equivalent to the equation $f(x) - w(x) = g(x)$, which is obtained from the first if we add the function $(-w(x))$ to both sides, i.e.

$$f(x) = g(x) + w(x) \Leftrightarrow f(x) - w(x) = g(x) + \underbrace{w(x) - w(x)}_{=0} \Rightarrow$$

$$f(x) = g(x) + w(x) \Leftrightarrow f(x) - w(x) = g(x) \qquad (*)$$

Formula (*) shows that, given an equation, we obtain an **equivalent** equation if we **transfer a term** from one side to the other, (from the left-hand side to the right-hand side or the other way around), provided that **we change the sign of the transferred term** (the plus becomes minus and the minus becomes plus).

For example, from $x^3 + 5x = 4x^2 + 7$ we obtain the **equivalent** equation $x^3 - 4x^2 + 5x = 7$, (by transferring $4x^2$ from the right-hand side to the left-hand side). From $x + 2 = 5 - 3x^5$ we obtain the equivalent equation $x + 2 - 5 + 3x^5 = 0$, etc.

Note: Any polynomial equation of the form $A(x) = B(x)$ can be written as $F(x) = 0$, **where $F(x)$ is a polynomial in x**. This is obvious, since from $A(x) = B(x)$ we obtain the equivalent equation $A(x) - B(x) = 0$, or if we call $F(x) \stackrel{\text{def}}{=} A(x) - B(x)$, $F(x) = 0$. **The degree of $F(x)$ is the degree of the equation**. For example:

1) $x^3 - 5x = 3x^2 + 1 \Leftrightarrow x^3 - 3x^2 - 5x - 1 = 0$, **Third** degree equation,

2) $x^5 + 6x^3 + 5x = 4x^3 - x + 2 \Leftrightarrow x^5 + 6x^3 + 5x - (4x^3 - x + 2) = 0 \Leftrightarrow x^5 + 2x^3 + 6x - 2 = 0$, **Fifth** degree equation,

3) $x^2 + 2x - 5 = x^2 + x - 9 \Leftrightarrow x^2 + 2x - 5 - (x^2 + x - 9) \Leftrightarrow x + 4 = 0$, **First** degree equation.

In the last equation, the term x^2 eventually disappears, and the resulting equation is a first degree equation.

c) Elimination of the denominators: In case the coefficients of an equation are rational numbers (**fractions**) then the equation can be transformed to an equivalent equation with **integer coefficients** if we multiply **both sides** of the equation by **a common multiple of the denominators** (application of Theorem 2-2-2). Let us for example consider the equation,

$$\frac{1}{2}x - 2 = \frac{3}{4}x + 5$$

If we multiply both sides by 4 we get an equivalent equation (Th. 2-2-2), i.e.

$$\frac{1}{2}x - 2 = \frac{3}{4}x + 5 \Leftrightarrow 4 \cdot \left(\frac{1}{2}x - 2\right) = 4 \cdot \left(\frac{3}{4}x + 5\right) \Leftrightarrow 2x - 8 = 3x + 20$$

$$\Leftrightarrow 2x - 3x = 20 + 8 \Leftrightarrow -x = 28 \stackrel{*(-1)}{\Leftrightarrow} x = -28$$

and this is the solution of the equation. Let the reader verify that $x = -28$ satisfies the original equation, (both sides of the original equation are equal to -16).

Another example: Solve the equation $\frac{1}{5}x + \frac{2}{3} = \frac{1}{2}x - 1$.

$$\frac{1}{5}x + \frac{2}{3} = \frac{1}{2}x - 1 \Leftrightarrow 30 \cdot \left(\frac{1}{5}x + \frac{2}{3}\right) = 30 \cdot \left(\frac{1}{2}x - 1\right) \Leftrightarrow$$

$$6x + 20 = 15x - 30 \Leftrightarrow 6x - 15x = -30 - 20 \Leftrightarrow -9x = -50 \Leftrightarrow$$

$$\left(-\frac{1}{9}\right) \cdot (-9x) = \left(-\frac{1}{9}\right) \cdot (-50) \Leftrightarrow x = \frac{50}{9}$$

2-3) First degree equations

A first degree equation is any equation of the form $ax + b = 0$, where a, b, are constant numbers (not depending on x). A first degree equation is also called **a linear equation**.

The usual procedure to solve any, first degree equation is the following:

1) If the equation contains **fractional coefficients**, we multiply both sides by a common multiple of the denominators (preferably **the least common multiple**), in order to get rid of the denominators. The coefficients of the equation thus obtained are now integers.

2) We transfer **all the unknowns** in one side (say the left-hand side) and all the known quantities (numbers) in the right-hand side. Each time we transfer a term (unknown or constant number) from one side to the other, **we change its sign**.

3) Divide both sides of the equation by **the coefficient** of x, to obtain the root (solution) of the equation.

Notice that these three steps are nothing else but repeated applications of Theorems 2-2-1 and 2-2-2, which warranty the equivalency of the equations from one step to the next.

Any first degree equation may have **only one** solution or **an infinite number** of solutions (identity) or even may have **no solution** at all (impossible equation). This depends, of course, on the coefficients a and b.

Indeed let us consider the equation $ax + b = 0$, (any first degree equation can be expressed in this form). We have:

$$ax + b = 0 \Leftrightarrow ax = -b \qquad (*)$$

There are three cases:

1) If $a \neq 0$ then the equation has a **unique solution:** $x = -\dfrac{b}{a}$

2) If $a = 0$ and $b = 0$, equation (*) becomes, $0 \cdot x = 0$ and is satisfied for all values of x, i.e. the solution set of (*) is the set of real numbers, (the equation is an **identity**, satisfied for all values of x).

3) If $a = 0$ and $b \neq 0$, equation (*) becomes, $0 \cdot x = b \neq 0$, and this is an **impossible** equation (since any number when multiplied by zero gives zero).

Literal equations: These are equations, in which the constants are represented by letters, called "**parameters**". For example, $kx + 3 = x - 5k$ is a literal equation, **where the parameter k is supposed to be a known number which does not depend on x**. The solution of this equation will be expressed in terms of k. Similarly, equation $k + \lambda x = (k - \lambda)x - k\lambda$ involves **two parameters** k and λ, and its solution will be expressed in terms of k and λ.

Example 2-3-1: Solve the equation $3x - 5 = x + 7$.

Solution

$$3x - 5 = x + 7 \Leftrightarrow 3x - x = 7 + 5 \Leftrightarrow 2x = 12 \Leftrightarrow x = \frac{12}{2} = 6$$

Example 2-3-2: Solve the equation $\frac{1}{3}x - 1 = \frac{1}{2}x + 2$.

Solution

$$\frac{1}{3}x - 1 = \frac{1}{2}x + 2 \Leftrightarrow 6 \cdot \left(\frac{1}{3}x - 1\right) = 6 \cdot \left(\frac{1}{2}x + 2\right) \Leftrightarrow$$

$$2x - 6 = 3x + 12 \Leftrightarrow 2x - 3x = 12 + 6 \Leftrightarrow -x = 18 \Leftrightarrow x = -18$$

Example 2-3-3: Solve the equation $2x + 3 = 2x + 3$.

Solution

$$2x + 3 = 2x + 3 \Leftrightarrow 2x - 2x = 3 - 3 \Leftrightarrow 0 \cdot x = 0$$

and this is **an identity**, satisfied by any number x.

Example 2-3-4: Solve the equation $3x - 7 = 3x + 2$.

Solution

$$3x - 7 = 3x + 2 \Leftrightarrow 3x - 3x = 2 + 7 \Leftrightarrow 0 \cdot x = 9$$

and this is **an impossible** equation, (since the multiplication of any number by zero is zero, and cannot be 9).

Example 2-3-5: Assuming that $x = 2$ is a root of the equation $cx + 2 = 9c - x$, find the constant c.

Solution

Since, by assumption, $x = 2$ is a root of the equation, it must satisfy the equation, i.e. $2c + 2 = 9c - 2$, and this is another equation where the **unknown is c**, i.e.

$$2c + 2 = 9c - 2 \Leftrightarrow 2c - 9c = -2 - 2 \Leftrightarrow -7c = -4 \Leftrightarrow c = \frac{4}{7}$$

Example 2-3-6: Solve the equation for x, $\frac{bx+2}{3} = \frac{x-1}{2}$, where b is considered to be a known real number (**a parameter**), and investigate for the various values of b.

Investigate means to find the values of b for which the equation has a unique solution or becomes an identity or becomes an impossible equation.

Solution

$$\frac{bx+2}{3} = \frac{x-1}{2} \Leftrightarrow 6 \cdot \left(\frac{bx+2}{3}\right) = 6 \cdot \left(\frac{x-1}{2}\right) \Leftrightarrow$$

$$2bx + 4 = 3x - 3 \Leftrightarrow 2bx - 3x = -3 - 4 \Leftrightarrow (2b-3)x = -7 \quad (*)$$

1) If $(2b-3) \neq 0$, i.e. if $b \neq \frac{3}{2}$ then the unique solution of the equation is $x = -\frac{7}{2b-3}$.

2) If $(2b-3) = 0$, i.e. if $b = \frac{3}{2}$ then equation (*) becomes $0 \cdot x = -7$ and this is an impossible equation.

Example 2-3-7: Assuming that a, b, c are known numbers (**parameters**) solve and investigate the equation $(a+b)x - c = (a-b)x + c$.

Solution

$$(a+b)x - c = (a-b)x + c \Leftrightarrow (a+b)x - (a-b)x = c + c \Leftrightarrow$$

$$ax + bx - ax + bx = 2c \Leftrightarrow 2bx = 2c \Leftrightarrow bx = c \quad (*)$$

1) If $b \neq 0$ equation (*) has a unique solution, $x = c/b$.

2) If $b = 0$ and $c = 0$, equation (*) becomes $0 \cdot x = 0$ and is an identity, (satisfied by all values of x).

3) If $b = 0$ and $c \neq 0$, equation (*) becomes $0 \cdot x = c \neq 0$, and this is an impossible equation.

Example 2-3-8: Assuming that a and b are known numbers (**parameters**) solve and investigate the equation $(1-x)(a-x) = (a-x)(1-b) - (1+x)(b-x)$.

Solution

We first simplify the given equation, as much as possible:

$$(1-x)(a-x) = (a-x)(1-b) - (1+x)(b-x) \Leftrightarrow$$

$$a - ax - x + x^2 = a - x - ab + bx - b - bx + x + x^2 \Leftrightarrow$$

$$-ax - x + x^2 + x - bx + bx - x - x^2 = a - ab - b - a \Leftrightarrow$$

$$-(a+1)x = -b(a+1) \Leftrightarrow (a+1)x = b(a+1) \qquad (*)$$

1) If $(a+1) \neq 0$ equation (*) has a unique solution, $x = \frac{b(a+1)}{a+1} = b$.

2) If $(a+1) = 0$, i.e. if $a = -1$, equation (*) becomes $0 \cdot x = 0$, and is an identity.

PROBLEMS

Solve the following equations:

2-3-1) $3x - 10 = 2x + 3$, (**Ans:** $x = 13$).

2-3-2) $2(x-1) + 3(x+4) = 2x + 5$.

2-3-3) $5(x-2) + 10(2-3x) + 10x = -(10+2x)$, (**Ans:** $x = 20/13$).

2-3-4) $(1+x)(2-3x) = (3x-1)(1-x)$

2-3-5) $\frac{5x-3}{2} - \frac{3x}{4} = x - 5$, (**Ans:** $x = -14/3$).

2-3-6) $\frac{8-x}{6} + \frac{x-1}{3} = \frac{x+6}{2} - \frac{x}{3}$, (**Ans:** Impossible).

2-3-7) $\frac{x+1}{4} - \frac{2x-1}{5} + \frac{3x+1}{2} = \frac{27x+19}{20}$, (**Ans:** Identity, true for all x).

Solve and investigate the following equations for the various values of the parameters involved:

2-3-8) $(a+x)(1+bx) = a(1+b) + a^2b^2 + bx^2$.

(**Ans:** If $(1+ab) \neq 0$, $x = ab$, if $(1+ab) = 0$ the equation becomes an identity, satisfied by all x).

2-3-9) $\frac{x}{a} + \frac{x}{b} = 1, (ab \neq 0)$,

(Ans: If $(a+b) \neq 0$, $x = ab/(a+b)$, if $(a+b) = 0$ impossible, no solution).

2-3-10) $\frac{ax}{2} - \frac{1}{3} = \frac{a}{3} + x$.

(Ans: If $a \neq 2$, $x = 2(a+1)/(3a-6)$, if $a = 2$ impossible).

2-3-11) $\frac{x}{(a-b)(a-c)} + \frac{x}{(b-a)(b-c)} + \frac{x}{(c-a)(c-b)} = 2$, **(Ans:** Impossible).

2-3-12) For what value of a the equation $\frac{ax-1}{3} + \frac{x+1}{2} = 4$ does not have a solution?

(Ans: $a = -3/2$).

2-3-13) If $F(x) = (5-3x)(x+4)$ and $G(x) = 2x^2 + 5$, find the value of x for which $2F(x) + 3G(x) = 0$.

(Ans: $x = 55/14$).

2-3-14) Assuming that $x = 1$ is a root of $kx^2 + 2kx + 8 = 5k + 2$, find k.

(Ans: $k = 3$).

Solve the following formulas for the indicated variables:

2-3-15) $\frac{1}{R} + \frac{1}{a} = \frac{1}{b}$, for R,   **(Ans:** $R = ab/(a-b)$).

2-3-16) $s = vt + \frac{1}{2}gt^2$, for v,   **(Ans:** $v = \left(s - \frac{1}{2}gt^2\right)/t$).

2-3-17) $V = \pi r^2 h$, for h,   **(Ans:** $h = V/(\pi r^2)$).

2-3-18) $C = \frac{5}{9}(F - 32)$, for F,   **(Ans:** $F = 32 + 9C/5$).

2-3-19) Solve for x and investigate the equation $(c-2)x = c^2 - 4$, (c is considered to be a known number).

(Ans: If $c \neq 2$, $x = c + 2$, if $c = 2$ the equation becomes an identity).

2-3-20) Solve for x and investigate the equation $(c^2 - 9)x = c^3 + 27$.

(**Ans:** If $c \neq \pm 3, x = (c^2 - 3c + 9)/(c - 3)$, if $c = -3$ the equation becomes an identity, if $c = 3$ the equation is impossible).

2-4) Factored equations

A factored equation is an equation of the form

$$f_1(x) \cdot f_2(x) \cdot f_3(x) \cdot \cdots \cdot f_n(x) = 0 \qquad (2-4-1)$$

where the functions $f_1(x), f_2(x), \cdots, f_n(x)$ are in general polynomials in x. In order equation (2-4-1) to be satisfied, it suffices that **at least one** of the factors in equation (2-4-1) be zero, since then the product will be zero. This means that the given equation **is equivalent** to the system of equations

$$f_1(x) = 0, \quad \text{or} \quad f_2(x) = 0, \quad \text{or} \quad \cdots \quad \text{or} \quad f_n(x) = 0 \qquad (2-4-2)$$

For example, the equation $(x - 2)(x + 1)(2x - 3) = 0$ is satisfied if $(x - 2) = 0$, i.e. $x = 2$, **or** if $(x + 1) = 0$, i.e. $x = -1$, **or** even if $(2x - 3) = 0$, i.e. if $x = 3/2$. The solution set of the equation is $\{2, -1, 3/2\}$.

As a matter of fact, this method (factoring) is a usual method of solving polynomial equations, i.e. we try to **factor out the given polynomial**, (with the aid of appropriate identities) and then **set each one of the factors equal to zero**.

In cases where one of the factors in (2-4-1) is impossible (let us for example assume that $f_1(x) = 0$ is impossible), then equation (2-4-1) is equivalent to the system $\{f_2(x) = 0, \cdots, f_n(x) = 0\}$. For example the roots of the equation

$$(x^2 + 10)(x - 5)(x + 3) = 0$$

are $x = 5$ and $x = -3$, since the expression $x^2 + 10$ is always **positive** for all values of the real variable x, i.e. $\forall x \in \mathbb{R}: x^2 + 10 \neq 0$.

A general remark: An equation $f(x) = 0$ may lead to different solution sets, depending on the set of numbers, within which we want to solve the equation.

For example, let us consider the equation

$$(x-1)(x+3)(2x-5)(x-\sqrt{2})(x^2+5) = 0$$

Within the set of **natural** numbers, the equation has **one** solution ($x = 1$), within the set of **integers** the same equation has **two** solutions ($x = 1, x = -3$), within the set of **rational** numbers the equation has **three** roots ($x = 1, x = -3, x = 5/2$) and within the set of **real** numbers the equation has **four** roots ($x = 1, x = -3, x = 5/2, x = \sqrt{2}$). Equation $x^2 + 5 = 0$ does not have a solution in the set of real numbers. However, as we shall see in Chapter 8, this equation does have solutions within a new set of numbers, called the set of **Complex numbers**.

For now, unless otherwise stated, we shall always assume that x is real.

Example 2-4-1: Solve the equation $(x+9)(x-1)(x+\sqrt{5}) = 0$.

Solution

$$(x+9)(x-1)(x+\sqrt{5}) = 0 \Leftrightarrow \begin{cases} x+9 = 0 \\ x-1 = 0 \\ x+\sqrt{5} = 0 \end{cases} \Rightarrow x \in \{-9, 1, -\sqrt{5}\}$$

Example 2-4-2: Factor the polynomial $x^2 - 16$ and then solve the equation $x^2 - 16 = 0$.

Solution

$$x^2 - 16 = x^2 - 4^2 = (x-4)(x+4)$$

$x^2 - 16 = 0 \Leftrightarrow (x-4)(x+4) \Leftrightarrow x = 4 \text{ or } x = -4$, (the roots of the equation).

Example 2-4-3: Solve the equation $x^2 - 6x + 5 = 0$ by factoring the left-hand side.

Solution

$$x^2 - 6x + 5 = 0 \Leftrightarrow x^2 \underbrace{-5x - x}_{-6x} + 5 = 0 \Leftrightarrow$$

$x(x-5) - (x-5) = 0 \Leftrightarrow (x-5)(x-1) = 0 \Leftrightarrow x = 5 \text{ or } x = 1$

Let the reader verify directly that both numbers 1 and 5 satisfy the original equation.

Example 2-4-4: Solve and investigate the literal equation $\{(a+b)x - c\}^2 = \{(a-b)x - c\}^2$, ($a, b, c$ are assumed to be known numbers).

Solution

$$\{(a+b)x - c\}^2 = \{(a-b)x - c\}^2 \Leftrightarrow$$

$$\{(a+b)x - c\}^2 - \{(a-b)x - c\}^2 = 0 \Leftrightarrow$$

$$\{(a+b)x - c + (a-b)x - c\} \cdot \{(a+b)x - c - (a-b)x + c\} = 0 \Leftrightarrow$$

$$(2ax - 2c)(2bx) = 0 \Leftrightarrow 4bx(ax - c) = 0 \Leftrightarrow bx(ax - c) = 0 \quad (*)$$

1) If $b = 0$ equation (*) is an identity, satisfied $\forall x \in \mathbb{R}$.

2) If $b \neq 0$ the equation in (*) is equivalent to $x(ax - c) = 0$.

a) If $a \neq 0$ and $c \neq 0$ we have two roots, $x = 0$ and $x = c/a$,

b) If $a = 0$ and $c \neq 0$ we have one root, $x = 0$,

c) If $a = 0$ and $c = 0$ the equation becomes an identity.

PROBLEMS

Solve the following equations:

2-4-1) $(x+3)(2x-1)(x+7) = 0$, (**Ans:** $-3, 1/2, -7$).

2-4-2) $(x-1)(x+2)(x-5)(x+3) = 0$.

2-4-3) $(x^2 + 1)(x + \sqrt{3})(x - \sqrt{5}) = 0$, (**Ans:** $-\sqrt{3}, \sqrt{5}$).

2-4-4) $x^2(x^2 + 5)(3x - 7) = 0$.

Solve the following equations by factoring:

2-4-5) $(x^2 - 9)(2x + 1) = 12(x - 3)$, (**Ans:** $1, 3, -9/2$).

2-4-6) $(2x + 4)(x - 3) = (2x + 4)(3x + 5)$.

2-4-7) $(x+2)^3 - (x-2)^3 = 32x + 16$, **(Ans: $0, 8/3$).**

2-4-8) $x^2 - 3x + 2 = 0$.

2-4-9) $(x^2 + 2x + 1)^2 = (x+1)^2(x+3)^2$, **(Ans: $-1, -2$).**

2-4-10) $3x^2 - 7x + 4 = 0$.

2-4-11) $x^2 + 13x + 42 = 0$, **(Ans: $-6, -7$).**

Hint: $13x = 7x + 6x$.

2-4-12) $2x^2 + 3x - 5 = 0$.

2-4-13) $(x^2 - 9)^2 - (x+3)^2 = 0$, **(Ans: $-3, 2, 4$).**

Solve for x and investigate the following equations, (a, b, c are constants):

2-4-14) $(x-a)^3 + (x-b)^3 + (x-c)^3 = 3(x-a)(x-b)(x-c)$

(Ans: If a, b, c not **all** equal to each other, $x = (a+b+c)/3$, if $a = b = c$ the equation is an identity, satisfied $\forall x \in \mathbb{R}$.

Hint: Apply **Cauchy's identity**,

$$A^3 + B^3 + C^3 - 3ABC = \frac{1}{2}(A+B+C)\{(A-B)^2 + (B-C)^2 + (C-A)^2\}$$

2-4-15) $(x-a)^3(x+a+2b) - (x+b)^3(x-2a-b) = 0$.

(Ans: If $(a+b) \neq 0$ the solution is $x = (a-b)/2$, if $a + b = 0$ the equation becomes an identity, satisfied by all numbers x.

Hint: Carrying out the calculations and simplifying, leads to the following expression, $(a+b)^3(2x+b-a) = 0$.

2-5) Equations with absolute values

Every equation of the form

$$cx + c_1|x - b_1| + c_2|x - b_2| + \cdots + c_n|x - b_n| = d \qquad (2-5-1)$$

where $c, c_1, \cdots, c_n, b_1, b_2, \cdots, b_n$ and d are **known constants**, (numbers), is called an **equation with absolute values**. (Absolute value of real numbers and related topics are thoroughly covered in my book, "**College Algebra, Vol.1**").

An absolute value equation **is not a first degree equation**. A first degree equation has one solution, while equations with absolute usually have more than one solution.

For example, let us consider the equation $|x - 3| = 2$. This means that either $(x - 3) = 2$, i.e. $x = 5$, or $(x - 3) = -2$ in which case $x = 1$. Indeed, note that both numbers, 1 and 5 satisfy $|x - 3| = 2$, ($|5 - 3| = 2$ and $|1 - 3| = |-2| = 2$).

The main idea to solve an equation with absolute values is **to get rid of the absolute values**. This can be achieved by making certain assumptions regarding the range of the values of x. The method is best illustrated by means of the following examples.

Example 2-5-1: Solve the equation $|x + 1| = -3$.

Solution

Since the absolute value of any number is **a non-negative number** (zero or positive), equation $|x + 1| = -3$ is impossible.

Example 2-5-2: Solve the equation $|2x + 3| = 5$.

Solution

$$|2x + 3| = 5 \Leftrightarrow \begin{matrix} 2x + 3 = 5, \text{ or} \\ 2x + 3 = -5 \end{matrix} \Leftrightarrow \begin{matrix} x = (5 - 3)/2 \Leftrightarrow x = 1 \\ x = (-5 - 3)/2 \Leftrightarrow x = -4 \end{matrix}$$

Check that both solutions satisfy the equation.

Example 2-5-3

Solve the equation $x + |x + 3| = 2$.

Solution

1) Assume that $(x + 3) \geq 0$, i.e. $x \geq -3$. Then $|x + 3| = x + 3$ and the original equation becomes,

$$x + (x + 3) = 2 \Leftrightarrow 2x + 3 = 2 \Leftrightarrow 2x = 2 - 3 = -1 \Leftrightarrow x = -1/2 \quad (*)$$

Since the found value of x ($-1/2$) is indeed bigger that -3, this is an **acceptable** solution.

2) Now assume that $(x + 3) < 0$, i.e. $x < -3$. Then $|x + 3| = -(x + 3)$ and the original equation becomes,

$$x - x - 3 = 2 \Leftrightarrow -3 = 2$$

which is impossible.

In summary: $x + |x + 3| = 2$ has **a unique** solution, $x = -1/2$.

Example 2-5-4: Solve the equation $2x - |x + 1| + |3x - 9| = 7$.

Solution

It helps if we consider the real axis as shown in Fig. 2-1.

$-\infty \qquad\qquad -1 \qquad\qquad 3 \qquad\qquad \infty$

Fig. 2-1: The real axis.

The numbers (-1) and 3, (at which **the absolute values vanish**), divide the set of real numbers in three subintervals, $-\infty < x \leq -1$, $-1 < x \leq 3$ and $3 \leq x < \infty$. **We consider each one of these subintervals separately**:

1) If $-\infty < x \leq -1$, then $(x + 1) \leq 0$ and $(3x - 9) < 0$, and therefore $|x + 1| = -(x + 1)$ and $|3x - 9| = -(3x - 9)$, and the original equation becomes,

$$2x - \{-(x + 1)\} + \{-(3x - 9)\} = 7 \Leftrightarrow 2x + x + 1 - 3x + 9 = 7 \Leftrightarrow$$

$$0 \cdot x + 10 = 7 \Leftrightarrow 0 \cdot x = 7 - 10 \Leftrightarrow 0 \cdot x = -3$$

and this is impossible. This means that there is no solution within the interval $-\infty < x \leq -1$.

2) If $-1 < x \leq 3$, then $(x + 1) > 0$ and $(3x - 9) \leq 0$, and therefore $|x + 1| = x + 1$ and $|3x - 9| = -(3x - 9)$, and the original equation becomes,

$$2x - (x + 1) - (3x - 9) = 7 \Leftrightarrow 2x - x - 1 - 3x + 9 = 7 \Leftrightarrow$$

$$-2x = 7 - 9 + 1 \Leftrightarrow -2x = -1 \Leftrightarrow x = \mathbf{1/2}$$

Since this solution lies in the interval $-1 < x \leq 3$, it is **an acceptable** solution.

3) If $3 < x < \infty$, then $(x + 1) > 0$ and $(3x - 9) > 0$, and therefore $|x + 1| = x + 1$ and $|3x - 9| = 3x - 9$, and the original equation becomes,

$$2x - (x + 1) + 3x - 9 = 7 \Leftrightarrow 2x - x - 1 + 3x - 9 = 7 \Leftrightarrow$$

$$4x = 7 + 1 + 9 \Leftrightarrow 4x = 17 \Leftrightarrow x = \mathbf{17/4}$$

Since this solution lies in the interval $3 < x < \infty$, it is an **acceptable** solution.

In summary: The given equation has two solutions, $x = 1/2$ and $x = 17/4$.

Let the reader check that both solutions satisfy the given equation.

PROBLEMS

Solve the following equations:

2-5-1) $|2x + 5| = 9$, (**Ans:** $-7, 2$).

2-5-2) $|7x + 2| = 0$.

2-5-3) $|x| + |x - 5| = 3$, (**Ans:** Impossible).

2-5-4) $2x + |3x + 10| = 7$.

2-5-5) $x|x - 2| - |x - 2||5x - 10| = -8|x - 2|$, (**Ans:** $2, 1/3, 9/2$).

2-6) Word problems

An important application of first degree equations is in the solution of word problems. The idea is **to convert a word problem into an equation** for the sought for quantity, and then solve the equation. The following problems illustrate the method.

Example 2-6-1: Find two numbers whose sum is 54 if one number is 8 larger than the other.

Solution

If we call x the smaller number the larger one will be $(x + 8)$. The sum of these two numbers is 54, i.e.

$$x + (x + 8) = 54 \iff 2x + 8 = 54 \iff x = (54 - 8)/2 = 46/2 = 23$$

This means that the smaller number is 23 and the larger one is $(23 + 8) = 31$. Check that 31 is 8 larger than 23 and that $23 + 31 = 54$.

Example 2-6-2: John is twice as old as Peter, but in 10 years he will be 11/8 as old as Peter. How old is John now?

Solution

If x years represents John's age now, then Peter's age now will be $x/2$. In 10 years from now, John's age will be $(x + 10)$ and Peter's age will be $\left(\frac{x}{2} + 10\right)$. According to the problem, we must have,

$$x + 10 = \frac{11}{8}\left(\frac{x}{2} + 10\right) \iff x + 10 = \frac{11x}{16} + \frac{110}{8} \iff$$

$$16(x + 10) = 11x + 220 \iff 16x + 160 = 11x + 220 \iff$$

$$16x - 11x = 220 - 160 \iff 5x = 60 \iff x = 60/5 = 12$$

John's age (now) is 12 years and Peter's age (today) is 6 years.

Example 2-6-3: The sides of two squares differ by 8 meters and their areas differ by 112 square meters. Find the lengths of the sides of the squares.

Solution

If x is the length of the side of the first square, then the side of the second square is $(x + 8)$. The area of the first square is x^2 and the area of the second square is $(x + 8)^2$. According to the problem we have,

$$(x + 8)^2 - x^2 = 112 \Leftrightarrow x^2 + 16x + 64 - x^2 = 112 \Leftrightarrow$$

$$16x + 64 = 112 \Leftrightarrow 16x = 112 - 64 = 48 \Leftrightarrow x = \frac{48}{16} = 3$$

The side of the first square is 3 meters and of the second square is 8+3=11 meters.

Example 2-6-4: One pipe can fill a tank in 10 hours and another pipe can fill the same tank in 18 hours. The drainpipe can empty the tank in 12 hours. With all pipes open, how long will it take to fill the tank?

Solution

In one hour the first pipe fills $1/10$ of the tank and the second pipe fills $1/18$ of the tank. Also, in one hour, the drainpipe empties $1/12$ of the tank. Let us now assume that, with all pipes open, it takes x hours to fill the tank. This means that

$$x\left(\frac{1}{10} + \frac{1}{18} - \frac{1}{12}\right) = 1 \Leftrightarrow x = \frac{1}{\frac{1}{10} + \frac{1}{18} - \frac{1}{12}} \cong 13.84 \; hours$$

Example 2-6-5: The sum of three consecutive integer numbers is 36. Find the numbers.

Solution

If x is the smallest number, then the second one will be $(x + 1)$ and the third one $(x + 2)$. According to the problem we must have,

$$x + (x + 1) + (x + 2) = 36 \Leftrightarrow 3x + 3 = 36 \Leftrightarrow x = \frac{36 - 3}{3} = 11$$

The sought for numbers are: 11, 12, 13.

PROBLEMS

2-6-1) Find two consecutive even integers whose sum is 54,   (**Ans:** 26, 28).

2-6-2) Find three consecutive odd integers whose sum is 27.

2-6-3) What number must be added to both terms of the fraction $\frac{3}{4}$ to make the result equal to $\frac{5}{6}$?   (**Ans:** 2).

2-6-4) Find three consecutive even integers such that the sum of the second and the third is 38.

2-6-5) Find four consecutive even integers such that the sum of the first and the last is 58,   (**Ans:** 26, 28, 30, 32).

2-6-6) Find a number such that if one-third of it is added to the number, the sum is 32.

2-6-7) John is 13 years older than his brother, and in 6 years from now he will be twice his brother's age then. How old is each now?   (**Ans:** 20, 7).

2-6-8) The length of a rectangle is three times its width. What are its dimensions if its perimeter is 48?

2-6-9) The length of a rectangle is 4 less than three times its width, and its perimeter is 112 meters. Find its dimensions,   (**Ans:** 15, 41).

2-6-10) Three more than twice a number is 29. Find the number.

2-6-11) Find two consecutive even integers such that 6 times the first exceeds 5 times the second by 26,   (**Ans:** 36, 38).

CHAPTER 3: FIRST DEGREE INEQUALITIES IN ONE UNKNOWN

3-1) Intervals of real numbers

Given two real numbers a and b with $a < b$ the **open interval** from a to b is the set of numbers greater than a and less than b, i.e. the set of real numbers x which satisfy the double inequality $\boldsymbol{a < x < b}$. (Inequalities, related properties and Theorems are covered in my book "**College Algebra, Vol. 1**").

The points a and b are called the **endpoints of the interval**. The **bracket** [or] indicates that the endpoint is included, while the **parenthesis** (or) indicates that the endpoint is not included. The **open** interval $a < x < b$ is also denoted as $(\boldsymbol{a, b})$, while the **closed** interval $a \leq x \leq b$ is denoted as $[\boldsymbol{a, b}]$. The intervals $[a, b)$ and $(a, b]$ are called **half-opened intervals**.

Given a and b with $a < b$, we have therefore, four cases:

$$
\begin{array}{lll}
a < x < b & \text{or} \quad (a, b) & \text{Open Interval} \\
a \leq x \leq b & \text{or} \quad [a, b] & \text{Closed Interval} \\
a < x \leq b & \text{or} \quad (a, b] & \text{Half-open Interval} \\
a \leq x < b & \text{or} \quad [a, b) & \text{Half-open Interval}
\end{array} \qquad (3-1-1)
$$

The set of real numbers x **greater than** a given number a, is expressed by the inequality, $a < x$, or $a < x < \infty$. Recall that ∞ (the symbol of the infinity) **is not a number**, it is a concept. We can approach infinity (through a limiting process) but we cannot reach infinity; in that sense, we **cannot** write $x = \infty$.

If we want to express the set of real numbers **greater than or equal** to a (i.e. the set of numbers **not less than** a), we may write $a \leq x$ or $a \leq x < \infty$. Note the difference between the two inequalities, $a < x$ and $a \leq x$. In the first one x is strictly greater than a, while in the second one x is greater than or equal to a, (i.e. x is not less than a).

Similarly the inequality $y < a$ or $-\infty < y < a$ means that y is **strictly less than** a, while $y \leq a$ means that y is **less than or equal** to a, (not greater than a).

Intervals can be described, using **set notation**. For example, we may express the open interval $a < x < b$ as $x \in (a, b)$ or the closed interval $a \leq x \leq b$ as $x \in [a, b]$, etc.

Let us consider the inequality $a < x \leq b$. The graph of this inequality is shown in Fig.3-1.

Fig. 3-1: Intervals on the real axis.

The set of points $x \in (a, b]$ is **a portion of the real axis**, as shown in Fig. 3-1. The circle at the point a has been left open to indicate that a **does not belong** to the interval. The circle at the point b has been filled to indicate that b **does belong** to the interval.

In various problems there is a need to determine the set of points that belong to two or more intervals, **simultaneously**. The common portion of two or more intervals can be found rather easily if we graph each interval separately and then find their common part (where they intersect).

Example 3-1-1: Find the common part of the intervals $-5 < x < 3$ and $1 < x < 6$.

Solution

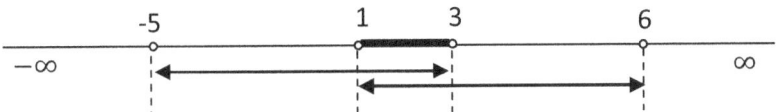

Fig. 3-2: Intersection of intervals.

We arrange the numbers on the real line, as shown in Fig. 3-2.

The interval which is common to the two given intervals is the interval from 1 to 3, (endpoints are not included). We may write,

$$(-5, 3) \cap (1, 6) = (1, 3)$$

or equivalently,

$$\{-5 < x < 3\} \cap \{1 < x < 6\} = \{1 < x < 3\}$$

If two intervals do not have points in common, then their intersection is the **null set**. For example, $(-1,3) \cap (5,10) = \emptyset$.

Example 3-1-2: Consider the intervals $I_1 = \{1 < x \leq 5\}$, $I_2 = \{5 \leq x < 7\}$ and $I_3 = \{7 < x < 9\}$. Find the intervals: $I_1 \cap I_2$, $I_1 \cup I_2$, $I_2 \cup I_3$, $I_2 \cap I_3$.

Solution

$$I_1 \cap I_2 = \{1 < x \leq 5\} \cap \{5 \leq x < 7\} = \{5\}$$

$$I_1 \cup I_2 = \{1 < x \leq 5\} \cup \{5 \leq x < 7\} = \{1 < x < 7\}$$

$$I_2 \cup I_3 = \{5 \leq x < 7\} \cup \{7 < x < 9\} = \{5 \leq x < 9\} - \{7\}$$

$$I_2 \cap I_3 = \{5 \leq x < 7\} \cap \{7 < x < 9\} = \emptyset$$

PROBLEMS

Find the common part of the intervals:

3-1-1) $0 \leq x \leq 5$ and $3 \leq x \leq 7$, (**Ans:** $3 \leq x \leq 5$).

3-1-2) $-3 < x < 5$ and $4 \leq x \leq 8$.

3-1-3) $x < 10$ and $x \geq 5$, (**Ans:** $5 \leq x < 10$).

3-1-4) $x \geq 2$ and $x \leq 7$.

3-1-5) $x > 10$ and $x < 9$, (**Ans:** $\emptyset$).

3-1-6) If $a < b$ show that

$$\frac{1}{2}||x-a| - |x-b|| = \begin{cases} |x - (a+b)/2| & \text{if } a \leq x \leq b \\ (b-a)/2 & \text{if } x \notin [a,b] \end{cases}$$

3-2) Inequalities in one unknown

a) If $F(x)$ and $G(x)$ are two functions of x, having the same domain of definition (i.e. defined over the same set of real numbers), then we say that we have to solve the inequality $F(x) < G(x)$ when we want to determine all the values of the independent variable x, which assign arithmetic values to $F(x)$ smaller than the values they assign to $G(x)$. Every number a which makes

$F(a) < G(a)$ is called **a solution** of the inequality. **The set of all the solutions of $F(x) < G(x)$ is called the solution set of the inequality.**

b) Equivalent inequalities: Two inequalities $F(x) < G(x)$ and $A(x) < B(x)$ are called **equivalent** if they have **the same solution set**, i.e. if every solution of the first is a solution of the second and conversely, every solution of the second is a solution of the first.

The following theorems are used in order to replace an inequality with an equivalent one, (which however is simpler than the given).

Theorem 3-2-1: **If to both sides of an inequality $F(x) < G(x)$ we add the same function $A(x)$, we obtain an inequality equivalent to the given. In symbols:**

$$F(x) < G(x) \Leftrightarrow F(x) + A(x) < G(x) + A(x) \qquad (3-2-1)$$

Proof: Let a be a solution of $F(x) < G(x)$. This means that $F(a) < G(a)$, i.e.

$$G(a) - F(a) > 0 \Leftrightarrow G(a) - F(a) + \underbrace{A(a) - A(a)}_{=0} > 0 \Leftrightarrow$$

$$\{G(a) + A(a)\} - \{F(a) + A(a)\} > 0 \Leftrightarrow F(a) + A(a) < G(a) + A(a)$$

and this means that a is a solution of $F(x) + A(x) < G(x) + A(x)$.

Conversely: Let b be a solution of $F(x) + A(x) < G(x) + A(x)$. This means that

$$F(b) + A(b) < G(b) + A(b) \Leftrightarrow \{F(b) + A(b)\} - \{G(b) + A(b)\} < 0 \Leftrightarrow$$

$$F(b) - G(b) < 0 \Leftrightarrow F(b) < G(b)$$

and this means that b is a solution of $F(x) < G(x)$ as well.

We have thus shown that the inequalities $F(x) < G(x)$ and $F(x) + A(x) < G(x) + A(x)$ have the same solution set, and therefore, **they are equivalent**.

For example, $3x^2 + 5x + 3 < 8x - 7$ is equivalent to $(3x^2 + 5x + 3) + (-3x^2) < (8x - 7) + (-3x^2)$, (we add $(-3x^2)$ to both sides), i.e.

$$3x^2 + 5x + 3 < 8x - 7 \Leftrightarrow 5x + 3 < -3x^2 + 8x - 7$$

Application of Theorem 3-2-1 implies that **we may transfer a term from one side of an inequality to the other, provided that we change the sign of the term transferred**. For instance,

$$10x - 7 > 3x + 2 \Leftrightarrow 10x - 3x > 2 - (-7) \Leftrightarrow 7x > 9$$

$$x^3 + 5 < 2x^2 - 3x + 8 \Leftrightarrow x^3 + 5 - 2x^2 + 3x - 8 < 0$$
$$\Leftrightarrow x^3 - 2x^2 + 3x - 3 < 0$$

Note: Any inequality $F(x) < G(x)$ can be put in the form $F(x) - G(x) < 0$.

Theorem 3-2-2: a) **If we multiply both sides of the inequality $F(x) < G(x)$ by a positive number we obtain an inequality equivalent to the given, i.e.**

$$\forall \lambda > 0: \quad F(x) < G(x) \Leftrightarrow \lambda F(x) < \lambda G(x) \qquad (3-2-2)$$

b) **If we multiply both sides of the inequality $F(x) < G(x)$ by a negative number we obtain an equivalent inequality, provided that we change the sense of the inequality**, i.e.

$$\forall k < 0: \quad F(x) < G(x) \Leftrightarrow kF(x) > kG(x) \qquad (3-2-3)$$

In simple terms: Multiplication of an inequality by a negative number changes the sense of the inequality.

Proof: The proof is simple and is left as an exercise for the reader, (see Problem 3-2-1).

For example:

$$2x - 3 < x + 5 \Leftrightarrow 7 \cdot (2x - 3) < 7 \cdot (x + 5)$$

$$2x + 3 < x^2 + 4x - 10 \Leftrightarrow (-2) \cdot (2x + 3) > (-2) \cdot (x^2 + 4x - 10)$$

In case the inequality involves **fractional coefficients**, we may multiply by a common multiple of the denominators to get **an equivalent inequality** with integer coefficients. For example:

$$\frac{1}{3}x + 2 < \frac{3}{2}x - 5 \Leftrightarrow 6 \cdot \left(\frac{1}{3}x + 2\right) < 6 \cdot \left(\frac{3}{2}x - 5\right) \Leftrightarrow$$

$$2x + 12 < 9x - 30$$

c) The same theorems apply for inequalities of the form $F(x) \leq G(x)$.

Example 3-2-1: Are the following inequalities equivalent?

a) $x^2(x-1) > 0$ and $x - 1 > 0$, and b) $x^2(x-1) \geq 0$ and $x - 1 \geq 0$.

Solution

a) The solution set of the first inequality is $x - 1 > 0$ and obviously coincides with the solution set of the second; the inequalities are equivalent.

b) The first inequality is satisfied either if $x = 0$ or $x - 1 \geq 0$, i.e. if $x = 0$ or $x \geq 1$ and since $x = 0$ is not a solution of the second inequality, the two inequalities are **not** equivalent.

Example 3-2-2: Show that $x + 3 < 2x - 7$ is equivalent to $-x^2 - x - 3 > -x^2 - 2x + 7$.

Solution

If we multiply both sides of the given inequality by (-1) we obtain the equivalent inequality, (Theorem 3-2-2), $-x - 3 > -2x + 7$, (note that **when multiplying an inequality by a negative number, the sense of the inequality is reversed**). If we now add in both sides $(-x^2)$ we obtain an equivalent inequality (by virtue of Theorem 3-2-1). In symbols:

$$x + 3 < 2x - 7 \Leftrightarrow (-1)(x+3) > (-1)(2x-7) \Leftrightarrow$$

$$-x - 3 > -2x + 7 \Leftrightarrow -x^2 - x - 3 > -x^2 - 2x + 7$$

PROBLEMS

3-2-1) Prove Theorem 3-2-2.

3-2-2) Show that $5x + 3 \geq 3x - 9$ is equivalent to $10x + 6 \geq 6x - 18$.

3-2-3) Show that $3x - 8 \leq 7 - x$ is equivalent to $-6x + 16 \geq -14 + 2x$.

3-2-4) Show that $\left(\frac{1}{2}x - \frac{2}{3}\right) \geq \left(\frac{1}{6}x + 1\right)$ is equivalent to $2x^2 - 3x + 4 \leq 2x^2 - x - 6$.

3-2-5) Are the following inequalities equivalent?

a) $x^4(x+5) < 0$ and $x + 5 < 0$, (**Ans:** Yes).

b) $x^4(x+5) \leq 0$ and $x + 5 \leq 0$, (**Ans:** No).

3-3) Solving first degree inequalities in one unknown

If **both sides** of the inequality $F(x) > G(x)$ are **first degree polynomials**, the inequality is called **a first degree inequality**. For example, $x + 3 > 5 - x$, $2 > 7 - x$, $3x + 1/2 \leq -5x + 7/8$ are **first degree** inequalities. The inequality $2x^2 + 5 > 3x - 1$ is **a second degree** inequality, since the left-hand side is a second degree polynomial, while $x^3 + 2x^2 \geq 5x - 2$ is **a third degree** inequality, etc.

The solution of first degree inequalities is achieved with the aid of Theorems 3-2-1 and 3-2-2. If there are fractional coefficients (fractions) we multiply through by a common multiple of the denominators (preferably by the LCD –Least Common Denominator) in order to get rid of the denominators, then we transfer all the unknowns in one side (say the left-hand side), simplify and then divide by the coefficient of the unknown. Remember that **when we divide (or multiply) by a negative number the sense of the inequality reverses**.

Example 3-3-1: Solve the inequality $\frac{x}{2} - 3 \geq \frac{x}{3} + 2$.

Solution

$$\frac{x}{2} - 3 \geq \frac{x}{3} + 2 \iff 6 \cdot \left(\frac{x}{2} - 3\right) \geq 6 \cdot \left(\frac{x}{3} + 2\right) \iff$$

$$3x - 18 \geq 2x + 12 \iff 3x - 2x \geq 18 + 12 \iff x \geq 30$$

Example 3-3-2: Solve the inequality $2 \cdot (3x - 5) > 9x + \frac{1}{2}$.

Solution

$$2 \cdot (3x - 5) > 9x + \frac{1}{2} \Leftrightarrow 6x - 10 > 9x + \frac{1}{2} \Leftrightarrow$$

$$2 \cdot (6x - 10) > 2 \cdot \left(9x + \frac{1}{2}\right) \Leftrightarrow 12x - 20 > 18x + 1 \Leftrightarrow$$

$$12x - 18x > 20 + 1 \Leftrightarrow -6x > 21 \Leftrightarrow x < -\frac{21}{6}$$

Note that in the last step, when we divide by (−6), we reverse the sense of the inequality.

Example 3-3-3: Find all the values of x which satisfy both inequalities simultaneously: $\frac{x-2}{3} - \frac{x-3}{4} \geq \frac{1}{2}$ and $\frac{x}{2} + \frac{x}{4} < \frac{x+8}{3}$.

Solution

First inequality:

$$\frac{x-2}{3} - \frac{x-3}{4} \geq \frac{1}{2} \Leftrightarrow 12 \cdot \left(\frac{x-2}{3} - \frac{x-3}{4}\right) \geq 12 \cdot \frac{1}{2} \Leftrightarrow$$

$$4 \cdot (x - 2) - 3 \cdot (x - 3) \geq 6 \Leftrightarrow 4x - 8 - 3x + 9 \geq 6 \Leftrightarrow$$

$$4x - 3x \geq 8 - 9 + 6 \Leftrightarrow x \geq 5 \qquad (*)$$

Second inequality:

$$\frac{x}{2} + \frac{x}{4} < \frac{x+8}{3} \Leftrightarrow 12 \cdot \left(\frac{x}{2} + \frac{x}{4}\right) < 12 \cdot \frac{x+8}{3} \Leftrightarrow$$

$$6x + 3x < 4x + 32 \Leftrightarrow 6x + 3x - 4x < 32 \Leftrightarrow 5x < 32 \Leftrightarrow x < \frac{32}{5} \qquad (**)$$

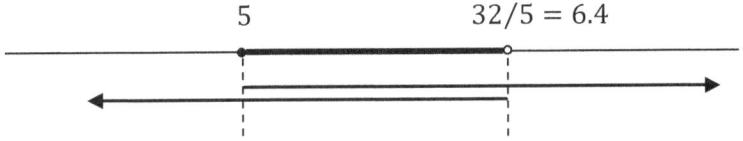

Fig. 3-3: The common solution of two inequalities.

The "**common solution**" of the two inequalities is the interval $5 \leq x < 6.4$. Any number within this interval satisfies both equations.

Note: Consider a slight modification of the problem; find **the integer values** of x which satisfy both inequalities, $\frac{x-2}{3} - \frac{x-3}{4} \geq \frac{1}{2}$ and $\frac{x}{2} + \frac{x}{4} < \frac{x+8}{3}$, Then, from Fig. 3-3 we see that there are **two integer** values of x which satisfy both inequalities simultaneously, $x = 5$ and $x = 6$.

Example 3-3-4: Solve and investigate the inequality $kx + 5 > 2kx + 3$, where k is considered to be a known real number (**parameter**).

Solution

$$kx + 5 > 2kx + 3 \Leftrightarrow kx - 2kx > 3 - 5 \Leftrightarrow -kx > -2 \stackrel{\cdot(-1)}{\Longleftrightarrow}$$

$$kx < 2$$

At this point, we **cannot just divide by k**, because we do not know whether k is positive (the sense of the inequality is retained) or negative (the sense of the inequality is reversed). We therefore proceed as follows:

$$kx < 2 \Leftrightarrow \begin{array}{l} If\ k > 0, then\ x < 2/k \\ If\ k = 0, then\ 0 \cdot x < 2\ (identity) \\ If\ k < 0, then\ x > 2/k \end{array}$$

PROBLEMS

Solve the following inequalities:

3-3-1) $2(x - 2) \geq 3(x + 3)$, (**Ans:** $x \leq -13$).

3-3-2) $3\left(x + \frac{1}{6}\right) + \frac{1}{3} \leq -2x + \frac{1}{2}$.

3-3-3) $\frac{x-1}{7} > \frac{2x+5}{3}$, (**Ans:** $x < -38/11$).

3-3-4) $\frac{2x-3}{2} + \frac{x+3}{5} < \frac{x-6}{10}$.

3-3-5) $\frac{2x-3}{3} + \frac{6x-1}{12} \geq \frac{3x}{4} + \frac{9}{2}$, (**Ans:** $x \geq 67/5$).

In the following problems find the **integer** values of x which satisfy the stated inequalities simultaneously:

3-3-6) $5 - 4(x - 3) > x - 2(x - 1)$ **and** $4(x + 3) < 5(x + 2) + 6$.

(**Ans:** $x \in \{-3, -2, -1, 0, 1, 2, 3, 4\}$).

3-3-7) $\frac{5x+2}{4} > \frac{x}{2} + 2$ and $\frac{x}{4} - \frac{5x+8}{6} < \frac{2x-9}{3}$ and $8 - \frac{7-x}{8} < \frac{23}{2} - \frac{3x}{4}$.

(**Ans:** $x \in \{3, 4\}$).

Solve and investigate the following inequalities:

3-3-8) $2kx + \frac{5x}{3} > 3x - k$.

(**Ans:** If $k > 2/3$, $x > 3k/(4 - 6k)$, if $k < 2/3$, $x < 3k/(4 - 6k)$, if $k = 2/3$ identity.

3-3-9) $(k^2 + 1)x + 9 \geq 2kx + 7k$.

(**Ans:** If $k \neq 1$, $x \geq (7k - 9)/(k - 1)^2$, if $k = 1$ identity).

3-4) The sign of the product $(x - a)(x - b)(x - c) \cdots (x - p)$

a) Let us start our analysis with the following problem:

For what values of x is the product $(x - 1)(x - 2)(x - 3)$ positive and for what values of x is negative? We arrange the values 1, 2, 3 (i.e. the **roots** of the product $(x - 1)(x - 2)(x - 3)$) on the real axis, **in ascending order**, as shown in Fig. 3-4.

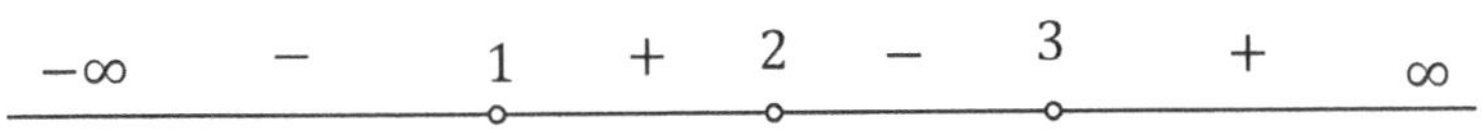

```
  -∞      −        1     +      2    −     3     +      ∞
─────────────────○──────────○─────────○──────────────
```

Fig. 3-4: Sign of the product $(x - 1)(x - 2)(x - 3)$.

These numbers divide the real axis in 4 subintervals, $(-\infty, 1), (1, 2), (2, 3)$, and $(3, \infty)$.

Assume that x lies in **the rightmost interval**, i.e. $x \in (3, \infty)$. Then, $x - 3 > 0$, $x - 2 > 0$ and $x - 1 > 0$ and this shows that if x takes on any value in the interval $(3, \infty)$ the product $(x - 1)(x - 2)(x - 3)$ is **positive**, since all its factors are positive.

If $x \in (2,3)$, then $x - 3 < 0$, $x - 2 > 0$ and $x - 1 > 0$, and this shows that when x takes on any value in the interval $(2,3)$ then the product $(x - 1)(x - 2)(x - 3)$ is **negative**, since it contains one negative factor and two positive ones.

If $x \in (1,2)$, then $x - 3 < 0$, $x - 2 < 0$ and $x - 1 > 0$, and this shows that when x takes on any value in the interval $(1,2)$ then the product $(x - 1)(x - 2)(x - 3)$ is **positive**, since it contains two negative factor and one positive.

Finally, if $x \in (-\infty, 1)$, then $x - 3 < 0$, $x - 2 < 0$ and $x - 1 < 0$, and this shows that when x takes on any value in the interval $(-\infty, 1)$ then the product $(x - 1)(x - 2)(x - 3)$ is **negative**, since it contains three negative factors.

In summary:

$$(x-1)(x-2)(x-3) = \begin{cases} Positive \ if \ 3 < x < \infty & (+) \\ Negative \ if \ 2 < x < 3 & (-) \\ Positive \ if \ 1 < x < 2 & (+) \\ Negative \ if \ -\infty < x < 1 & (-) \end{cases} \quad (*)$$

Notice that the sign of the product $(x - 1)(x - 2)(x - 3)$ is positive in the rightmost interval $(3, \infty)$, and then alternates ($+, -, +, -$).

Suppose now that we want to solve the inequality $(x - 1)(x - 2)(x - 3) > 0$. From formula (*) we find that the solution is $(1,2) \cup (3, \infty)$, i.e. either $1 < x < 2$ or $3 < x < \infty$. These intervals constitute **the solution set** of the given inequality.

Similarly, the solution of $(x - 1)(x - 2)(x - 3) < 0$ is $(-\infty, 1) \cup (2,3)$.

Note: A product of **first degree polynomials in x** is in its **"canonical form"** if the coefficients of x in each factor of the product is equal to $(+1)$. If the product is not in its canonical form, it is easily **converted to its canonical form**.

For example the product $(x - 1)(x - 2)(x - 3)$ is in its canonical form. The product $(2x + 6)(3x + 6)(x - 5)$ is not in canonical form, since the coefficients of x in the first and the second factor are 2 and 3 respectively, and not 1. However, we may convert the product to its canonical form, since

$$(2x + 6)(3x + 6)(x - 5) = 2 \cdot (x + 3) \cdot 3 \cdot (x + 2)(x - 5)$$
$$= 6 \underbrace{(x + 3)(x + 2)(x - 5)}_{Canonical\ form}$$

Similarly,

$$(-5x + 2)(3x + 7) = (-5) \cdot \left(x - \frac{2}{5}\right) \cdot 3 \cdot \left(x + \frac{7}{3}\right) = -15 \underbrace{\left(x - \frac{2}{5}\right)\left(x + \frac{7}{3}\right)}_{Canonical\ form}$$

b) We may now outline **the general method** of solving an inequality of the form $(a_1x - b_1)(a_2x - b_2) \cdots (a_nx - b_n) > 0$, (or < 0).

1) Express the product $(a_1x - b_1)(a_2x - b_2) \cdots (a_nx - b_n)$ in its **canonical form** $c(x - d_1)(x - d_2) \cdots (x - d_n)$, where (without loss of generality) we may assume that $d_1 < d_2 < \cdots < d_n$.

2) Arrange the roots of the product $d_1, d_2, \cdots, d_n$ on the real line, in **ascending** order $d_1, d_2, \cdots, d_n$.

3) The sign of the product $(x - d_1)(x - d_2) \cdots (x - d_n)$ is positive (+) when x lies in **the rightmost interval** (d_n, ∞), in the immediately preceding interval (d_{n-1}, d_n) is negative (−), etc, (**alternates**).

c) One method to solve a **polynomial inequality** $P(x) > 0$ is to factor $P(x)$ and then apply the method illustrated in (b), (see Example 3-4-3).

Example 3-4-1: Solve the inequality $(2x - 4)(-3x + 9)(x + 1) > 0$.

Solution

First of all, we have to express the inequality in **its equivalent canonical form**:

$$(2x - 4)(-3x + 9)(x + 1) > 0 \Leftrightarrow 2 \cdot (x - 2) \cdot (-3)(x - 3)(x + 1) > 0 \Leftrightarrow$$

$$(-6)(x + 1)(x - 2)(x - 3) > 0 \Leftrightarrow (x + 1)(x - 2)(x - 3) < 0 \quad (*)$$

So the given inequality **is equivalent** to the inequality in (*).

We arrange the roots $(-1, 2, 3)$ on the real line, as shown in Fig. 3-5.

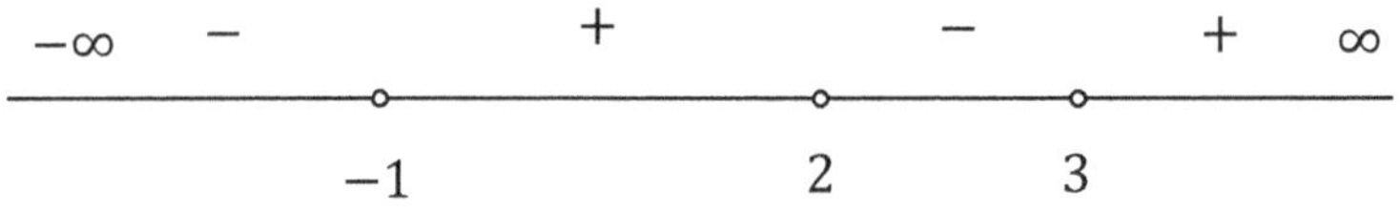

```
 -∞    −        +        −        +   ∞
─────────○────────○────────○──────────
        −1        2        3
```

Fig. 3-5: Solution of the inequality $(x+1)(x-2)(x-3) < 0$.

As we see from Fig. 3-5, the solution of the inequality $(x+1)(x-2)(x-3) < 0$, (which is **equivalent** to the given inequality) is

$$x \in (-\infty, -1) \cup (2,3), \quad or \quad (-\infty < x < -1) \cup (2 < x < 3) \qquad (**)$$

Note: The solution of $(2x-4)(-3x+9)(x+1) \geq 0$, (**greater than or equal to zero**), will be

$$x \in (-\infty, -1] \cup [2,3], \quad or \quad (-\infty < x \leq -1) \cup (2 \leq x \leq 3) \qquad (***)$$

This is so because the product vanishes at $x = -1, 2, 3$, (the roots of the product).

Example 3-4-2: Solve the inequality
$$(x^2 + 1)(2x - 4)(-3x + 9)(x + 1) > 0$$

Solution

Since the factor $(x^2 + 1)$ is positive for all real values of x, ($\forall x \in \mathbb{R}: x^2 + 1 > 0$), the given inequality is **equivalent** to $(2x - 4)(-3x + 9)(x + 1) > 0$, which was solved in Example 3-4-1, and its solution set is given by equation (**), (in Ex. 3-4-1).

Example 3-4-3: Solve the polynomial inequality $x^3 - 5x^2 + 6x < 0$.

Solution

We first try to factor the polynomial $x^3 - 5x^2 + 6x$.

$$x^3 - 5x^2 + 6x = x^3 - 3x^2 - 2x^2 + 6x = x^2(x-3) - 2x(x-3)$$
$$= (x-3)(x^2 - 2x) = x(x-2)(x-3)$$

So, the problem reduces to solving the inequality $x(x-2)(x-3) < 0$, (note that the left-hand side of the inequality is in its canonical form).

The roots of the product are 0, 2 and 3. We arrange these roots on the real line, as shown in Fig. 3-6.

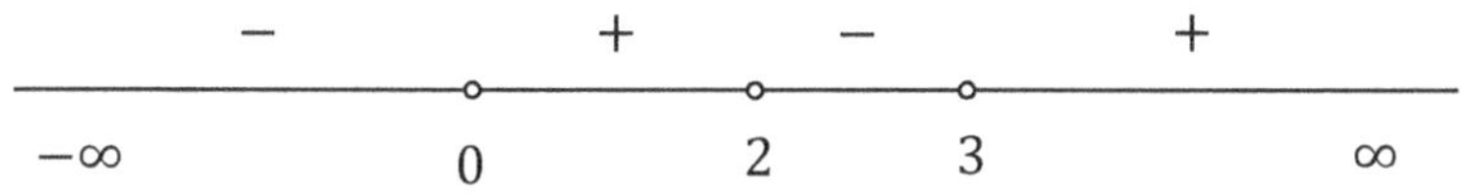

```
      −        +       −       +
────────o──────o──────o──────────
  −∞    0      2      3       ∞
```

Fig. 3-6: Solution of the inequality $x(x-2)(x-3) < 0$.

The solution of $x(x-2)(x-3) < 0$ is: $(-\infty < x < 0) \cup (2 < x < 3)$.

Example 3-4-4: Solve the inequality $x^5(x-2)^7(x-3)^{11} < 0$.

Solution

$$x^5(x-2)^7(x-3)^{11} < 0 \Leftrightarrow x^4 x(x-2)^6(x-2)(x-3)^{10}(x-3) < 0 \Leftrightarrow$$

$$\underbrace{\{x^4(x-2)^6(x-3)^{10}\}}_{Positive\ quantity} x(x-2)(x-3) < 0 \Leftrightarrow \qquad (*)$$

$$x(x-2)(x-3) < 0 \qquad (**)$$

For every $x \in \mathbb{R} - \{0, 2, 3\}$ the term $x^4(x-2)^6(x-3)^{10}$ is **a positive term**, (due to the even powers), and therefore, inequality (*) is **equivalent** to the inequality $x(x-2)(x-3) < 0$, which was solved in Ex. 3-4-3. Therefore, the solution of the inequality $x^5(x-2)^7(x-3)^{11} < 0$ is: $(-\infty < x < 0) \cup (2 < x < 3)$.

PROBLEMS

Solve the following inequalities:

3-4-1) $(2x+1)(3x-9)(x-5) > 0$, (**Ans:** $x \in (-1/2, 3) \cup (5, \infty)$).

3-4-2) $(x-1)^3(3x-5)(x+7) \leq 0$.

3-4-3) $x^4(x+5)(x-4)(x+3) > 0$, (**Ans:** $x \in (-5, -3) \cup (4, \infty)$).

3-4-4) $x^2 - 5x + 6 \geq 0$.

3-4-5) $x^3 - x^2 - 10x \leq 8$, (**Ans:** $x \in (-\infty, -2] \cup [-1, 4]$).

3-4-6) $x^3 - 9x^2 < 15 - 23x$.

3-4-7) $-2x^3 - x^2 + 2x + 1 > 0$,   (**Ans:** $x \in (-\infty, -1) \cup (-1/2, 1)$).

3-4-8) For what real values of x is the square root $\sqrt{x^3 - 8x^2 + 12x}$ defined? (The square root is defined for all values of x which make the radicand **a positive number**; square roots of negative numbers do not exist in the set of real numbers).

3-4-9) For what values of x are the inequalities $(x-1)(x-3)(x-5) \leq 0$ and $(x+1)^3(x-2)(x-4) > 0$ satisfied simultaneously?

(**Ans:** $(-1 < x \leq 1) \cup (4 < x \leq 5)$).

3-4-10) Find the natural numbers for which the expression $\sqrt{x^3 - 4x^2 - 11x + 30}$ is not defined, (i.e. the natural numbers making the radicand a negative number).

(**Ans:** $x = 3$ or 4).

Hint: Show that $x^3 - 4x^2 - 11x + 30 = (x+3)(x-2)(x-5)$.

3-5) Inequalities with absolute values

Inequalities involving absolute values of the unknown x, are called **inequalities with absolute values**. Te main idea to solve such inequalities is **to get rid of the absolute values**. In order to do this, we have to make certain assumptions regarding the range of x, get rid of the absolute values, find the solution and **then verify that this is a valid solution of the original inequality**, (the one involving the absolute values). The approach is analogues to the approach used in solving equations with absolute values, (see Sec. 2-5).

The following two inequalities are important and are used quite often, (see "College Algebra, Vol. 1"),

$$\begin{cases} |x| < a \ (a > 0) \Leftrightarrow -a < x < a \\ |x| > a \ (a > 0) \Leftrightarrow -\infty < x < -a \ \text{ or } \ a < x < \infty \end{cases} \quad (3-5-1)$$

For example,

$$|x| < 3 \Leftrightarrow -3 < x < 3$$

$$|x - 2| < 5 \Leftrightarrow -5 < x - 2 < 5 \Leftrightarrow -3 < x < 7$$

$$|x| > 1 \Leftrightarrow -\infty < x < -1 \ \ or \ \ 1 < x < \infty$$

The following Examples illustrate the method.

Example 3-5-1: Solve the inequality $x - 2|x| < 8$.

Solution

1) We assume that $x \geq 0$. In this case $|x| = x$ and the original inequality is equivalent to the following system of inequalities:

$$x - 2|x| < 8 \Leftrightarrow \begin{cases} x - 2x < 8 \\ and \\ x \geq 0 \end{cases} \Leftrightarrow \begin{cases} -x < 8 \\ and \\ x \geq 0 \end{cases} \Leftrightarrow \begin{cases} x > -8 \\ and \\ x \geq 0 \end{cases} \Rightarrow x \geq 0 \quad (*)$$

2) Now, assuming that $x < 0$, $|x| = -x$ and we have:

$$x - 2|x| < 8 \Leftrightarrow \begin{cases} x - 2(-x) < 8 \\ and \\ x < 0 \end{cases} \Leftrightarrow \begin{cases} 3x < 8 \\ and \\ x < 0 \end{cases} \Leftrightarrow$$

$$\begin{cases} x < 8/3 \\ and \\ x < 0 \end{cases} \Rightarrow x < 0 \quad (**)$$

From (*) and (**) it follows that the solution set of $x - 2|x| < 8$ is the set of real numbers, $(-\infty < x < \infty)$.

Example 3-5-2: Solve the inequality $|x - 2| + |x - 5| < 2x$.

Solution

First of all we notice that any real number x which might satisfy the given inequality must be **positive**, i.e. $x > 0$. The numbers 2 and 5 divide the set of **positive** real numbers into **three subintervals**, $0 < x < 2$, $2 \leq x < 5$ and $5 \leq x < \infty$.

1) We first assume that $0 < x < 2$. In this case $(x - 2) < 0, (x - 5) < 0$ and therefore, $|x - 2| = -(x - 2) = 2 - x$ and $|x - 5| = -(x - 5) = 5 - x$. The given inequality is equivalent to the following system:

$$|x-2|+|x-5|<2x \Leftrightarrow \begin{cases} 2-x+5-x<2x \\ \text{and} \\ 0<x<2 \end{cases} \Leftrightarrow$$

$$\begin{cases} -4x<-7 \\ \text{and} \\ 0<x<2 \end{cases} \Leftrightarrow \begin{cases} x>7/4 \\ \text{and} \\ 0<x<2 \end{cases} \Rightarrow \frac{7}{4}<x<2 \quad (*)$$

2) We now assume that $2 \leq x < 5$. In this case $(x-2) \geq 0, (x-5) < 0$ and therefore, $|x-2| = x-2$ and $|x-5| = -(x-5) = 5-x$. The original inequality is equivalent to the following system:

$$|x-2|+|x-5|<2x \Leftrightarrow \begin{cases} x-2+5-x<2x \\ \text{and} \\ 2 \leq x < 5 \end{cases} \Leftrightarrow$$

$$\begin{cases} 3<2x \\ \text{and} \\ 2 \leq x < 5 \end{cases} \Leftrightarrow \begin{cases} \frac{3}{2}<x \\ \text{and} \\ 2 \leq x < 5 \end{cases} \Rightarrow 2 \leq x < 5 \quad (**)$$

3) Finally, we assume that $5 \leq x < \infty$. Then $(x-2) > 0, (x-5) \geq 0$ and therefore, $|x-2| = x-2$ and $|x-5| = x-5$. The original equation is equivalent to the following system:

$$|x-2|+|x-5|<2x \Leftrightarrow \begin{cases} x-2+x-5<2x \\ \text{and} \\ 5 \leq x < \infty \end{cases} \Leftrightarrow$$

$$\begin{cases} 0 \cdot x < 7 \text{ (identity)} \\ \text{and} \\ 5 \leq x < \infty \end{cases} \Rightarrow 5 \leq x < \infty \quad (***)$$

The solution set of the inequality is the union of the sets in (*), (**) and (***), which as we easily verify is the interval $(7/4) < x < \infty$. In other words, any number in the interval $(7/4, \infty)$ satisfies the inequality.

Example 3-5-3: Solve the inequality $x^2 - 7|x| + 6 < 0$.

Solution

Since $\forall x \in \mathbb{R}, x^2 = |x|^2$, the given inequality becomes $|x|^2 - 7|x| + 6 < 0$, or if we set $y = |x| \geq 0$,

$$y^2 - 7y + 6 < 0 \Leftrightarrow (y-1)(y-6) < 0 \Leftrightarrow 1 < y < 6 \qquad (*)$$

(see Sec. 3-4), or, going back to the variable x,

$$1 < |x| < 6 \Leftrightarrow \begin{cases} 1 < |x| \\ \text{and} \\ |x| < 6 \end{cases} \overset{(3\text{-}5\text{-}1)}{\Longleftrightarrow}$$

$$\begin{cases} -\infty < x < -1 \text{ or } 1 < x < \infty \\ \text{and} \\ -6 < x < 6 \end{cases} \Rightarrow -6 < x < -1 \text{ or } 1 < x < 6$$

We have thus found that the solution of the inequality is $(-6, -1) \cup (1, 6)$.

PROBLEMS

Solve the following inequalities:

3-5-1) $|x + 7| < 13$, (**Ans:** $-20 < x < 6$).

3-5-2) $|x + 3| > 5$.

3-5-3) $|x + 4| + 2x \leq 10$, (**Ans:** $-\infty < x \leq 2$).

3-5-4) $x^2 - 3 \leq 0$, (**Ans:** $-\sqrt{3} \leq x \leq \sqrt{3}$).

3-5-5) $|x - 1| + |x - 2| < 2x + 1$, (**Ans:** $(1/2) < x < \infty$).

3-5-6) $x^2 - |x| - 12 \leq 0$.

3-5-7) $2x^2 - 9|x| - 5 \leq 0$, (**Ans:** $-5 \leq x \leq 5$).

3-5-8) $2x - 3|x| < 9$.

3-5-9) $x^4 - x^2 - 2 \leq 0$, (**Ans:** $-\sqrt{2} \leq x \leq \sqrt{2}$).

Hint: $x^4 - x^2 - 2 = (x^2 + 1)(x^2 - 2)$.

3-5-10) $|x + 1| + 2|x - 3| < 7x$.

CHAPTER 4: SYSTEMS OF LINEAR EQUATIONS

4-1) Equations with many unknowns

a) Let us consider two functions $F(x, y)$ and $G(x, y)$ of the real variables x and y. We assume that both functions have the same domain of definition, which could be the set $\mathbb{R} \times \mathbb{R}$ or a subset of it.

Solving the equation $F(x, y) = G(x, y)$ means to find **all ordered pairs** (a, b) of real numbers which designate equal arithmetic values to the functions $F(x, y) = G(x, y)$. The variables x and y are called the "**unknowns**" of the equation.

In other words, (a, b) will be a solution of the equation $F(x, y) = G(x, y)$ if $F(a, b) = G(a, b)$. We say that the solution (a, b) satisfies the equation $F(x, y) = G(x, y)$. **The solution set of the equation is the set of all ordered pairs which satisfy the equation**.

Usually, an equation with two variables has **an infinite number of solutions**, (an infinite number of ordered pairs which satisfy the equation). For example, let us consider the equation,

$$\underbrace{2x + y + 1}_{F(x,y)} = \underbrace{x - 3y + 5}_{G(x,y)} \Leftrightarrow 2x - x = -y - 3y + 5 - 1 \Leftrightarrow x = 4(1 - y)$$

We see that if we choose (**arbitrarily** a value for y) we can determine the corresponding x. For instance we easily check that the following pairs are solutions of the equation $F(x, y) = G(x, y)$.

$$\{(12, -2), (8, -1), (4,0), (0,1), (-4,2), (-8,3),\}$$

In general, if we assign a value b to the variable y, i.e. if we set $y = b$ then the corresponding $x = 4(1 - b)$. This means that the solution of $2x + y + 1 = x - 3y + 5$ is any ordered pair $(4(1 - b), b)$, $b \in \mathbb{R}$. For each b we have one solution (ordered pair) and since b can be any real number we have **an infinite number of solutions**.

If $F(x, y) = G(x, y) \; \forall (x, y) \in \mathbb{R} \times \mathbb{R}$, then this equality is called an **identity**. An identity holds true for all the values of the variables involved. For example, $(x - y)^2 = x^2 - 2xy + y^2$, $x^2 - y^2 = (x + y)(x - y)$ are identities, (each side of an identity is the other side written in a different but equivalent form). In order to show that $F(x, y)$ is **identically equal** to $G(x, y)$ we write $\boldsymbol{F(x, y) \equiv G(x, y)}$.

b) Similarly, we may consider equations in three unknowns,

$$F(x, y, z) = G(x, y, z)$$

Any **ordered triad** of real numbers (a, b, c) which satisfies the equation is a **solution** of the equation, i.e. (a, b, c) is a solution if $\boldsymbol{F(a, b, c) = G(a, b, c)}$. The totality of the solutions is called the **solution set** of the equation.

An equation in three unknowns x, y, z admits an infinite number of solutions, (ordered triads). For example, let us consider the equation $2x + 3y - z = 5$, which can be cast in the equivalent form $z = 2x + 3y - 5$. If we set $x = a, y = b$, then $z = 2a + 3b - 5$, and this means that the solutions of the equation $2x + 3y - z = 5$ are all the ordered triads $(a, b, 2a + 3b - 5), (a, b) \in \mathbb{R} \times \mathbb{R}$.

For instance, for $a = 1, b = 0$ we obtain the solution $(1, 0, -3)$; for $a = 0, b = 2$ we obtain the solution $(0, 2, 1)$; for $a = 2, b = 3$ we obtain the solution $(2, 3, 8)$, etc.

If $F(x, y, z) = G(x, y, z)$, $\forall (x, y, z) \in \mathbb{R} \times \mathbb{R} \times \mathbb{R}$ then this equality is called an **identity** and we write $F(x, y, z) \equiv G(x, y, z)$. For example $(x + y + z)^2 = x^2 + y^2 + z^2 + 2xy + 2yz + 2zx$ is an identity, since it is true for **all the values** of the variables x, y and z.

In a similar fashion we may define equations in four, five, etc. unknowns.

c) Equivalent equations: Two equations are said to be **equivalent if they have the same solution set**, i.e. if any solution of the first is a solution of the second, and vice versa, every solution of the second is a solution of the first.

Theorems 2-2-1 and 2-2-2 are used in order to replace an equation with an equivalent one.

d) Every equation of the form $F(x, y, z, \cdots) = G(x, y, z, \cdots)$ can be cast in the form $\boldsymbol{f(x, y, z, \cdots) = 0}$, (obviously, if we define $f(x, y, z, \cdots) = F(x, y, z, \cdots) - G(x, y, z, \cdots)$). In case $f(x, y, z, \cdots)$ is **a polynomial** in $x, y, z, \cdots$, then $f(x, y, z, \cdots) = 0$ is called **a polynomial equation**. The degree of this polynomial (see Sec. 1-5(c)) is the degree of the equation. For instance, $x^2y^3 - 3x^2y + xy - 10 = 0$ is a fifth degree polynomial equation in x and y, $x^4y^3 + 3x^3y^2 + 5x^2y + 15x - 3y + 7 = 0$ is a seventh degree polynomial equation in x and y, etc.

e) The following two theorems are sometimes useful in solving some special types of equations.

Theorem 4-1-1: If the product $A_1 \cdot A_2 \cdot \cdots \cdot A_n = 0$, where $A_1, A_2, \cdots, A_n$ are functions of $x, y, z, \cdots$ then $A_1 = 0$, or $A_2 = 0$, or..., or $A_n = 0$, i.e.

$$A_1 \cdot A_2 \cdot \cdots \cdot A_n = 0 \iff A_1 = 0, \text{ or } A_2 = 0, \text{ or } \cdots, \text{ or } A_n = 0$$

Proof: Obvious.

As an example, the equation $(2x + 3y - 1)(x + y + z + 5) = 0$ is equivalent to $2x + 3y - 1 = 0$ **or** $x + y + z + 5 = 0$.

Before stating the second theorem, we need the following definition.

A function $A(x, y, z, \cdots)$ is called **non-negative** if
$$A(x, y, z, \cdots) \geq 0, \qquad \forall (x, y, z, \cdots) \in \mathbb{R} \times \mathbb{R} \times \mathbb{R} \ldots$$

In other words, $A(x, y, z, \cdots)$ is either zero or positive (**never negative**) for all values of the variables $x, y, z, \cdots$. For example $|x|, |x + y|, (x - y)^2, |x + y - z|, (x + y + z - 1)^2$ are all **non-negative quantities**, they can be either zero or positive. The function $x^2 + 1$ is **strictly positive**, $\forall x \in \mathbb{R}$. The function $(x - y)^3$ could be either positive (if $x - y > 0$) or negative (if $x - y < 0$).

Theorem 4-1-2: If the sum of non-negative terms is zero, then each term in the sum must be zero, i.e. assuming that $A_1, A_2, \cdots, A_n$ are non-negative quantities, then

$$A_1 + A_2 + \cdots + A_n = 0 \iff A_1 = 0 \text{ and } A_2 = 0 \text{ and } \cdots \text{ and } A_n = 0$$

Proof: If we assume that **at least one** of the terms is not zero, say for example $A_1 \neq 0$, then A_1 must be **positive** (since A_1 is a non-negative quantity), so even in the "**worse case**" where all the other terms $A_2, \cdots, A_n$ are zero, the sum $A_1 + A_2 + \cdots + A_n$ would be positive and not zero, and this contradicts our assumption (that the sum is zero). This forces us to conclude that $A_1 = 0$. Reasoning similarly, we conclude that $A_2 = 0,..., A_n = 0$ and this completes the proof.

For an application of this theorem, see Example 4-1-3. The important thing to remember when applying this theorem is that **all the terms in the summation must be non-negative quantities**.

Example 4-1-1: Show that the pair $(a, (7 - 2a)/3)$ is a solution of the equation $2x + 3y = 7$, for all values of the number a.

Solution

It suffices to show that $(a, (7 - 2a)/3)$ satisfies the equation $\forall a \in \mathbb{R}$. Indeed, we have, $2a + 3(7 - 2a)/3 = 2a + 7 - 2a = 7$, and this completes the proof.

Example 4-1-2: Show that the triad $(a, b, (15 - 2a + b)/3)$ is a solution of $2x - y + 3z = 15$ for all values of the numbers a and b.

Solution

It suffices to show that $(a, b, (15 - 2a + b)/3)$ satisfies the equation $\forall (a, b) \in \mathbb{R} \times \mathbb{R}$. Indeed, we have,

$$2a - b + 3 \cdot \frac{15 - 2a + b}{3} = 2a - b + 15 - 2a + b = 15$$

and this completes the proof.

Example 4-1-3: Solve the equation $|x - 1| + |y + 2| + |2z - 3| = 0$.

Solution

Since all the terms of the equation are **non-negative quantities**, application of Theorem 4-1-2 implies that

$$\begin{cases} x - 1 = 0, \textbf{and} \\ y + 2 = 0, \textbf{and} \\ 2z - 3 = 0 \end{cases} \Leftrightarrow \begin{cases} x = 1 \\ y = -2 \\ z = 3/2 \end{cases}$$

and this is the **unique** solution of the equation.

Example 4-1-4: Solve the equation $(x + y - 5)(2x - 3y - 1) = 0$.

Solution

$$(x + y - 5)(2x - 3y - 1) = 0 \Leftrightarrow \begin{cases} x + y - 5 = 0 \text{ or} & (*) \\ 2x - 3y - 1 = 0 & (**) \end{cases}$$

Equation (*):

$$x + y - 5 = 0 \Leftrightarrow y = 5 - x \Rightarrow Sol.: (a, 5 - a), a \in \mathbb{R} \quad (***)$$

Equation (**):

$$2x - 3y - 1 = 0 \Leftrightarrow y = (2x - 1)/3 \Rightarrow Sol.: (b, (2b - 1)/3), b \in \mathbb{R} (****)$$

The solutions of the original equation are all the ordered pairs

$$(a, 5 - a), a \in \mathbb{R} \text{ and } (b, (2b - 1)/3), b \in \mathbb{R} \quad (*****)$$

For example, for $a = 1$ we obtain one solution (1,4) and for $b = 2$ we obtain another solution (2,1), etc.

PROBLEMS

4-1-1) Show that each one of the following equations does not have real solutions: $x^4 + 5 = 0$, $|x + 1| + |y - 5| = -5$, $x^2 + y^4 + |z - x - y| = -1$.

Use Theorem 4-1-2 to solve the following three problems:

4-1-2) Find the unique solution of the equation $|x + 1| + |y - 3| + |z| = 0$.

4-1-3) Find the unique solution of the equation $(2x - 1)^2 + (y + 3)^8 + (3z - 4)^{10} = 0$, **(Ans:** $x = 1/2, y = -3, z = 4/3$).

4-1-4) Find the unique solution of the equation $(x + 2y)^2 = 4xy - 12y + 2x - 10$, **(Ans:** $x = 1, y = -3/2$).

4-1-5) Solve the equation $(2x - 2y + z - 7)(x - y + 3z + 2) = 0$.

(Ans: $(a, b, (7 - 2a + 2b)), a, b \in \mathbb{R}$ or $(c, d, (-2 - c + d)/3), c, d \in \mathbb{R}$).

4-1-6) Show that for every number a the pair $\left(\frac{4(1-a^2)}{1+a^2}, \frac{8a}{1+a^2}\right)$ is a solution of the equation $x^2 + y^2 = 16$.

4-1-7) Show the identity:

$$(x-y)(x+y-z) + (y-z)(y+z-a) + (z-x)(z+x-y) \equiv 0$$

4-1-8) Show the identity, (known as **Cauchy's identity**)

$$x^3 + y^3 + z^3 - 3xyz \equiv (x+y+z)(x^2+y^2+z^2-xy-yz-zx)$$
$$\equiv \frac{1}{2}(x+y+z)\{(x-y)^2 + (y-z)^2 + (z-x)^2\}$$

4-1-9) Based on Cauchy's identity, show that if the sum of the cubes of three numbers is equal to their triple product, then, either the three numbers are equal to each other, or their sum is equal to zero.

4-2) Linear equations in two unknowns

Every equation of the form

$$ax + by = c \qquad (4-2-1)$$

is called **a first degree** equation or **a linear equation** in two unknowns, x and y. The numbers a, b and c are the **coefficients** of the equation. Every such equation admits an infinite number of solutions. For example, assuming that $b \neq 0$, if $x = k \in \mathbb{R}$, then the corresponding y (as obtained from (4-2-1)) is $y = (c - ax)/b$, i.e. the solution of (4-2-1) is

$$(x = k, \qquad y = (c - ak)/b \qquad k \in \mathbb{R} \qquad (4-2-2)$$

Similarly, assuming that $a \neq 0$, if $y = \lambda \in \mathbb{R}$, then the corresponding x (as obtained from (4-2-1)) is

$$(x = (c - b\lambda)/a, \qquad y = \lambda) \quad \lambda \in \mathbb{R} \qquad (4-2-3)$$

Example 4-2-1: Find the positive and integers solutions of the linear equation $3x + 5y = 13$.

Solution

We want to find the positive and integer solutions of $3x + 5y = 13$, (if there are any). We have,

$$3x + 5y = 13 \Leftrightarrow x = \frac{13 - 5y}{3} \quad (*)$$

and since $x \geq 1$, equation (*) implies

$$\frac{13 - 5y}{3} \geq 1 \Leftrightarrow 13 - 5y \geq 3 \Leftrightarrow -5y \geq -10 \Leftrightarrow y \leq \frac{-10}{-5} = 2$$

and since also $y \geq 1$, we conclude that **the allowed values** of y are, $y = 1$ and $y = 2$.

If $y = 1$, then the corresponding value of x (as obtained from (*)) is $x = (13 - 5 \cdot 1)/3 = 8/3$, (not integer).

If $y = 2$, then the corresponding value of x is $x = (13 - 5 \cdot 2)/3 = 1$, (integer). Thus the sought for solution is the pair $(x = 1, y = 2)$.

PROBLEMS

4-2-1) Find the positive and integers solutions of $5x + 7y = 24$.

(Ans: $(x = 2, y = 2)$).

4-2-2) Assuming that $(x = 1, y = 3)$ is a solution of $ax + by = 3$, show that $a^2 + 6ab = 9(1 - b^2)$.

Hint: Since $(x = 1, y = 3)$ is a solution of $ax + by = 3$, we must have, $a + 3b = 3$, etc.

4-3) Systems of equations and equivalent systems

a) Let us consider two equations $F(x, y) = 0$ and $G(x, y) = 0$. If (x_0, y_0) is a solution of the first equation, (i.e. if $F(x_0, y_0) = 0$), then in general, we do not expect this solution to be also a solution of the second equation.

When we want to determine all the pairs (x, y) which satisfy both equations simultaneously, then we say that we have to solve the system of equations $F(x, y) = 0$ and $G(x, y) = 0$.

Every pair (x_0, y_0) which satisfies **both equations simultaneously** is called a **solution of the system**. The totality of the solutions is called **the solution set of the system**.

The system $\{F(x,y) = 0, G(x,y) = 0\}$ is **a system of two equations with two unknowns**.

Similarly, we may have **a system of three equations with three unknowns** x, y and z,

$$\begin{cases} F(x,y,z) = 0 \\ G(x,y,z) = 0 \\ R(x,y,z) = 0 \end{cases}$$

Any order triad of numbers (x_0, y_0, z_0) which satisfies all three equations **simultaneously** is called **a solution of the system**, i.e. (x_0, y_0, z_0) is a solution if

$$\begin{cases} F(x_0, y_0, z_0) = 0 & and \\ G(x_0, y_0, z_0) = 0 & and \\ R(x_0, y_0, z_0) = 0 & \end{cases}$$

Of course, we may have systems of four equations with four unknowns, five equations with five unknowns, etc.

b) Equivalent systems: Two systems are equivalent if they have the same solutions, i.e. if every solution of the first is a solution of the second and conversely, every solution of the second is a solution of the first.

Theorem 4-3-1: **The system of two equations $\{F = 0, G = 0\}$ is equivalent to the system $\{F = 0, kF + \lambda G = 0\}$, where k and λ are constant numbers different from zero, (for brevity we write F instead of $F(x,y)$ and G instead of $G(x,y)$).**

Proof: a) Let (x_0, y_0) is a solution of the first system. Then

$$\begin{cases} F(x_0, y_0) = 0 \\ G(x_0, y_0) = 0 \end{cases} \Rightarrow \begin{cases} F(x_0, y_0) = 0 \\ kF(x_0, y_0) + \lambda G(x_0, y_0) = 0 \end{cases}$$

and this shows that (x_0, y_0) is a solution of the system $\{F = 0, kF + \lambda G = 0\}$.

b) Let (x_1, y_1) be a solution of $\{F = 0, kF + \lambda G = 0\}$. Then

$$\begin{cases} F(x_1, y_1) = 0 \\ kF(x_1, y_1) + \lambda G(x_1, y_1) = 0 \end{cases} \Rightarrow \begin{cases} kF(x_1, y_1) = 0 \\ kF(x_1, y_1) + \lambda G(x_1, y_1) = 0 \end{cases} \Rightarrow$$

$$kF(x_1, y_1) + \lambda G(x_1, y_1) - kF(x_1, y_1) = 0 \Rightarrow \lambda G(x_1, y_1) = 0 \overset{(\lambda \neq 0)}{\Longrightarrow}$$

$$G(x_1, y_1) = 0 \Rightarrow \{F(x_1, y_1) = 0, \quad G(x_1, y_1) = 0\}$$

and this shows that (x_1, y_1) is a solution of $\{F = 0, G = 0\}$.

The equation $kF + \lambda G = 0$ is called **a linear combination** of the equations $F = 0$ and $G = 0$.

Example 4-3-1: a) Verify that $(x = 1, y = 2)$ is a solution of the system

$$\{\underbrace{x^2 + y^2 - 2x + 3y - 9 = 0}_{F(x,y)}, \quad \underbrace{3x + 2y - 7 = 0}_{G(x,y)}\}$$

b) Is $(x = 2, y = 1)$ a solution of the same system?

Solution

a) $F(1,2) = 1^2 + 2^2 - 2 \cdot 1 + 3 \cdot 2 - 9 = 1 + 4 - 2 + 6 - 9 = 0$

$G(1,2) = 3 \cdot 1 + 2 \cdot 2 - 7 = 3 + 4 - 7 = 0$

and this verifies that $(1,2)$ is a solution of the system.

b) $F(2,1) = 2^2 + 1^2 - 2 \cdot 2 + 3 \cdot 1 - 9 = 4 + 1 - 4 + 3 - 9 = -5 \neq 0$.

Since $(2,1)$ is **not** a solution of the first equation, we do not even have to check the second equation. The pair $(2,1)$ is **not** a solution to the system, since a solution has to satisfy both equations simultaneously.

4-4) Solving a linear system of two equations with two unknowns

In this section we shall consider **linear systems of two equations with two unknowns**, which is a system consisting of two linear equations in two unknowns, (see Sec. 4-2). Such systems are known **as linear 2 × 2 systems**. The general form of such systems is

$$\begin{cases} a_1 x + b_1 y = c_1 \\ a_2 x + b_2 y = c_2 \end{cases} \qquad (4-4-1)$$

where the coefficients $a_1, b_1, c_1, a_2, b_2, c_2$ are considered to be **known** numbers or quantities **not** depending on the unknowns x and y. Solving the system means to find all the pairs (x_0, y_0) which satisfy both equations **simultaneously**. The methods developed in this section, can be easily generalized to solve 3×3 linear systems (3 linear equations in three unknowns,), 4×4 linear systems (4 linear systems in four unknowns), etc.

a) **Solving by substitution:**

Let us for definiteness consider the following system

$$\begin{cases} (a): & 3x + 2y = 9 \\ (b): & 5x - 4y = 37 \end{cases} \qquad (*)$$

We may solve the first equation for x, (i.e. express x in terms of y) and substitute this expression into the second equation, which now becomes an equation for y. Solving this equation we find y and then from the expression of x in terms of y, we find x.

In our problem:

From (a) we get, $x = (9 - 2y)/3$ and substituting in (b) we have:

$$5 \cdot \frac{9 - 2y}{3} - 4y = 37 \Leftrightarrow 5 \cdot (9 - 2y) - 12y = 111 \Leftrightarrow$$

$$45 - 10y - 12y = 111 \Leftrightarrow -22y = 111 - 45 = 66 \Leftrightarrow$$

$$y = -\frac{66}{22} \Leftrightarrow y = -3$$

Having found y the corresponding $x = (9 - 2y)/3 = (9 + 6)/3 = 5$.

The solution of the system is the pair $(x = 5, y = -3)$. Let the reader verify that (3,5) satisfies both equations.

Of course, we could have solved, for example, (b) for y and substitute in (a). The solution would be the same. Indeed, in this case we would have, $y = (5x - 37)/4$ and substituting in (a) yields,

$$3x + 2 \cdot \frac{5x - 37}{4} = 9 \Leftrightarrow 3x + \frac{5x - 37}{2} = 9 \Leftrightarrow$$

$$6x + 5x - 37 = 18 \Leftrightarrow 11x = 37 + 18 = 55 \Leftrightarrow x = \frac{55}{11} = 5$$

and then $y = (5 \cdot 5 - 37)/4 = -12/4 = -3$. Again the solution of the system is $(x = 5, y = -3)$.

b) Solving by elimination (linear combination):

In order to illustrate this method let us consider again the system of part (a), i.e.

$$\begin{cases} (a): & 3x + 2y = 9 \\ (b): & 5x - 4y = 37 \end{cases}$$

If we multiply (a) by 4, (b) by 2 and then add together the resulting two equations, we obtain,

$$\begin{cases} 12x + 8y = 36 \\ 10x - 8y = 74 \end{cases} \stackrel{(add)}{\Longrightarrow} 22x = 110 \Leftrightarrow x = 110/2 = 5$$

and then from (a), $3 \cdot 5 + 2y = 9 \Leftrightarrow 2y = 9 - 15 = -6 \Leftrightarrow y = -6/2 = -3$. The solution of the system is $(x = 5, y = -3)$, identical to the solution found in (a), as it ought to be.

This method is called the **elimination method (or the linear combination method)**, since by a judicious choice of the numbers with which we multiply the equations (before adding) we **may eliminate one of the unknowns**, leaving thus us with **one equation in one unknown**. This method is actually based on Theorem 4-3-1 (let the reader think about this).

Example 4-4-1: Solve the system $\{2x - y = 5, \quad 3x - 4y = 7\}$.

Solution

a) By substitution:

Solving the first equation for y we get, $y = 2x - 5$ and substituting into the second, yields,

$$3x - 4(2x - 5) = 7 \Leftrightarrow 3x - 8x + 20 = 7 \Leftrightarrow -5x = 7 - 20 \Leftrightarrow$$

$$-5x = -13 \Leftrightarrow x = 13/5$$

The corresponding y is: $y = 2x - 5 = 2 \cdot (13/5) - 5 = (26/5) - 5 = 1/5$. The solution of the system is $\left(x = \frac{13}{5}, y = \frac{1}{5}\right)$.

b) By elimination:

$$\begin{cases}(a) & 2x - y = 5 \\ (b) & 3x - 4y = 7\end{cases}$$

The linear combination $(-3) \cdot (a) + 2 \cdot (b)$ eliminates x. Indeed, if we multiply the first equation by (-3), the second the 2 and then **add** together, we obtain,

$$(-3)(2x - y) + 2(3x - 4y) = (-3) \cdot 5 + 2 \cdot 7 \Leftrightarrow$$

$$-6x + 3y + 6x - 8y = -15 + 14 \Leftrightarrow -5y = -1 \Leftrightarrow y = \frac{1}{5}$$

and then from (a), we can find x: $2x - \frac{1}{5} = 5 \Leftrightarrow 2x = 5 + \frac{1}{5} = \frac{26}{5} \Leftrightarrow x = \frac{13}{5}$. Again, the solution of the system is $\left(x = \frac{13}{5}, y = \frac{1}{5}\right)$.

Of course, we could have eliminated y instead of x. The linear combination $(-4) \cdot (a) + 1 \cdot (b)$ leads to the following equation:

$$(-4)(a) + 1 \cdot (b) \Rightarrow (-4) \cdot (2x - y) + 1 \cdot (3x - 4y) = -20 + 7 \Leftrightarrow$$

$$-8x + 4y + 3x - 4y = -13 \Leftrightarrow -5x = -13 \Leftrightarrow x = \frac{13}{5}$$

and then from (a), $2 \cdot \frac{13}{5} - y = 5 \Leftrightarrow y = \frac{26}{5} - 5 = \frac{1}{5}$.

Example 4-4-2: Solve the system $\left\{\frac{x-2}{3} + \frac{y+1}{4} = 3, \frac{2x+1}{5} - \frac{3y-4}{2} = 5\right\}$.

Solution

We first eliminate the denominators. The least common denominator in the first equation is 12, while in the second equation is 10. We have,

$$\begin{cases} \dfrac{x-2}{3} + \dfrac{y+1}{4} = 3 \\ \dfrac{2x+1}{5} - \dfrac{3y-4}{2} = 5 \end{cases} \Leftrightarrow \begin{cases} 12 \cdot \left(\dfrac{x-2}{3} + \dfrac{y+1}{4}\right) = 12 \cdot 3 \\ 10 \cdot \left(\dfrac{2x+1}{5} - \dfrac{3y-4}{2} = 5\right) = 10 \cdot 5 \end{cases} \Leftrightarrow$$

$$\begin{cases} 4(x-2) + 3(y+1) = 36 \\ 2(2x+1) - 5(3y-4) = 50 \end{cases} \Leftrightarrow \begin{cases} 4x - 8 + 3y + 3 = 36 \\ 4x + 2 - 15y + 20 = 50 \end{cases} \Leftrightarrow$$

$$\begin{cases} 4x + 3y = 36 + 8 - 3 \\ 4x - 15y = 50 - 2 - 20 \end{cases} \Leftrightarrow \begin{cases} 4x + 3y = 41 \\ 4x - 15y = 28 \end{cases}$$

This system is solved easily if we just subtract the second equation from the first, (since by doing this **we eliminate** the unknown x). We have,

$$(4x + 3y) - (4x - 15y) = 41 - 28 \Leftrightarrow 18y = 13 \Leftrightarrow y = 13/18.$$

The unknown x can now be found from the first equation, i.e.

$$4x + 3 \cdot \dfrac{13}{18} = 41 \Leftrightarrow 4x = 41 - \dfrac{39}{18} = \dfrac{41 \cdot 18 - 39}{18} \Leftrightarrow$$

$$4x = \dfrac{699}{18} \Leftrightarrow x = \dfrac{699}{18 \cdot 4} = \dfrac{699}{72} = \dfrac{233}{24}$$

The solution of the system is $\left\{x = \dfrac{233}{24}, y = \dfrac{13}{18}\right\}$.

PROBLEMS

Solve the following linear systems with both methods: **substitution** and **elimination**.

4-4-1) $\{x - 2y = 1, \quad 2x + y = 3\}$, (**Ans:** $(x = 7/5, y = 1/5)$).

4-4-2) $\{3x - y = 7, \quad 4x + 9y = 2\}$.

4-4-3) $\{x - 3(2 - y) = 5, \quad x + 2(y - 4) = 1\}$, (**Ans:** $x = 5, y = 2$).

4-4-4) $\{7(1 - x) = 3 - y, \quad 13x - 8y = 5 + y\}$.

4-4-5) $\{4(2x - 3y) + 3(x + 2y) = 3, \quad 2x + 7y = 6\}$.

(**Ans:** $x = 57/89, y = 60/89$).

4-4-6) $\left\{\frac{x-1}{2} = \frac{y+3}{5}, \quad 25x - 32y = 45\right\}$.

4-4-7) $\left\{\frac{2x+3}{7} - \frac{y-1}{2} = \frac{9}{28}, \quad \frac{x-1}{3} + \frac{5y+2}{5} = \frac{7}{15}\right\}$.

(Ans: $x = -171/190, y = 133/190$).

4-4-8) $\{3x + 2y = 7 - 3(2x + y), \quad 5x - 4y = 2 + 7(x - y)\}$.

4-4-9) $\left\{3x - \frac{y-5}{7} = \frac{4x-3}{2}, \quad \frac{3y+4}{5} - \frac{2x-5}{3} = y\right\}$.

(Ans: $x = -28/26, y = 207/26$).

4-4-10) $\left\{\frac{x-3y+4}{x+y} = \frac{7}{2}, \quad \frac{x+y+5}{x-y-3} = \frac{1}{2}\right\}$.

4-4-11) Consider the function $F(x,y) = ax + by$. If $F(5,-2) = 14$ and $F(-3,5) = 8$, find a and b.

(Ans: $a = 86/19, b = 82/19$).

4-4-12) If $\frac{x+1}{y} = \frac{1}{3}$ and $\frac{x}{y+1} = \frac{1}{4}$, find the ratio $\frac{x}{y}$, (Ans: $4/15$).

4-4-13) Determine the constants a and b so that $(x = 2, y = -3)$ is the solution of the system:

$$(2a + 3b)x - (a + 1)y = 5a + 4b - 1$$
$$(3a - 5b)x + (a + b - 2)y = 2a + 3b + 7$$

(Ans: $a = -31/17, b = -3/17$).

4-5) Linear 2 × 2 systems, the general case-Investigation

Any 2 × 2 linear system can eventually (after some possible simplifications) be expressed in the following form:

$$\begin{cases} a_1 x + b_1 y = c_1 \\ a_2 x + b_2 y = c_2 \end{cases} \quad (4-5-1)$$

We assume, without loss of generality, that the coefficients a_2, b_2, a_2, b_2 are different from zero. We may express this fact by writing $\boldsymbol{a_1 b_1 a_2 b_2 \neq 0}$. The case where one of the coefficients is zero, for example $a_1 = 0$, is a trivial case of a system, since in this case the first equation is an equation for y, ($b_1 y = c_1$), the y is determined, and then from the second equation we may easily find x.

Multiplying the first equation by b_2, the second equation by b_1 and subtracting term wise we obtain:

$$b_2(a_1 x + b_1 y) - b_1(a_2 x + b_2 y) = b_2 c_1 - b_1 c_2 \Leftrightarrow$$

$$b_2 a_1 x + b_2 b_1 y - b_1 a_2 x - b_1 b_2 y = b_2 c_1 - b_1 c_2 \Leftrightarrow$$

$$(a_1 b_2 - a_2 b_1) x = b_2 c_1 - b_1 c_2 \tag{*}$$

Notice that the unknown y has been **eliminated**. Equation (*) is a first degree equation for the unknown x, expressed in terms of the coefficients of the system.

Similarly, multiplying the first equation of the system in (4-5-1) by a_2, the second by a_1 and subtracting term wise, the unknown x is **eliminated** and the result is the following first degree equation for y,

$$(a_1 b_2 - a_2 b_1) y = a_1 c_2 - a_2 c_1 \tag{**}$$

Summarizing:

$$\begin{cases} a_1 x + b_1 y = c_1 \\ a_2 x + b_2 y = c_2 \end{cases} \Rightarrow \begin{cases} (a_1 b_2 - a_2 b_1) x = b_2 c_1 - b_1 c_2 \\ (a_1 b_2 - a_2 b_1) y = a_1 c_2 - a_2 c_1 \end{cases} \tag{4-5-2}$$

Case 1: If $(a_1 b_2 - a_2 b_1) \neq 0$ then the system has **a unique solution**,

$$\left\{ x = \frac{b_2 c_1 - b_1 c_2}{a_1 b_2 - a_2 b_1}, \quad y = \frac{a_1 c_2 - a_2 c_1}{a_1 b_2 - a_2 b_1} \right\} \tag{4-5-3}$$

Case 2: If $(a_1 b_2 - a_2 b_1) = 0$ and **at least one** of the numbers $(b_2 c_1 - b_1 c_2)$ or $(a_1 c_2 - a_2 c_1)$ is **different from zero**, then the system does not have any solutions, it is **an impossible system**.

Indeed, suppose that $(a_1 b_2 - a_2 b_1) = 0$ and that $(b_2 c_1 - b_1 c_2) \neq 0$. Then the first equation in (4-5-2) becomes $0 \cdot x = b_2 c_1 - b_1 c_2 \neq 0$, and this cannot happen, since any number when multiplied by zero gives zero. The first equation in (4-5-2) is impossible, and therefore the system is impossible.

Case 3: If $(a_1 b_2 - a_2 b_1) = 0$ and $(b_2 c_1 - b_1 c_2) = 0$ and $(a_1 c_2 - a_2 c_1) = 0$, then the two equations in (4-5-2) become, $0 \cdot x = 0$ and $0 \cdot y = 0$ and is satisfied $\forall (x, y) \in \mathbb{R} \times \mathbb{R}$. The system has **an infinite number of solutions**.

We have thus shown that any 2×2 linear system either has **a unique solution**, or it has no solutions at all (**impossible system**) or even it may have an **infinite number of solutions**, (depending on the coefficients of the system).

For instance, let the reader verify that the system $(2x + 3y = 5, \quad 6x + 9y = 11)$ has no solutions at all, while the system $(4x - 7y = 10, \quad 20x - 35y = 50)$ has an infinite number of solutions.

The solution and the investigation of a 2×2 linear system are facilitated when we express its solution using the symbolism **of determinants**. This is developed in the next section.

4-6) Solving 2×2 linear systems using determinants: Cramer's rule

a) Determinants of second order (or 2×2 determinants):

Various mathematical formulas are expressed in a compact and easy to remember form if we express the difference $(ad - bc)$ by the symbol

$$\begin{vmatrix} a & b \\ c & d \end{vmatrix}$$

which is called **a second order determinant**, and which is composed of **two horizontal rows and two vertical columns**.

So by definition we have:

$$\begin{vmatrix} a & b \\ c & d \end{vmatrix} \stackrel{\text{def}}{=} ad - bc \qquad (4-6-1)$$

For example,

$$\begin{vmatrix} 3 & 4 \\ 2 & 6 \end{vmatrix} = 3 \cdot 6 - 4 \cdot 2 = 18 - 8 = 10$$

$$\begin{vmatrix} 7 & -4 \\ 5 & 3 \end{vmatrix} = 7 \cdot 3 - (-4) \cdot 5 = 21 + 20 = 41$$

$$\begin{vmatrix} 4 & 2 \\ 8 & 4 \end{vmatrix} = 4 \cdot 4 - 2 \cdot 8 = 16 - 16 = 0$$

$$\begin{vmatrix} -3 & -7 \\ -y & x \end{vmatrix} = -3x - (-7)(-y) = -3x - 7y$$

b) Let us again consider the system

$$\begin{cases} a_1 x + b_1 y = c_1 \\ a_2 x + b_2 y = c_2 \end{cases}$$

Using determinants, equation (4-5-2) is written as follows:

$$\begin{vmatrix} a_1 & b_1 \\ a_2 & b_2 \end{vmatrix} x = \begin{vmatrix} c_1 & b_1 \\ c_2 & b_2 \end{vmatrix} \quad \text{and} \quad \begin{vmatrix} a_1 & b_1 \\ a_2 & b_2 \end{vmatrix} y = \begin{vmatrix} a_1 & c_1 \\ a_2 & c_2 \end{vmatrix} \quad (4-6-2)$$

The determinant

$$\begin{vmatrix} a_1 & b_1 \\ a_2 & b_2 \end{vmatrix} \stackrel{\text{def}}{=} a_1 b_2 - a_2 b_1$$

which multiplies both x and y (in formula (4-6-2)) is called the **determinant of the coefficients**. The other two determinants which appear in our analysis are,

$$\begin{vmatrix} c_1 & b_1 \\ c_2 & b_2 \end{vmatrix} \quad \text{and} \quad \begin{vmatrix} a_1 & c_1 \\ a_2 & c_2 \end{vmatrix} \quad (4-6-3)$$

Using determinants the three cases studied in section 4-5, can now be expressed in the following, easy to remember form:

Case 1: If the determinant of the coefficients is **not zero**, i.e. if

$$\begin{vmatrix} a_1 & b_1 \\ a_2 & b_2 \end{vmatrix} \neq 0$$

then the system has **a unique solution** given by the formulas

$$x = \frac{\begin{vmatrix} c_1 & b_1 \\ c_2 & b_2 \end{vmatrix}}{\begin{vmatrix} a_1 & b_1 \\ a_2 & b_2 \end{vmatrix}}, \quad y = \frac{\begin{vmatrix} a_1 & c_1 \\ a_2 & c_2 \end{vmatrix}}{\begin{vmatrix} a_1 & b_1 \\ a_2 & b_2 \end{vmatrix}} \quad (4-6-4)$$

Case 2: If the determinant of the coefficients is zero and **at least one** of the determinants in (4-6-3) is different from zero, then the system is impossible, i.e.

$$If \begin{vmatrix} a_1 & b_1 \\ a_2 & b_2 \end{vmatrix} = 0 \quad and \quad \left\{ \begin{vmatrix} c_1 & b_1 \\ c_2 & b_2 \end{vmatrix} \neq 0 \quad or \quad \begin{vmatrix} a_1 & c_1 \\ a_2 & c_2 \end{vmatrix} \neq 0 \right\} \quad (4-6-5)$$

then the system is **impossible**.

Case 3: If the determinant of the coefficients is zero and **the two** determinants in (4-6-3) are zero, then the system has an infinite number of solutions, i.e.

$$If \begin{vmatrix} a_1 & b_1 \\ a_2 & b_2 \end{vmatrix} = 0 \quad and \quad \left\{ \begin{vmatrix} c_1 & b_1 \\ c_2 & b_2 \end{vmatrix} \neq 0 \quad and \quad \begin{vmatrix} a_1 & c_1 \\ a_2 & c_2 \end{vmatrix} \neq 0 \right\} \quad (4-6-6)$$

then the system has an **infinite number** of solutions.

c) Cramer's rule:

Equation (4-6-4) is known as the **Cramer's rule**. According to this rule, each unknown is expressed as **the quotient of two determinants**. The denominator (for both x and y) is the same and equal to the determinant of the coefficients. The numerator in the formula expressing x is the determinant obtained from the determinant of the coefficients if the "$x - column$", i.e. the column $\begin{pmatrix} a_1 \\ a_2 \end{pmatrix}$ is replaced by the column of constants $\begin{pmatrix} c_1 \\ c_2 \end{pmatrix}$. Similarly, the numerator in the formula expressing y is the determinant obtained from the determinant of the coefficients if the "$y - column$", i.e. the column $\begin{pmatrix} b_1 \\ b_2 \end{pmatrix}$ is replaced by the column of constants $\begin{pmatrix} c_1 \\ c_2 \end{pmatrix}$.

Provided that **the determinant of the coefficients is not zero**, the system has **a unique solution**, expressed by the two formulas in (4-6-4). However, if **the determinant is zero, then the system may have either no solution or an infinite number of solutions**.

Cramer's rule is easily generalized in solving linear systems 3×3, or 4×4, etc.

Example 4-6-1: Using Cramer's rule solve the system

$$\begin{cases} 7x - 3y = 13 \\ -5x + 4y = -8 \end{cases}$$

Solution

$$x = \frac{\begin{vmatrix} 13 & -3 \\ -8 & 4 \end{vmatrix}}{\begin{vmatrix} 7 & -3 \\ -5 & 4 \end{vmatrix}} = \frac{13 \cdot 4 - (-3) \cdot (-8)}{7 \cdot 4 - (-3) \cdot (-5)} = \frac{52 - 24}{28 - 15} = \frac{28}{13}$$

$$y = \frac{\begin{vmatrix} 7 & 13 \\ -5 & -8 \end{vmatrix}}{\begin{vmatrix} 7 & -3 \\ -5 & 4 \end{vmatrix}} = \frac{7 \cdot (-8) - 13 \cdot (-5)}{7 \cdot 4 - (-3) \cdot (-5)} = \frac{-56 + 65}{28 - 15} = \frac{9}{13}$$

Example 4-6-2: Solve an investigate the following system for the various values of the real parameter λ,

$$\{(\lambda + 3)x - (\lambda + 1)y = 2, \quad (\lambda - 2)x - (\lambda + 2)y = 5\}$$

Solution

The determinant D of the coefficients is,

$$D \overset{\text{def}}{=} \begin{vmatrix} \lambda + 3 & -(\lambda + 1) \\ \lambda - 2 & -(\lambda + 2) \end{vmatrix} = -(\lambda + 3)(\lambda + 2) + (\lambda + 1)(\lambda - 2) \Rightarrow$$

$$D = -\lambda^2 - 5\lambda - 6 + \lambda^2 - \lambda - 2 \Rightarrow D = -6\lambda - 8 \Rightarrow \boldsymbol{D = -2(3\lambda + 4)}$$

1) If $D \neq 0$, i.e. if $-2(3\lambda + 4) \neq 0 \Leftrightarrow \lambda \neq -\frac{4}{3}$, then the system has a unique solution,

$$\forall \lambda \in \mathbb{R} - \left\{-\frac{4}{3}\right\}, \quad \begin{aligned} x &= \frac{\begin{vmatrix} 2 & -(\lambda + 1) \\ 5 & -(\lambda + 2) \end{vmatrix}}{\begin{vmatrix} \lambda + 3 & -(\lambda + 1) \\ \lambda - 2 & -(\lambda + 2) \end{vmatrix}} = -\frac{3\lambda + 1}{2(3\lambda + 4)} \\ y &= \frac{\begin{vmatrix} \lambda + 3 & 2 \\ \lambda - 2 & 5 \end{vmatrix}}{\begin{vmatrix} \lambda + 3 & -(\lambda + 1) \\ \lambda - 2 & -(\lambda + 2) \end{vmatrix}} = -\frac{3\lambda + 19}{2(3\lambda + 4)} \end{aligned}$$

2) At $\lambda = -4/3$, the original system becomes, (check it!)

$$\begin{cases} \dfrac{5}{3}x + \dfrac{1}{3}y = 2 \\ \dfrac{5}{3}x + \dfrac{1}{3}y = -\dfrac{5}{2} \end{cases}$$

which is impossible, (one and the same quantity $(\dfrac{5}{3}x + \dfrac{1}{3}y)$ cannot be equal to 2, and at the same time equal to $-5/2$).

In summary, for every $\lambda \in \mathbb{R} - \left\{-\dfrac{4}{3}\right\}$ the system has a unique solution, if $\lambda = -\dfrac{4}{3}$ the system is impossible.

Example 4-6-3: Solve and investigate the following system for the various values of the real parameter a,

$$\begin{cases} ax - 6y = 5a - 3 \\ 2x + (a-7)y = -7a + 29 \end{cases}$$

Solution

The determinant of the coefficients is

$$D = \begin{vmatrix} a & -6 \\ 2 & (a-7) \end{vmatrix} = a^2 - 7a + 12 = (a-4)(a-3) \qquad (*)$$

1) If $D \neq 0 \Leftrightarrow (a-4)(a-3) \neq 0$, i.e. if $a \neq 4$ **and** $a \neq 3$, the system has a unique solution, (let the reader verify it),

$$x = \dfrac{\begin{vmatrix} 5a-3 & -6 \\ -7a+29 & (a-7) \end{vmatrix}}{\begin{vmatrix} a & -6 \\ 2 & (a-7) \end{vmatrix}} = \dfrac{5(a-13)}{a-4}$$

$\forall a \in \mathbb{R} - \{3, 4\}$, $\qquad (**)$

$$y = \dfrac{\begin{vmatrix} a & 5a-3 \\ 2 & -7a+29 \end{vmatrix}}{\begin{vmatrix} a & -6 \\ 2 & (a-7) \end{vmatrix}} = -\dfrac{(7a+2)}{a-4}$$

2) If $a = 4$ the original system becomes,

$$\begin{cases} 4x - 6y = 17 \\ 2x - 3y = 1 \end{cases} \Leftrightarrow \begin{cases} 2x - 3y = 17/2 \\ 2x - 3y = 1 \end{cases}$$

and this shows that the system, in this case ($a = 4$) is impossible.

3) If $a = 3$ the original system becomes,

$$\begin{cases} 3x - 6y = 12 \\ 2x - 4y = 8 \end{cases} \Leftrightarrow \begin{cases} x - 2y = 4 \\ x - 2y = 4 \end{cases}$$

and this system has an infinite number of solutions. For example, if $x = x_0 \in \mathbb{R}$ then $y_0 = (x_0 - 4)/2$.

In summary, if $D = (a - 3)(a - 4) \neq 0$, the system has **a unique** solution given in (**), if $a = 4$ the system has no solutions at all, while if $a = 3$ the system has an infinite number of solutions $(x_0, (x_0 - 4)/2)$, $x_0 \in \mathbb{R}$.

PROBLEMS

Evaluate the following determinants:

4-6-1) $\begin{vmatrix} 2 & 5 \\ 4 & 1 \end{vmatrix}, \begin{vmatrix} -3 & 7 \\ -2 & 7 \end{vmatrix}, \begin{vmatrix} 2 & x \\ -5 & 2y \end{vmatrix}, \begin{vmatrix} 2 & 6 \\ 4 & 12 \end{vmatrix}, \begin{vmatrix} -1 & 7 \\ 2y - 3 & x + 5 \end{vmatrix}.$

(**Ans:** $-18, -7, 5x + 4y, 0, -x - 14y + 16$).

Solve the following systems using Cramer's rule:

4-6-2) $\{3x - 5y = 10, \ 2x + y = 5\}$.

4-6-3) $\left\{ \frac{x}{3} - \frac{2y}{5} = \frac{4}{15}, \ \frac{3x}{2} + \frac{5y}{4} = -3 \right\}$, (**Ans:** $x = -\frac{52}{61}, y = -\frac{84}{61}$).

4-6-4) $\{2(x - 4) + 3(y + 11) = 0, \ -3x + 7(y - 2) = 10\}$.

4-6-5) $\left\{ \frac{3x+2}{3} + \frac{y-x}{4} = 2x + 1, \ \frac{5y-2x}{3} + 3y = 3(x - y) - 2 \right\}$,

(**Ans:** $x = -55/156, \ y = -67/156$).

Solve and investigate the following systems for the real values of the parameters involved:

4-6-6) $\begin{cases} (a - 1)x + (a - 2)y = 2a - 3 \\ (a + 1)x + (a + 2)y = 2a + 3 \end{cases}$.

4-6-7) $\begin{cases} ax + 4y = 3a - 2 \\ x + ay = a - 2 \end{cases}$.

(**Ans:** $a^2 - 4 \neq 0$, i.e. if $a \neq 2$ and $a \neq -2$, unique solution $x = (3a^2 - 6a + 8)/(a^2 - 4)$, $y = (a^2 - 5a + 2)/(a^2 - 4)$, if $a = \pm 2$ impossible).

4-6-8) $\begin{cases} (\lambda + 2)x + (\mu - 2)y = 3\lambda\mu \\ (\lambda - 1)x + (\mu + 1)y = 2\lambda\mu \end{cases}$.

4-6-9) $\begin{cases} x + \lambda y = 1 \\ \lambda x - 3\lambda y = 2\lambda + 3 \end{cases}$.

(**Ans:** $\lambda(\lambda + 3) \neq 0$, unique solution $x = 2, y = -1/\lambda$, if $\lambda = 0$ impossible, if $\lambda = -3$ infinite number of solutions).

4-6-10) $\begin{cases} bx + b^2 y = b - 1 \\ 2x + 4y = 4b - 7 \end{cases}$.

4-6-11) $\begin{cases} (a + b)x + (a - b)y = 2ab \\ (a + c)x + (a - c)y = 2ac \end{cases}$, $bc \neq 0$.

(**Ans:** $a(b - c) \neq 0$, unique solution, $x = a, y = -a$, if $a = 0$ an infinite number of solutions $x = y = arbitrary$, if $b = c$ infinite number of solutions.

4-6-12) $\begin{cases} (a - b)x + (a + b)y = a \\ (a^2 - b^2)x + (a^2 + b^2)y = a^2 \end{cases}$.

4-6-13) $\begin{cases} (a - 1)x + (a + 1)y = 2(a^3 - 1) \\ (a^2 - 1)x + (a^2 + 1)y = 2(a^3 - 1) \end{cases}$.

(**Ans:** $a(1 - a) \neq 0$, unique solution $x = 1 - a^3, y = a^3 - 1$, if $a = 0$ infinite number of solutions, if $a = 1$ infinite number of solutions).

4-6-14) Determine a and b in terms of λ and μ so that the system

$$\begin{cases} (3a - 5b + \lambda)x + (8a - 3b - \mu)y = 1 \\ (2a - 3b + \lambda)x + (4a - b)y = 2 \end{cases}$$

has an infinite number of solutions.

(**Ans:** $a = (5\lambda + 14\mu)/64$, $b = (3\lambda + 2\mu)/16$).

4-6-15) Consider the two triads of numbers (A, B, C) and (a, b, c) and show the identity,

$$(a^2 + b^2 + c^2)(A^2 + B^2 + C^2) - (aA + bB + cC)^2$$
$$\equiv \begin{vmatrix} a & b \\ A & B \end{vmatrix}^2 + \begin{vmatrix} a & c \\ A & C \end{vmatrix}^2 + \begin{vmatrix} b & c \\ B & C \end{vmatrix}^2.$$

This is known as **Lagrange's identity**.

4-6-16) a) In Problem 4-6-15, show the **permanent inequality**,

$$(a^2 + b^2 + c^2)(A^2 + B^2 + C^2) \geq (aA + bB + cC)^2$$

b) If

$$(a^2 + b^2 + c^2)(A^2 + B^2 + C^2) = (aA + bB + cC)^2$$

show that $\frac{a}{A} = \frac{b}{B} = \frac{c}{C}$.

Hint: Use Theorem 4-1-2.

4-7) Linear systems with more than two unknowns

Let us consider the following systems:

$$\begin{cases} 2x - y + z = 5 \\ 3x + 2z - 7z = 1 \\ x + y + 2z = 7 \end{cases}, \quad \begin{cases} x + y + 3z + 2w = 4 \\ 2x + 3y - 4z + 5w = 2 \\ -3x + 6y - 2z + 8w = -1 \\ 4x - 8y + 3z - 2w = 5 \end{cases}$$

The first system is **a linear system** (since each equation is a linear equation in the unknowns x, y and z) **in three unknowns**, or as we say is a 3×3 linear system, (three linear equations in three unknowns). The second system is a **linear 4×4 system**, four linear equations in four unknowns x, y, z and w. Similarly we may have linear systems 5×5, 6×6, etc.

The following theorem is used often in order to replace a linear system by an equivalent one.

Theorem 4-7-1: Given a system $\{F = 0, G = 0, R = 0, ...\}$ we obtain an equivalent system if any one of the equations is replaced by a linear combination of this equation and some of the remaining ones.

Here, for brevity, we write $F, G, R,...$instead of $F(x, y, z, ...), G(x, y, z, ...), R(x, y, z, ...),....$

This theorem is actually a generalization of theorem 4-3-1. The proof of theorem 4-7-1 is similar to the proof of theorem 4-3-1 and is therefore omitted.

For example, application of theorem 4-7-1 implies that

$$\{F = 0, G = 0, R = 0, ...\} \Leftrightarrow \{F = 0, kF + \lambda G + \mu R = 0, R = 0, ...\}$$

where k, λ, μ are real numbers $\neq 0$. In the second system, the equation $G = 0$ has been replaced by the **linear combination** of the first three equations.

a) Gauss's elimination method: Repeated application of this theorem leads to the so called "**Gauss's elimination method**" which is a powerful method in solving linear systems.

As an application let us consider the following 3×3 linear system,

$$\begin{array}{ll} (a): & 3x + 2y + z = 11 \\ (b): & x - y + 2z = 7 \\ (c): & 2x + 4y - 3z = -1 \end{array} \qquad (S)$$

We replace the **second** equation by the linear combination $1 \cdot (a) + (-3) \cdot (b)$ and the **third** equation by the linear combination $2 \cdot (a) + (-3) \cdot (c)$. The new system thus obtained, is equivalent to the system (S), by virtue of theorem 4-7-1. Notice that the numbers 1 and (-3) in the first linear combination, are chosen so that x is eliminated.

$$\begin{array}{ll} 1 \cdot (a): & 3x + 2y + z = 11 \\ (-3) \cdot (b): & -3x + 3y - 6z = -21 \\ \mathbf{1 \cdot (a) + (-3) \cdot (b):} & 5y - 5z = -10 \end{array} \qquad (*)$$

The second linear combination $2 \cdot (a) + (-3) \cdot (c)$ also leads to the elimination of x in the third equation, i.e.

$$\begin{array}{ll} 2 \cdot (a): & 6x + 4y + 2z = 22 \\ (-3) \cdot (c): & -6x - 12y + 9z = 3 \\ \mathbf{2 \cdot (a) + (-3) \cdot (c):} & -8y + 11z = 25 \end{array} \qquad (**)$$

By virtue of theorem 4-7-1, the original system (S) is **equivalent** to the system

$$(a'): \quad 3x + 2y + z = 11$$
$$(b'): \quad 5y - 5z = -10 \qquad (S')$$
$$(c'): \quad -8y + 11z = 25$$

We notice that in the last two equations the unknown x does not appear, **it has been eliminated**, while at the same time, (S') is equivalent to (S).

We now may replace the last equation (c') in (S') by the linear combination $8 \cdot (b') + 5 \cdot (c')$. Notice that this combination eliminates the unknown y, i.e.

$$8 \cdot (b'): \qquad 40y - 40z = -80$$
$$5 \cdot (c'): \qquad -40y + 55z = 125$$
$$8 \cdot (b') + 5 \cdot (c'): \qquad 15z = 45$$

We now see that the last equation contains the unknown z only, i.e. **it is an equation in the unknown z.**

Summarizing, we have:

$$\underbrace{\begin{cases} 3x + 2y + z = 11 \\ x - y + 2z = 7 \\ 2x + 4y - 3z = -1 \end{cases}}_{(S)} \Leftrightarrow \underbrace{\begin{cases} 3x + 2y + z = 11 \\ 5y - 5z = -10 \\ -8y + 11z = 25 \end{cases}}_{(S')} \Leftrightarrow \underbrace{\begin{cases} 3x + 2y + z = 11 \\ 5y - 5z = -10 \\ 15z = 45 \end{cases}}_{(S'')}$$

From the third equation in (S'') we get $z = 3$, and then from the second equation in (S'') we get $y = (5z - 10)/5 = 5/5 = 1$ and finally from the first equation in (S'') we get $x = (11 - 2y - z)/3 = 6/3 = 2$.

We have thus found that the solution of the system (S'') is $(x = 2, y = 1, z = 3)$ and this is also **the solution of the original system** (S) which is equivalent to (S''), since (S'') was obtained from (S) by repeated application of theorem 4-7-1.

This example outlines the general procedure followed when applying **Gauss's elimination method**. In each step we **eliminate one variable** (unknown), (by applying suitable linear combinations) and eventually we end up with **one equation in one unknown**. This unknown is easily found, and then going backwards, we find the second unknown, then the third, etc.

This method can obviously be applied to solve higher order systems (4 × 4, 5 × 5), etc.

b) The method of substitution: This is another method of solving linear systems. The method is best illustrated by means of an example. Let us consider again the same system (S) we used in the previous example,

$$\begin{aligned} (a): & \quad 3x + 2y + z = 11 \\ (b): & \quad x - y + 2z = 7 \\ (c): & \quad 2x + 4y - 3z = -1 \end{aligned} \quad (S)$$

We may solve, for example, the first equation (a) for x, **find x in terms of y and z**, and **substitute** this expression in the other two equations. From the first equation in (S) we obtain,

$$x = \frac{11 - 2y - z}{3} \quad (*)$$

and substituting this expression in (b) and (c) we get,

$$\begin{cases} \dfrac{11 - 2y - z}{3} - y + 2z = 7 \\ 2 \cdot \dfrac{11 - 2y - z}{3} + 4y - 3z = -1 \end{cases} \Leftrightarrow \begin{cases} -5y + 5z = 10 \\ 8y - 11z = -25 \end{cases}$$

This system is a 2 × 2 linear system in the unknowns y and z. Solving the system (using for example **Cramer's rule**) we find $y = 1, z = 3$ and from equation $(*)$ we find $x = 2$, i.e. the solution of the original system is $(x = 2, y = 1, z = 3)$.

c) A generalized method of substitution: The method of substitution presented in (b) may be generalized as follows: Instead of solving one equation for one unknown (say the unknown x) and then substitute in the remaining equations, **we may solve two equations for two unknowns, say x and y**, and then substitute these expressions in the remaining equations. The thus obtained system has **two unknowns fewer** than the original system. The method is illustrated in Example 4-7-1.

d) Cramer's rule: Cramer's rule developed in Section 4-6 for linear 2 × 2 systems can be generalized to linear systems 3 × 3, 4 × 4, etc. In order to be able to apply this method we need to define **higher order determinants**. This is done in section 4-9.

Example 4-7-1: Solve the system (**generalized method of substitution**):

$$\begin{cases} 2x - y + 3z = -2 \\ x + 2y + z = 6 \\ 3x + 2y + 5z = 8 \end{cases}$$

Solution

We may write the system as follows:

$$\begin{cases} 2x - y = -2 - 3z \\ x + 2y = 6 - z \\ 3x + 2y + 5z = 8 \end{cases} \quad (*)$$

We consider the first two equations as a linear 2 × 2 system, with unknowns x and y. The solution of this system will give x and y in terms of z, i.e.

$$x = \frac{\begin{vmatrix} -2-3z & -1 \\ 6-z & 2 \end{vmatrix}}{\begin{vmatrix} 2 & -1 \\ 1 & 2 \end{vmatrix}} = \frac{-4 - 6z + 6 - z}{5} = \frac{-7z + 2}{5} \quad (**)$$

$$y = \frac{\begin{vmatrix} 2 & -2-3z \\ 1 & 6-z \end{vmatrix}}{\begin{vmatrix} 2 & -1 \\ 1 & 2 \end{vmatrix}} = \frac{12 - 2z + 2 + 3z}{5} = \frac{z + 14}{5} \quad (***)$$

and substituting in the third equation in (*) we get,

$$3 \cdot \frac{-7z + 2}{5} + 2 \cdot \frac{z + 14}{5} + 5z = 8 \Leftrightarrow$$

$$-21z + 6 + 2z + 28 + 25z = 40 \Leftrightarrow 6z = 40 - 6 - 28 = 6 \Leftrightarrow z = 1$$

and from (**) and (***) we get $x = (-7 + 2)/5 = -1$ and $y = (1 + 14)/5 = 3$. The solution of the system is $(x = -1, y = 3, z = 1)$.

Example 4-7-2: Solve the 4 × 4 system

$$\begin{cases} 2x + 3y + z + w = 10 \\ x - 2y + 3z - 2w = -22 \\ 3x + 4y + z + 3w = 23 \\ 2x - 3y + 4z + w = -11 \end{cases}$$

Solution

We solve the first two equations for x and y and express these two unknowns in terms of z and w.

$$x = \frac{\begin{vmatrix} 10 - z - w & 3 \\ -22 - 3z + 2w & -2 \end{vmatrix}}{\begin{vmatrix} 2 & 3 \\ 1 & -2 \end{vmatrix}} = \frac{-11z + 4w - 46}{7} \quad (*)$$

$$y = \frac{\begin{vmatrix} 2 & 10 - z - w \\ 1 & -22 - 3z + 2w \end{vmatrix}}{\begin{vmatrix} 2 & 3 \\ 1 & -2 \end{vmatrix}} = \frac{5z - 5w + 54}{7} \quad (**)$$

By virtue of (*) and (**) the third and fourth equations in the original system become,

$$\begin{cases} 3 \cdot \dfrac{-11z + 4w - 46}{7} + 4 \cdot \dfrac{5z - 5w + 54}{7} + z + 3w = 23 \\ 2 \cdot \dfrac{-11z + 4w - 46}{7} - 3 \cdot \dfrac{5z - 5w + 54}{7} + 4z + w = -11 \end{cases} \Leftrightarrow$$

$$\begin{cases} -6z + 13w = 83 \\ -9z + 30w = 177 \end{cases} \quad (***)$$

System (***) is a system of two equations in two unknowns, and can be solved, for example, using Cramer's rule.

$$z = \frac{\begin{vmatrix} 83 & 13 \\ 177 & 30 \end{vmatrix}}{\begin{vmatrix} -6 & 13 \\ -9 & 30 \end{vmatrix}} = \frac{189}{-63} = -3, \quad w = \frac{\begin{vmatrix} -6 & 83 \\ -9 & 177 \end{vmatrix}}{\begin{vmatrix} -6 & 13 \\ -9 & 30 \end{vmatrix}} = \frac{-315}{-63} = 5$$

Having found z and w, we may now determine x and y from equations (*) and (**) respectively,

$$x = \frac{-11z + 4w - 46}{7} = \frac{33 + 20 - 46}{7} = \frac{7}{7} = 1$$

$$y = \frac{5z - 5w + 54}{7} = \frac{-15 - 25 + 54}{7} = \frac{14}{7} = 2$$

The solution of the system is $(x = 1, y = 2, z = -3, w = 5)$.

Example 4-7-3: Solve the system

$$\begin{cases} x + y + z = a \\ y + z + w = b \\ z + w + x = c \\ w + x + y = d \end{cases}$$

where a, b, c, d are considered to be known constants.

Solution

The given system is **a linear 4×4 system**, (unknowns x, y, z, w). It can be solved either by Gauss's elimination method or by substitution. However, we can solve the system by a much easier method, (this is due to the symmetry involved- all coefficients are equal to 1). If we **add the four equations together**, we find,

$$3(x + y + z + w) = a + b + c + d \Leftrightarrow x + y + z + w = \frac{a + b + c + d}{3} \quad (*)$$

and if we subtract the first equation of the system from (*) we find,

$$w = \frac{a + b + c + d}{3} - a \Leftrightarrow w = \frac{b + c + d - 2a}{3}$$

and similarly we find x, y and z.

PROBLEMS

Solve the following systems using: **a)** Gauss's method of elimination and **b)** the method of substitution.

4-7-1) $\begin{cases} 12x + 5y + 2z = -1 \\ 7x + y - 3z = 2 \\ 4x + 3y + 2z = 1 \end{cases}$ **(Ans:** $x = -1, y = 3, z = -2$).

4-7-2) $\begin{cases} x+y-z=2 \\ x+z-w=2 \\ z+w-x=7 \\ w+x-y=8 \end{cases}$

4-7-3) $\begin{cases} 2x-3y+5z=12 \\ 3x+y+z=-4 \\ 4x-4y+13z=16 \end{cases}$ (**Ans:** $x=0, y=-4, z=0$).

4-7-4) $\{\frac{x}{2}+\frac{y}{3}+\frac{z}{4}=62, \quad \frac{x}{3}+\frac{y}{4}+\frac{z}{5}=47, \quad \frac{x}{4}+\frac{y}{5}+\frac{z}{6}=38\}$.

4-7-5) $\begin{cases} x+3y+z+5w=24 \\ x+y+2z+w=1 \\ x+2y+5z+4w=6 \\ x+3y+5z+3w=4 \end{cases}$ (**Ans:** $x=1, y=2, z=-3, w=4$).

4-7-6) $\{\frac{x+1}{y+2}=3, \quad \frac{z+4}{y+3}=2, \quad \frac{z+5}{x+2}=1\}$.

4-7-7) $\begin{cases} a^2x+ay+z=-a^3 \\ b^2x+by+z=-b^3 \\ c^2x+cy+z=-c^3 \end{cases}$ where $(a-b)(b-c)(c-a) \neq 0$

The parameters a, b and c are considered to be known numbers, x, y and z are the unknowns.

(**Ans:** $x=-(a+b+c), \ y=ab+bc+ca, \ z=-abc$).

4-7-8) Solve the system,

$$\begin{cases} x+y+z+w=10 \\ x+y-z-w=16 \\ x-y+z-w=30 \\ x-y-z+w=40 \end{cases}$$

Hint: See Example 4-7-3.

4-7-9) Solve the system,

$$\begin{cases} 3x+2y=711 \\ 3y-5w+z=-29 \\ x+y-2w=90 \\ w-z=22 \end{cases}$$

(**Ans:** $x = 155, y = 123, w = 94, z = 72$).

4-7-10) Solve the system,

$$\{3x + 2y + 4z = 17, \quad 3x - y + z = 8, \quad 2x + 5y + 3z = 19\}$$

4-8) Third order determinants, evaluation and properties

a) In section 4-6 we defined second order determinants by means of the formula,

$$\begin{vmatrix} a & b \\ c & d \end{vmatrix} \stackrel{\text{def}}{=} ad - bc$$

We may now define the **determinant of order 3** as follows:

$$\begin{vmatrix} a_1 & a_2 & a_3 \\ b_1 & b_2 & b_3 \\ c_1 & c_2 & c_3 \end{vmatrix} \stackrel{\text{def}}{=} a_1 \begin{vmatrix} b_2 & b_3 \\ c_2 & c_3 \end{vmatrix} - a_2 \begin{vmatrix} b_1 & b_3 \\ c_1 & c_3 \end{vmatrix} + a_3 \begin{vmatrix} b_1 & b_2 \\ c_1 & c_2 \end{vmatrix} \qquad (4-8-1)$$

The left-hand side symbol in (4-8-1) is called **a third order determinant**. A third order determinant contains **9 elements**, arranged in **3 rows** $((a_1 \ a_2 \ a_3), (b_1 \ b_2 \ b_3), (c_1 \ c_2 \ c_3)$ and in **3 columns** $((a_1 \ b_1 \ c_1), (a_2 \ b_2 \ c_2), (a_3 \ b_3 \ c_3))$. The right-hand side in (4-8-1) is called "**the expansion of the determinant along the first row**".

For example,

$$\begin{vmatrix} 2 & 5 & 3 \\ -1 & 4 & 2 \\ 7 & 1 & 6 \end{vmatrix} = 2 \begin{vmatrix} 4 & 2 \\ 1 & 6 \end{vmatrix} - 5 \begin{vmatrix} -1 & 2 \\ 7 & 6 \end{vmatrix} + 3 \begin{vmatrix} -1 & 4 \\ 7 & 1 \end{vmatrix}$$

$$= 2 \cdot (24 - 2) - 5 \cdot (-6 - 14) + 3 \cdot (-1 - 28)$$

$$= 44 + 100 - 87 = 57$$

The **value** of the determinant is 57.

b) Algebraic cofactors: Let us consider the following third order determinant, and the adjacent arrangement of "plus" and "minus" signs.

$$\begin{vmatrix} a_1 & a_2 & a_3 \\ b_1 & b_2 & b_3 \\ c_1 & c_2 & c_3 \end{vmatrix} \qquad \begin{pmatrix} + & - & + \\ - & + & - \\ + & - & + \end{pmatrix} \qquad (4-8-2)$$

The algebraic cofactor of an element of the third order determinant is the second order determinant obtained if we delete the row and the column in which the element lies, with **the appropriate + or − sign in front**, as obtained from the table of signs in (4-8-2).

For example, the cofactor A_1 of the element a_1 is $A_1 \stackrel{\text{def}}{=} + \begin{vmatrix} b_2 & b_3 \\ c_2 & c_2 \end{vmatrix}$; the cofactor A_2 of the element a_2 is $A_2 \stackrel{\text{def}}{=} - \begin{vmatrix} b_1 & b_3 \\ c_1 & c_3 \end{vmatrix}$; the cofactor A_3 of the element a_3 is $A_3 \stackrel{\text{def}}{=} + \begin{vmatrix} b_1 & b_2 \\ c_1 & c_2 \end{vmatrix}$; the cofactor B_3 of the element b_3 is $B_3 \stackrel{\text{def}}{=} - \begin{vmatrix} a_1 & a_2 \\ c_1 & c_2 \end{vmatrix}$, etc.

We may now express (4-8-1) in the following form,

$$\begin{vmatrix} a_1 & a_2 & a_3 \\ b_1 & b_2 & b_3 \\ c_1 & c_2 & c_3 \end{vmatrix} = a_1 A_1 + a_2 A_2 + a_3 A_3 \qquad (4-8-3)$$

Let us, for instance, consider the determinant $\begin{vmatrix} 3 & 5 & -2 \\ -1 & 1 & 2 \\ 4 & -4 & 7 \end{vmatrix}$. The cofactor of the element 3 is $+ \begin{vmatrix} 1 & 2 \\ -4 & 7 \end{vmatrix} = 15$, the cofactor of the element 5 is $- \begin{vmatrix} -1 & 2 \\ 4 & 7 \end{vmatrix} = 15$ and the cofactor of the element (-2) is $+ \begin{vmatrix} -1 & 1 \\ 4 & -4 \end{vmatrix} = 0$.

According to (4-8-3) we may find the value of the determinant as follows:

$$\begin{vmatrix} 3 & 5 & -2 \\ -1 & 1 & 2 \\ 4 & -4 & 7 \end{vmatrix} = 3 \cdot 15 + 5 \cdot 15 + (-2) \cdot 0 = 45 + 75 = 120$$

c) Properties of determinants: When the elements of a determinant are large numbers, the computation of the determinant may become very cumbersome. The following properties of determinants usually facilitate the calculations.

1) The value of a determinant does not change if we interchange rows with columns, (i.e. if the rows become columns and the columns become rows), i.e.

$$\begin{vmatrix} a_1 & a_2 & a_3 \\ b_1 & b_2 & b_3 \\ c_1 & c_2 & c_3 \end{vmatrix} = \begin{vmatrix} a_1 & b_1 & c_1 \\ a_2 & b_2 & c_2 \\ a_3 & b_3 & c_3 \end{vmatrix} \qquad (4-8-4)$$

To prove (4-8-4) expand both sides according to (4-8-1) and verify that both expansions are identical.

As an immediate consequence of this property, a determinant can be evaluated **by expansion along the first column**, i.e.

$$\begin{vmatrix} a_1 & a_2 & a_3 \\ b_1 & b_2 & b_3 \\ c_1 & c_2 & c_3 \end{vmatrix} = a_1 \begin{vmatrix} b_2 & b_3 \\ c_2 & c_3 \end{vmatrix} - b_1 \begin{vmatrix} a_2 & a_3 \\ c_2 & c_3 \end{vmatrix} + c_1 \begin{vmatrix} a_2 & a_3 \\ b_2 & b_3 \end{vmatrix} \qquad (4-8-5)$$

2) If we interchange two rows (or two columns) the determinant changes sign, i.e.

$$\begin{vmatrix} a_1 & a_2 & a_3 \\ b_1 & b_2 & b_3 \\ c_1 & c_2 & c_3 \end{vmatrix} = - \begin{vmatrix} b_1 & b_2 & b_3 \\ a_1 & a_2 & a_3 \\ c_1 & c_2 & c_3 \end{vmatrix} \qquad (4-8-6)$$

This property can be shown by direct computation of the determinants.

3) If two rows (or two columns) of a determinant are the same, the determinant is zero.

Because if we call D the value of a given determinant, if we interchange two rows the new determinant must be $(-D)$, according to property 2. However, assuming that the two rows are the same, then we must have $D = -D$, i.e. $D = 0$.

4) If all the elements of some row (or some column) of a determinant are multiplied by one and the same number, then the whole determinant is multiplied by that number, i.e.

$$\begin{vmatrix} ka_1 & ka_2 & ka_3 \\ b_1 & b_2 & b_3 \\ c_1 & c_2 & c_3 \end{vmatrix} = k \begin{vmatrix} a_1 & a_2 & a_3 \\ b_1 & b_2 & b_3 \\ c_1 & c_2 & c_3 \end{vmatrix} \quad (k \neq 0) \qquad (4-8-7)$$

The proof is obtained by direct computation of the determinants.

5) If all the elements of one row (or column) are proportional to the corresponding elements of another row (or column), then the determinant is equal to zero, i.e.

$$\begin{vmatrix} a_1 & a_2 & a_3 \\ ka_1 & ka_2 & ka_3 \\ c_1 & c_2 & c_3 \end{vmatrix} = k \cdot \underbrace{\begin{vmatrix} a_1 & a_2 & a_3 \\ a_1 & a_2 & a_3 \\ c_1 & c_2 & c_3 \end{vmatrix}}_{=0 \ (Prop. \ 3 \ and \ 4)} = k \cdot 0 = 0 \qquad (4-8-8)$$

6) The value of a determinant does not change if to the elements of one row we add the corresponding elements of another row multiplied by the same number (the same applies to columns), i.e.

$$\underbrace{\begin{vmatrix} a_1 & a_2 & a_3 \\ b_1 & b_2 & b_3 \\ c_1 & c_2 & c_3 \end{vmatrix}}_{D_1} = \underbrace{\begin{vmatrix} a_1 + ka_2 & a_2 & a_3 \\ b_1 + kb_2 & b_2 & b_3 \\ c_1 + kc_2 & c_2 & c_3 \end{vmatrix}}_{D_2} = \underbrace{\begin{vmatrix} a_1 + ka_2 + \lambda a_3 & a_2 & a_3 \\ b_1 + kb_2 + \lambda b_3 & b_2 & b_3 \\ c_1 + kc_2 + \lambda c_3 & c_2 & c_3 \end{vmatrix}}_{D_3}$$

Indeed, if we expand D_2 along the elements of the first column, we have:

$$D_2 = (a_1 + ka_2)\begin{vmatrix} b_2 & b_3 \\ c_2 & c_3 \end{vmatrix} - (b_1 + kb_2)\begin{vmatrix} a_2 & a_3 \\ c_2 & c_3 \end{vmatrix} + (c_1 + kc_2)\begin{vmatrix} a_2 & a_3 \\ b_2 & b_2 \end{vmatrix} \Rightarrow$$

$$D_2 = \underbrace{a_1\begin{vmatrix} b_2 & b_3 \\ c_2 & c_3 \end{vmatrix} - b_1\begin{vmatrix} a_2 & a_3 \\ c_2 & c_3 \end{vmatrix} + c_1\begin{vmatrix} a_2 & a_3 \\ b_2 & b_2 \end{vmatrix}}_{D_1}$$

$$+ k\left\{ a_2\begin{vmatrix} b_2 & b_3 \\ c_2 & c_3 \end{vmatrix} - b_2\begin{vmatrix} a_2 & a_3 \\ c_2 & c_3 \end{vmatrix} + c_2\begin{vmatrix} a_2 & a_3 \\ b_2 & b_2 \end{vmatrix} \right\} \Rightarrow$$

$$D_2 = D_1 + k\underbrace{\begin{vmatrix} a_2 & a_2 & a_3 \\ b_2 & b_2 & b_3 \\ c_2 & c_2 & c_3 \end{vmatrix}}_{=0 \ (Property \ 3)} \Rightarrow D_2 = D_1$$

Similarly we may show that $D_2 = D_3$.

7) The value of a determinant can be evaluated by expanding along the elements of any row or any column.

This is a consequence of the aforesaid properties.

A general remark: The properties stated for third order determinants apply for determinants of any order, 4×4, 5×5, etc. Thus we may find the value of a fourth order determinant by expanding along the elements of the first column or the elements of the third row, etc. As a rule, **we expand along the row or the column which contains the greatest number of zeros**, (for obvious reasons). If none of the elements is zero, then we try to "create" some zeros, using property 6, and then expand along the row or column containing the zero (or zeros).

Example 4-8-1: Evaluate the determinant $D = \begin{vmatrix} 1 & 2 & -1 \\ 3 & 5 & -2 \\ -3 & 4 & 6 \end{vmatrix}$.

Solution

a) Expansion along the elements of the **first column**:

$$D = 1 \cdot \begin{vmatrix} 5 & -2 \\ 4 & 6 \end{vmatrix} - 3 \cdot \begin{vmatrix} 2 & -1 \\ 4 & 6 \end{vmatrix} + (-3) \cdot \begin{vmatrix} 2 & -1 \\ 5 & -2 \end{vmatrix} \Rightarrow$$

$$D = 38 - 3 \cdot 16 - 3 \cdot 1 = -13$$

b) Of course, we could have expanded along any row or any column (property 7). Let us, for example, expand along the **third row**:

$$D = (-3) \cdot \begin{vmatrix} 2 & -1 \\ 5 & -2 \end{vmatrix} - 4 \cdot \begin{vmatrix} 1 & -1 \\ 3 & -2 \end{vmatrix} + 6 \cdot \begin{vmatrix} 1 & 2 \\ 3 & 5 \end{vmatrix} \Rightarrow$$

$$D = (-3) \cdot 1 - 4 \cdot 1 + 6 \cdot (-1) = -3 - 4 - 6 = -13$$

Example 4-8-2: Evaluate the determinant $D = \begin{vmatrix} 86 & 19 & 21 \\ 52 & 13 & 14 \\ 75 & 24 & 26 \end{vmatrix}$.

Solution

One possible way to evaluate D is to expand directly, along the elements of one row or one column. However, since the elements are rather large numbers, we prefer to apply some of the properties, in order to work with smaller numbers. We have,

$$D = \begin{vmatrix} 86 & 19 & 21 \\ 52 & 13 & 14 \\ 105 & 24 & 26 \end{vmatrix} = \underbrace{\begin{vmatrix} 86 - 4 \cdot 19 & 19 & 21 - 19 \\ 52 - 4 \cdot 13 & 13 & 14 - 13 \\ 105 - 4 \cdot 24 & 24 & 26 - 24 \end{vmatrix}}_{Property\ 6} = \begin{vmatrix} 10 & 19 & 2 \\ 0 & 13 & 1 \\ 9 & 24 & 2 \end{vmatrix}$$

and expanding along the first column we find

$$D = 10 \cdot \begin{vmatrix} 13 & 1 \\ 24 & 2 \end{vmatrix} + 9 \cdot \begin{vmatrix} 19 & 2 \\ 13 & 1 \end{vmatrix} = 10 \cdot (26 - 24) + 9 \cdot (19 - 26) = -43$$

Example 4-8-3: Evaluate the determinant $D = \begin{vmatrix} x & 5 & y+z \\ y & 5 & z+x \\ z & 5 & x+y \end{vmatrix}$.

Solution

By virtue of properties 4 and 6, we have:

$$D = \begin{vmatrix} x & 5 & y+z \\ y & 5 & z+x \\ z & 5 & x+y \end{vmatrix} = 5 \cdot \begin{vmatrix} x & 1 & y+z \\ y & 1 & z+x \\ z & 1 & x+y \end{vmatrix} = 5 \cdot \begin{vmatrix} x+(y+z) & 1 & y+z \\ y+(z+x) & 1 & z+x \\ z+(x+y) & 1 & x+y \end{vmatrix} \Rightarrow$$

$$D = 5 \cdot (x+y+z) \cdot \underbrace{\begin{vmatrix} 1 & 1 & y+z \\ 1 & 1 & z+x \\ 1 & 1 & x+y \end{vmatrix}}_{=0 \ (Property\ 3)} = 0$$

Example 4-8-4: Solve the equation $\begin{vmatrix} x & x+6 & x+3 \\ x+5 & x & x+2 \\ x+3 & x+4 & x \end{vmatrix} = 0$.

Solution

$$\begin{vmatrix} x & x+6 & x+3 \\ x+5 & x & x+2 \\ x+3 & x+4 & x \end{vmatrix} = 0 \xRightarrow{(Property\ 6)}$$

$$\begin{vmatrix} x-(x+6) & (x+6)-(x+3) & x+3 \\ (x+5)-x & x-(x+2) & x+2 \\ (x+3)-(x+4) & (x+4)-x & x \end{vmatrix} = 0 \Leftrightarrow$$

$$\begin{vmatrix} -6 & 3 & x+3 \\ 5 & -2 & x+2 \\ -1 & 4 & x \end{vmatrix} = 0 \Leftrightarrow \begin{vmatrix} -6+2\cdot 3 & 3 & x+3 \\ 5+2\cdot(-2) & -2 & x+2 \\ -1+2\cdot 4 & 4 & x \end{vmatrix} = 0 \Leftrightarrow$$

$$\begin{vmatrix} 0 & 3 & x+3 \\ 1 & -2 & x+2 \\ 7 & 4 & x \end{vmatrix} = 0$$

and expanding along the elements of the first column, we have,

$$-1 \cdot \begin{vmatrix} 3 & x+3 \\ 4 & x \end{vmatrix} + 7 \cdot \begin{vmatrix} 3 & x+3 \\ -2 & x+2 \end{vmatrix} = 0 \Leftrightarrow$$

$$-(3x - 4x - 12) + 7(3x + 6 + 2x + 6) = 0 \Leftrightarrow$$

$$x + 12 + 35x + 84 = 0 \Leftrightarrow 36x = -96 \Leftrightarrow x = -\frac{96}{36} = -\frac{8}{3}$$

PROBLEMS

Evaluate the following determinants:

4-8-1) $\begin{vmatrix} 3 & 4 & -3 \\ -2 & 5 & 1 \\ 1 & 6 & -1 \end{vmatrix}$, (**Ans:** 14).

4-8-2) $\begin{vmatrix} 4 & 6 & 3 \\ 5 & 8 & 5 \\ 6 & 10 & 7 \end{vmatrix}$

4-8-3) $\begin{vmatrix} 15 & 10 & 18 \\ 10 & 1 & 18 \\ 25 & 16 & 54 \end{vmatrix}$, (**Ans:** -1980).

4-8-4) $\begin{vmatrix} -3 & 2 & 1 \\ 5 & 8 & -1 \\ 4 & 7 & 9 \end{vmatrix}$

4-8-5) $\begin{vmatrix} 1 & 1 & 1 \\ 4 & 5 & 9 \\ 16 & 25 & 81 \end{vmatrix}$, (**Ans:** 20).

4-8-6) $\begin{vmatrix} 7 & 11 & -5 \\ 14 & 21 & 6 \\ 2 & -3 & 4 \end{vmatrix}$

4-8-7) $\begin{vmatrix} 4 & -3 & 5 \\ 3 & -2 & 8 \\ 1 & -7 & -5 \end{vmatrix}$, (**Ans:** 100).

4-8-8) Show that

$$\begin{vmatrix} 1+x & y & z \\ x & 1+y & z \\ x & y & 1+z \end{vmatrix} = 1 + x + y + z$$

Hint: To simplify the calculations, subtract the second row from the first and the third row from the second, (property 6).

4-8-9) Solve the equation

$$\begin{vmatrix} 3 & x+4 & 5 \\ x+2 & 3 & 3 \\ 3 & 5 & x+4 \end{vmatrix} = 0$$

(**Ans:** $0, 1, -11$).

4-8-10) Show that

$$\begin{vmatrix} (x+y)(x^2+y^2) & x+y & 1 \\ (y+z)(y^2+z^2) & y+z & 1 \\ (z+x)(z^2+x^2) & z+x & 1 \end{vmatrix} = 0$$

Hint: To simplify the calculations, subtract the second row from the first and the third row from the second, (property 6).

4-8-11) Without expanding the determinants show that

$$\begin{vmatrix} a_1 & a_2 & a_3 \\ b_1 & b_2 & b_3 \\ c_1 & c_2 & c_3 \end{vmatrix} = \begin{vmatrix} b_2 & a_2 & c_2 \\ b_1 & a_1 & c_1 \\ b_3 & a_3 & c_3 \end{vmatrix} = \begin{vmatrix} b_1 & b_2 & b_3 \\ c_1 & c_2 & c_3 \\ a_1 & a_2 & a_3 \end{vmatrix}$$

4-8-12) If $\begin{vmatrix} a & b & c \\ c & a & b \\ b & c & a \end{vmatrix} = 0$, show that either $a = b = c$ or $(a+b+c) = 0$.

Hint: See Problems 4-1-8 and 4-1-9.

4-8-13) Solve for x the equation

$$\begin{vmatrix} x+a & x & x \\ x & x+b & x \\ x & x & x+c \end{vmatrix} = 0$$

(**Ans:** $x = -abc/(ab + bc + ca)$, assuming that the denominator is not zero).

4-8-14) Show that $\begin{vmatrix} 1+a^2 & ab & ac \\ ba & 1+b^2 & bc \\ ca & cb & 1+c^2 \end{vmatrix} = 1 + a^2 + b^2 + c^2$.

4-8-15) If x and y are any two angles, show that

$$\begin{vmatrix} \cos x & \sin x \cos y & \sin x \sin y \\ -\sin x & \cos x \cos y & \cos x \sin y \\ 0 & -\sin y & \cos y \end{vmatrix} = 1$$

4-8-16) Show that from the equality

$$\begin{vmatrix} 1 & \cos x & \cos y \\ \cos x & 1 & \cos z \\ \cos y & \cos z & 1 \end{vmatrix} = \begin{vmatrix} 0 & \cos x & \cos y \\ \cos x & 0 & \cos z \\ \cos y & \cos z & 0 \end{vmatrix}$$

it follows that $(\cos x)^2 + (\cos y)^2 + (\cos z)^2 = 1$.

4-9) Solving linear 3×3 systems using determinants: Cramer's rule

a) Let us consider the following 3×3 linear system,

$$\begin{cases} a_1 x + a_2 y + a_3 z = d_1 \\ b_1 x + b_2 y + b_3 z = d_2 \\ c_1 x + c_2 y + c_3 z = d_3 \end{cases} \qquad (4-9-1)$$

This systems consists of three equations with three unknowns x, y and z. The coefficients of the unknowns and d_1, d_2, d_3 are assumed to be known numbers. **Solving the system means to find the values of the unknowns x, y and z which satisfy simultaneously all the equations of the system.**

In solving this system with the aid of determinants, we shall find that the following four determinants play a decisive role.

$$D = \begin{vmatrix} a_1 & a_2 & a_3 \\ b_1 & b_2 & b_3 \\ c_1 & c_2 & c_3 \end{vmatrix} \qquad (4-9-2)$$

The determinant D is **the determinant of the coefficients of the system**.

$$D_x = \begin{vmatrix} d_1 & a_2 & a_3 \\ d_2 & b_2 & b_3 \\ d_3 & c_2 & c_3 \end{vmatrix} \quad D_y = \begin{vmatrix} a_1 & d_1 & a_3 \\ b_1 & d_2 & b_3 \\ c_1 & d_3 & c_3 \end{vmatrix} \quad D_z = \begin{vmatrix} a_1 & a_2 & d_1 \\ b_1 & b_2 & d_2 \\ c_1 & c_2 & d_3 \end{vmatrix} \quad (4-9-3)$$

We notice that the determinant D_x is obtained from D when the "**column**" **of the coefficients of the unknown x**, i.e. the column $\begin{pmatrix} a_1 \\ b_1 \\ c_1 \end{pmatrix}$ is replaced by the

column of the constants $\begin{pmatrix} d_1 \\ d_2 \\ d_3 \end{pmatrix}$; similarly D_y is obtained from D when the column of the coefficients of y is replaced by the column of constants and D_z is obtained from D when the column of the coefficients of z is replaced by the column of constants.

In order to solve the system (4-9-1) we may, for instance, solve the second and third equation for y and z, i.e. find y and z **in terms of** x and then replace in the first equation, which now becomes **an equation in one unknown**, the unknown x, (see section 4-7(c)). We have:

$$\begin{cases} b_2 y + b_3 z = d_2 - b_1 x \\ c_2 y + c_3 z = d_3 - c_1 x \end{cases} \Rightarrow$$

$$y = \frac{\begin{vmatrix} d_2 - b_1 x & b_3 \\ d_3 - c_1 x & c_3 \end{vmatrix}}{\begin{vmatrix} b_2 & b_3 \\ c_2 & c_3 \end{vmatrix}}, \quad z = \frac{\begin{vmatrix} b_2 & d_2 - b_1 x \\ c_2 & d_3 - c_1 x \end{vmatrix}}{\begin{vmatrix} b_2 & b_3 \\ c_2 & c_3 \end{vmatrix}} \quad (*)$$

and substituting in the first equation in (4-9-1) yields,

$$a_1 x + a_2 y + a_3 z = d_1 \stackrel{(*)}{\Rightarrow}$$

$$a_1 x + a_2 \cdot \frac{\begin{vmatrix} d_2 - b_1 x & b_3 \\ d_3 - c_1 x & c_3 \end{vmatrix}}{\begin{vmatrix} b_2 & b_3 \\ c_2 & c_3 \end{vmatrix}} + a_3 \cdot \frac{\begin{vmatrix} b_2 & d_2 - b_1 x \\ c_2 & d_3 - c_1 x \end{vmatrix}}{\begin{vmatrix} b_2 & b_3 \\ c_2 & c_3 \end{vmatrix}} = d_1$$

which, after a few simple simplifications leads to the following equation:

$$\underbrace{\left\{ a_1 \begin{vmatrix} b_2 & b_3 \\ c_2 & c_3 \end{vmatrix} - a_2 \begin{vmatrix} b_1 & b_3 \\ c_1 & c_3 \end{vmatrix} + a_3 \begin{vmatrix} b_1 & b_2 \\ c_1 & c_2 \end{vmatrix} \right\}}_{\text{Determinant } D} x$$

$$= \underbrace{\left\{ d_1 \begin{vmatrix} b_2 & b_3 \\ c_2 & c_3 \end{vmatrix} - d_2 \begin{vmatrix} a_2 & a_3 \\ c_2 & c_3 \end{vmatrix} + d_3 \begin{vmatrix} a_2 & a_3 \\ b_2 & b_3 \end{vmatrix} \right\}}_{\text{Determinant } D_x} \quad (**)$$

We notice that the coefficient of x in the left-hand side in (**) is nothing else but the expansion of the determinant in (4-9-2) along the elements of the first row, and similarly the right-hand side in (**) is the expansion of the determinant D_x in (4-9-3) along the first column. In terms of the determinants

D and D_x equation (**) takes the simple form, $x \cdot D = D_x$, i.e. (assuming that $D \neq 0$), $x = D_x/D$. Quite similarly we obtain, $y \cdot D = D_y$ and $z \cdot D = D_z$.

Summarizing: The solution of linear 3×3 the system

$$\begin{cases} a_1 x + a_2 y + a_3 z = d_1 \\ b_1 x + b_2 y + b_3 z = d_2 \\ c_1 x + c_2 y + c_3 z = d_3 \end{cases}$$

is given by the formulas

$$\{x \cdot D = D_x, \quad y \cdot D = D_y, \quad z \cdot D = D_z\} \qquad (4-9-4)$$

where the determinants D, D_x, D_y, D_z are defined in formulas (4-9-2) and (4-9-3). Provided that **the determinant of the coefficients $D \neq 0$**, we find,

$$\begin{cases} x = \dfrac{D_x}{D} = \dfrac{\begin{vmatrix} d_1 & a_2 & a_3 \\ d_2 & b_2 & b_3 \\ d_3 & c_2 & c_3 \end{vmatrix}}{\begin{vmatrix} a_1 & a_2 & a_3 \\ b_1 & b_2 & b_3 \\ c_1 & c_2 & c_3 \end{vmatrix}} \\[2em] y = \dfrac{D_y}{D} = \dfrac{\begin{vmatrix} a_1 & d_1 & a_3 \\ b_1 & d_2 & b_3 \\ c_1 & d_3 & c_3 \end{vmatrix}}{\begin{vmatrix} a_1 & a_2 & a_3 \\ b_1 & b_2 & b_3 \\ c_1 & c_2 & c_3 \end{vmatrix}} \\[2em] z = \dfrac{D_z}{D} = \dfrac{\begin{vmatrix} a_1 & a_2 & d_1 \\ b_1 & b_2 & d_2 \\ c_1 & c_2 & d_3 \end{vmatrix}}{\begin{vmatrix} a_1 & a_2 & a_3 \\ b_1 & b_2 & b_3 \\ c_1 & c_2 & c_3 \end{vmatrix}} \end{cases} \quad (\textbf{\textit{Cramer's rule}}) \quad (4-9-5)$$

Formulas (4-9-5) are known as **"the Cramer's rule"**.

b) Investigation of a linear, 3×3 system: Assuming that the determinant of the coefficients $D \neq 0$, the system has **a unique solution** for x, y, z given by equation (4-9-5).

If $D = 0$, and $D_x = 0$ and $D_y = 0$ and $D_z = 0$, then it follows from (4-9-4) that the system has **an infinite number of solutions**, (every triad of numbers x, y, z satisfies the system).

If $D = 0$ and **at least one** of D_x or D_y or D_z is **not zero**, then the system is **impossible**; for example if $D = 0$ and $D_x \neq 0$, then the first equation in (4-9-4) implies that $0 \cdot x = D_x \neq 0$ and this is impossible.

c) Cramer's rule applies also to higher order systems. However in such cases we have to deal with higher order determinants.

Example 4-9-1: Using Cramer's rule solve the system

$$2x - y + 3z = -2$$
$$x + 3y + 7z = 4$$
$$3x + 2y + 4z = 8$$

Solution

We have to find the determinants D, D_x, D_y, and D_z.

$$D = \begin{vmatrix} 2 & -1 & 3 \\ 1 & 3 & 7 \\ 3 & 2 & 4 \end{vmatrix} = 2 \cdot \begin{vmatrix} 3 & 7 \\ 2 & 4 \end{vmatrix} - 1 \cdot \begin{vmatrix} -1 & 3 \\ 2 & 4 \end{vmatrix} + 3 \cdot \begin{vmatrix} -1 & 3 \\ 3 & 7 \end{vmatrix} \Rightarrow$$

$$D = 2 \cdot (12 - 14) - 1 \cdot (-4 - 6) + 3 \cdot (-7 - 9) = -42$$

$$D_x = \begin{vmatrix} -2 & -1 & 3 \\ 4 & 3 & 7 \\ 8 & 2 & 4 \end{vmatrix} = (-2) \cdot \begin{vmatrix} 3 & 7 \\ 2 & 4 \end{vmatrix} - 4 \cdot \begin{vmatrix} -1 & 3 \\ 2 & 4 \end{vmatrix} + 8 \cdot \begin{vmatrix} -1 & 3 \\ 3 & 7 \end{vmatrix} = -84$$

and similarly we find,

$$D_y = \begin{vmatrix} 2 & -2 & 3 \\ 1 & 4 & 7 \\ 3 & 8 & 4 \end{vmatrix} = -126, \quad D_z = \begin{vmatrix} 2 & -1 & -2 \\ 1 & 3 & 4 \\ 3 & 2 & 8 \end{vmatrix} = 42$$

and from formula (4-9-5) we obtain,

$$\left\{ x = \frac{D_x}{D} = \frac{-84}{-42} = 2, \quad y = \frac{D_y}{D} = \frac{-126}{-42} = 3, \quad z = \frac{D_z}{D} = \frac{42}{-42} = -1 \right\}$$

Example 4-9-2: Solve and investigate the system (k is a real parameter),

$$kx + 3y = 1$$
$$kx + 8y + z = 0$$
$$5x + 7y + 2z = 0$$

Solution

Expanding the determinant of the coefficients along the elements of the first row we find,

$$D = \begin{vmatrix} k & 3 & 0 \\ k & 8 & 1 \\ 5 & 7 & 2 \end{vmatrix} = k \begin{vmatrix} 8 & 1 \\ 7 & 2 \end{vmatrix} - 3 \begin{vmatrix} k & 1 \\ 5 & 2 \end{vmatrix} = 3(k+5)$$

Also, we find easily that $D_x = 9, D_y = 5 - 2k$ and $D_z = 7k - 40$, (let the reader verify the results).

1) When $D \neq 0$, i.e. $k \neq -5$, the system has a unique solution (see equation (4-9-5)),

$$x = \frac{D_x}{D} = \frac{9}{3(k+5)}, \quad y = \frac{D_y}{D} = \frac{5-2k}{3(k+5)}, \quad z = \frac{D_z}{D} = \frac{7k-40}{3(k+5)}$$

2) When $D = 0$, i.e. $k = -5$, the first equation in (4-9-4) becomes $x \cdot 0 = D_x = 9$ and since this equation is impossible, the system is impossible.

PROBLEMS

Solve the following systems with the aid of Cramer's rule:

4-9-1) $\begin{cases} x + 4y - 8z = -8 \\ 4x + 8y - z = 76 \\ 8x - y - 4z = 110 \end{cases}$, (Ans: $x = 16, y = 2, z = 4$).

4-9-2) $\begin{cases} 2x - 3y + 4z = 8 \\ 3x + 7y - z = -1 \\ 5z - y - 2z = 10 \end{cases}$.

4-9-3) $\begin{cases} x + 2y + 3z = 8 \\ 2x + 3y + 4z = 12 \\ 3x + 5y + 7z = 20 \end{cases}$, (Ans: $x = \lambda, y = 4 - 2\lambda, z = \lambda, \ \lambda \in \mathbb{R}$).

4-9-4) $\begin{cases} 3x - 2y + 6z = 0 \\ x + y + 2z = 7 \\ 4x + 11y - 3z = 5 \end{cases}$.

4-9-5) $\begin{cases} x + y + z = -2 \\ 2x + 5y + z = -1 \\ 5x - 2y + 3z = -4 \end{cases}$, (Ans: $x = 1, y = 0, z = -3$).

4-9-6) $\begin{cases} 2x + 3y + 7z = -9 \\ -x + y - z = 7 \\ 5x + 4y - 5z = -2 \end{cases}$.

4-9-7) $\begin{cases} 2x + 3y + 4z = 53 \\ 3x + 5y - 4z = 2 \\ 4x + 7y - 2z = 31 \end{cases}$, (Ans: $x = 3, y = 5, z = 8$).

4-9-8) Solve the system: $\begin{cases} ax + cy = b \\ cz + bx = a \\ by + az = c \end{cases}$, ($a, b, c$ are known numbers).

4-9-9) Solve the system: $\begin{cases} (b+c)x + (c+a)y - (a+b)z = 2c^3 \\ (b+c)x - (c+a)y + (a+b)z = 2b^3 \\ -(b+c)x + (c+a)y + (a+b)z = 2a^3 \end{cases}$.

(Assume $(a+b)(b+c)(c+a) \neq 0$).

(Ans: $x = b^2 - bc + c^2$, $y = a^2 - ac + c^2$, $z = a^2 - ab + b^2$).

4-9-10) Solve and investigate the system: $\begin{cases} x + 2y + (k+3)z = 8 \\ 2x + 3y + (k+4)z = 12 \\ 3x + (6k+5)y + 7z = 20 \end{cases}$.

(Ans: If $k(3k+7) \neq 0$, unique solution $x = \frac{12(k+1)}{3k+7}, y = \frac{4}{3k+7}, z = \frac{12}{3k+7}$, if $k = 0$ an infinite number of solutions, if $k = -7/3$ the system is impossible).

4-9-11) Solve and investigate the system: $\begin{cases} kx + y + z = 1 \\ x + ky + z = k \\ x + y + kz = k^2 \end{cases}$, $k \in \mathbb{R}$.

(**Ans:** If $(k-1)(k+2) \neq 0$, unique solution $x = -\frac{k+1}{k+2}, y = \frac{1}{k+2}, z = \frac{(k+1)^2}{k+2}$, if $k = 1$ infinite number of solutions, if $k = -2$ impossible.

4-9-12) Solve and investigate the system: $\begin{cases} x + y + \lambda z = 1 \\ x + \lambda y + z = 2 \\ x - y + z = 3 \end{cases}$, $\lambda \in \mathbb{R}$.

4-9-13) Solve the system for x, y, z: $\begin{cases} x + y + z = a + b + c \\ bx + cy + az = a^2 + b^2 + c^2 \\ cx + ay + bz = a^2 + b^2 + c^2 \end{cases}$.

(**Ans:** $x = b + c - a$, $y = c + a - b$, $z = a + b - c$).

4-10) Consistent equations

Let us consider a system consisting of three equations with two unknowns, or a system of four equations with three unknowns, etc. **The characteristic feature of these systems is that the number of equations exceeds the number of the unknowns.**

Let us for definiteness, consider the following system, (three equations and two unknowns),

$$\begin{cases} 2x + 3y = 13 \\ 5x - 2y = 4 \\ ax + by = 9 \end{cases}$$

and let us investigate **under what conditions** this system has a solution. Solving the system of the first two equations, we find $(x = 2, y = 3)$, and this solution will satisfy the third equation **if and only** if $2a + 3b = 9$. **Otherwise, the system does not have a solution.** So, if a and b are such that they satisfy the equation $2a + 3b = 9$, we say that the three equations of the system are **consistent or compatible**, (i.e. the system does have a solution), otherwise they are **inconsistent**, (the system does not have a solution).

In general, for a system where the number of equations **exceeds** the number of unknowns to be **consistent**, a suitable relation should exist between the coefficients of the unknowns.

Theorem 4-10-1: **The necessary and sufficient condition for the system**

$$\{a_1x + b_1y = c_1 \quad a_2x + b_2y = c_2 \quad a_3x + b_3y = c_3\} \qquad (*)$$

to be consistent, is

$$\begin{vmatrix} a_1 & b_1 & c_1 \\ a_2 & b_2 & c_2 \\ a_3 & b_3 & c_3 \end{vmatrix} = 0 \qquad (4-10-1)$$

For **a proof** see Example 4-10-1.

Let us consider the following, illustrative examples.

Example 4-10-1: Prove equation (4-10-1).

Solution

Assuming that $x = x_0, y = y_0$ is a solution of the system (*), in Th. 4-10-1, we have,

$$\{a_1x_0 + b_1y_0 = c_1 \quad a_2x_0 + b_2y_0 = c_2 \quad a_3x_0 + b_3y_0 = c_3\}$$

Solving the first two equations for x_0, y_0 we find,

$$\left\{ x_0 = \frac{\begin{vmatrix} c_1 & b_1 \\ c_2 & b_2 \end{vmatrix}}{\begin{vmatrix} a_1 & b_1 \\ a_2 & b_2 \end{vmatrix}} \quad y_0 = \frac{\begin{vmatrix} a_1 & c_1 \\ a_2 & c_2 \end{vmatrix}}{\begin{vmatrix} a_1 & b_1 \\ a_2 & b_2 \end{vmatrix}} \right\} \qquad (*)$$

and assuming that the system is consistent, the expressions for x_0 and y_0 in (*) should also satisfy the third equation of the system, i.e.

$$a_3x_0 + b_3y_0 = c_3 \Longleftrightarrow$$

$$a_3 \cdot \frac{\begin{vmatrix} c_1 & b_1 \\ c_2 & b_2 \end{vmatrix}}{\begin{vmatrix} a_1 & b_1 \\ a_2 & b_2 \end{vmatrix}} + b_3 \cdot \frac{\begin{vmatrix} a_1 & c_1 \\ a_2 & c_2 \end{vmatrix}}{\begin{vmatrix} a_1 & b_1 \\ a_2 & b_2 \end{vmatrix}} = c_3 \Longleftrightarrow$$

$$a_3 \cdot \begin{vmatrix} c_1 & b_1 \\ c_2 & b_2 \end{vmatrix} + b_3 \cdot \begin{vmatrix} a_1 & c_1 \\ a_2 & c_2 \end{vmatrix} - c_3 \cdot \begin{vmatrix} a_1 & b_1 \\ a_2 & b_2 \end{vmatrix} = 0 \Longleftrightarrow$$

$$\begin{vmatrix} a_1 & c_1 & b_1 \\ a_2 & c_2 & b_2 \\ a_3 & c_3 & b_3 \end{vmatrix} = 0 \Leftrightarrow \begin{vmatrix} a_1 & b_1 & c_1 \\ a_2 & b_2 & c_2 \\ a_3 & b_3 & c_3 \end{vmatrix} = -\begin{vmatrix} a_1 & c_1 & b_1 \\ a_2 & c_2 & b_2 \\ a_3 & c_3 & b_3 \end{vmatrix} = 0$$

(see properties of determinants, equation (4-8-6)), and this completes the proof.

Example 4-10-2: For what value of the parameter k are the following three equations consistent?

$$\{2x + ky = 8, \quad 2x + (k+1)y = 3 \quad 4x - 5ky = 9\}$$

Solution

Method 1: We may solve the first two equations for x and y,

$$x = \frac{\begin{vmatrix} 8 & k \\ 3 & k+1 \end{vmatrix}}{\begin{vmatrix} 2 & k \\ 2 & k+1 \end{vmatrix}} = \frac{5k+8}{2}, \quad y = \frac{\begin{vmatrix} 2 & 8 \\ 2 & 3 \end{vmatrix}}{\begin{vmatrix} 2 & k \\ 2 & k+1 \end{vmatrix}} = -5$$

If the system is to be **consistent**, these values of x and y **should also satisfy the third equation**, i.e.

$$4 \cdot \frac{5k+8}{2} - 5k \cdot (-5) = 9 \Leftrightarrow 2 \cdot (5k+8) + 25k = 9 \Leftrightarrow$$

$$10k + 16 + 25k = 9 \Leftrightarrow 35k = 9 - 16 = -7 \Leftrightarrow k = -\frac{7}{35}$$

Method 2 (Theorem 4-10-1): According to Th. 4-10-1, the given system will be consistent provided that

$$\begin{vmatrix} 2 & k & 8 \\ 2 & k+1 & 3 \\ 4 & -5k & 9 \end{vmatrix} = 0 \Leftrightarrow k = -\frac{7}{35}$$

(Let the reader verify the calculations, for example expand the determinant along the first row).

Example 4-10-3: Show that the following system of equations is compatible:

$$\begin{cases} \dfrac{x}{a} + \dfrac{y}{b} = 1 \\ \dfrac{x^2}{a} + \dfrac{y^2}{b} = \dfrac{ab}{a+b} \\ \dfrac{x^{n+1}}{a} + \dfrac{y^{n+1}}{b} = \left(\dfrac{ab}{a+b}\right)^n, \quad n = 2,3,4,\cdots \end{cases}$$

Solution

It suffices to show that the solution of the first two equations, satisfies also the third equation.

From the first equation of the system we obtain, $y = b\left(1 - \dfrac{x}{a}\right)$ and substituting in the second equation we find the following equation for x,

$$\dfrac{x^2}{a} + \dfrac{1}{b} \cdot b^2 \left(1 - \dfrac{x}{a}\right)^2 - \dfrac{ab}{a+b} = 0$$

which after a few, straightforward simplifications leads to the equation,

$$(a+b)^2 x^2 - 2ab(a+b)x + a^2 b^2 = 0 \Leftrightarrow$$

$$\{(a+b)x - ab\}^2 = 0 \Leftrightarrow (a+b)x - ab = 0 \Leftrightarrow x = \dfrac{ab}{a+b}$$

and the corresponding $y = b\left(1 - \dfrac{x}{a}\right) = b\left(1 - \dfrac{b}{a+b}\right) = \dfrac{ab}{a+b}$.

In summary, the solution obtained from the first two equations of the system is

$$\left\{ x_0 = \dfrac{ab}{a+b}, \quad y_0 = \dfrac{ab}{a+b} \right\} \qquad (*)$$

The system will be compatible, if x_0 and y_0 satisfy the third equation as well. We verify that this is indeed the case, since

$$\dfrac{x_0^{n+1}}{a} + \dfrac{y_0^{n+1}}{b} = \dfrac{(ab)^{n+1}}{a(a+b)^{n+1}} + \dfrac{(ab)^{n+1}}{b(a+b)^{n+1}} = \dfrac{a^n b^{n+1}}{(a+b)^{n+1}} + \dfrac{a^{n+1} b^n}{(a+b)^{n+1}}$$

$$= \dfrac{a^n b^n (a+b)}{(a+b)^{n+1}} = \dfrac{a^n b^n}{(a+b)^n} = \left(\dfrac{ab}{a+b}\right)^n$$

and this completes the proof.

PROBLEMS

4-10-1) Determine λ for the following system to be consistent:

$$\begin{cases} x + (\lambda + 1)y = 0 \\ (1 - \lambda)x + \lambda y = \lambda + 1 \\ (\lambda + 1)x + (12 - \lambda)y = -(\lambda + 1) \end{cases}$$

(Ans: $\lambda = -1$, or $\lambda = 5$).

4-10-2) For which values of λ is the following system consistent?

$$\{(\lambda - 1)x + y = 1, \quad x + (\lambda - 1)y = 2, \quad 2x - y = 3\}$$

4-10-3) For which values of λ is the following system consistent? What is the solution of the system?

$$\{x = y - 1 = \lambda - z, \quad x - y - z = -\lambda, \quad 3x + y - z = 2\}$$

(Ans: $\lambda = 4$, $x = 1$, $y = 2$, $z = 3$).

4-10-4) Find the equation that should be satisfied by k, λ, μ if the system

$$\{x - \lambda y = k, \quad y - \lambda x = \mu, \quad kx + \mu y = 1\}$$

is to be consistent.

Hint: Apply Theorem 4-10-1.

4-10-5) Provided that the following system is consistent, show that $a + b = c$, (assume $a + b + c \neq 0$).

$$\begin{cases} x + y + z = a + b + c \\ ax + by + cz = a^2 + b^2 + c^2 \\ bx + cy + az = a^2 + b^2 + c^2 \\ cx + ay + bz = 4ab \end{cases}$$

4-10-6) If the linear equations $F(x, y) \equiv ax + by + cz - d = 0$ and $G(x, y) \equiv Ax + By + Cz - D = 0$ are equivalent, show that there exists a constant number k, such that $F(x, y) \equiv kG(x, y)$, (assume $abcd \neq 0$).

Hint: A solution of $F(x, y) = 0$ is given by $x = (d - bk - c\lambda)/a$, $y = k$, $z = \lambda$, where k and λ are arbitrary real numbers. Since $F(x, y) = 0$ and $G(x, y) = 0$ are equivalent, every solution of the former is a solution of the later, and vice versa.

4-10-7) Consider the system of the four linear equations $\{A = 0, B = 0, C = 0, D = 0\}$. If the equations $A - B = 0$ and $C - D = 0$ are equivalent, show that one of the equations of the system results as a linear combination of the other three.

Hint: See Problem 4-10-6.

4-10-8) Provided that the system $\{ax + by = 1, bx + ay = ab, x + y = a + b\}$ is consistent, show that $a^2 + ab + b^2 = 1$, (assume $ab \neq 0$).

4-10-9) Provided that the system $\{\lambda x + \mu y = 0, x + y = xy, x^2 + y^2 = 1\}$ is consistent, show that $\dfrac{1}{\lambda^2} + \dfrac{1}{\mu^2} = \dfrac{1}{(\lambda-\mu)^2}$, ($\lambda, \mu$ are real parameters).

4-11) Linear and homogeneous systems

If the constant term in a linear equation is zero, the equation is called **homogeneous**. For example the equations $ax + by = 0$, $ax + by + cz = 0$, $ax + by + cz + dw = 0$ are homogeneous. **If all the equations in a linear system are homogeneous, the system is called a linear, homogeneous system.**

a) Linear homogeneous system, 3 equations with 3 unknowns.

Let us consider the following 3×3 linear, homogeneous system:

$$\begin{cases} a_1 x + a_2 y + a_3 z = 0 \\ b_1 x + b_2 y + b_3 z = 0 \\ c_1 x + c_2 y + c_3 z = 0 \end{cases} \qquad (4-11-1)$$

This system, has the obvious (**trivial**) solution ($x = 0, y = 0, z = 0$). For every linear and homogeneous system there are two possibilities:

1) The system has **a unique solution** ($x = 0, y = 0, z = 0$) or

2) Has an **infinite number of solutions**.

If the determinant of the coefficients $D \neq 0$, then the system has a unique solution, (see Section 4-9), which is the trivial solution $(x = 0, y = 0, z = 0)$.

If the determinant of the coefficients $D = 0$, then the system has an infinite number of non-trivial solutions. Indeed, if (x_0, y_0, z_0) is a nontrivial solution of the system (4-11-1) then (kx_0, ky_0, kz_0), $k \neq 0$ will also be a solution.

We may thus state the following theorem:

Theorem 4-11-1: **Let D be the determinant of the coefficients of the linear, homogeneous system (4-11-1). Then,**

$$\text{If } D = \begin{vmatrix} a_1 & a_2 & a_3 \\ b_1 & b_2 & b_3 \\ c_1 & c_2 & c_3 \end{vmatrix} = 0 \qquad (4-11-2)$$

the system has an infinite number of non-trivial solutions. If $D \neq 0$ the system has a unique solution $(x = 0, y = 0, z = 0)$.

b) Linear homogeneous system, 2 equations with 3 unknowns.

Let us consider the system

$$\begin{cases} a_1 x + a_2 y + a_3 z = 0 \\ b_1 x + b_2 y + b_3 z = 0 \end{cases} \qquad (4-11-3)$$

For this system the following theorem holds true:

Theorem 4-11-2:

Assuming that, $\begin{vmatrix} a_1 & a_2 \\ b_1 & b_2 \end{vmatrix} \neq 0$, $\begin{vmatrix} a_1 & a_3 \\ b_1 & b_3 \end{vmatrix} \neq 0$, $\begin{vmatrix} a_2 & a_3 \\ b_2 & b_3 \end{vmatrix} \neq 0$, then

$$\begin{cases} a_1 x + a_2 y + a_3 z = 0 \\ b_1 x + b_2 y + b_3 z = 0 \end{cases} \Leftrightarrow \frac{x}{\begin{vmatrix} a_2 & a_3 \\ b_2 & b_3 \end{vmatrix}} = \frac{y}{\begin{vmatrix} a_3 & a_1 \\ b_3 & b_1 \end{vmatrix}} = \frac{z}{\begin{vmatrix} a_1 & a_2 \\ b_1 & b_2 \end{vmatrix}} \qquad (4-11-4)$$

Proof:

$$\begin{cases} a_1 x + a_2 y + a_3 z = 0 \\ b_1 x + b_2 y + b_3 z = 0 \end{cases} \Leftrightarrow \begin{cases} a_1 x + a_2 y = -a_3 z \\ b_1 x + b_2 y = -b_3 z \end{cases} \Leftrightarrow$$

$$\left\{x = \frac{\begin{vmatrix}-a_3z & a_2\\-b_3z & b_2\end{vmatrix}}{\begin{vmatrix}a_1 & a_2\\b_1 & b_2\end{vmatrix}} = z\frac{\begin{vmatrix}a_2 & a_3\\b_2 & b_3\end{vmatrix}}{\begin{vmatrix}a_1 & a_2\\b_1 & b_2\end{vmatrix}}, \quad y = \frac{\begin{vmatrix}a_1 & -a_3z\\b_1 & -b_3z\end{vmatrix}}{\begin{vmatrix}a_1 & a_2\\b_1 & b_2\end{vmatrix}} = z\frac{\begin{vmatrix}a_3 & a_1\\b_3 & b_1\end{vmatrix}}{\begin{vmatrix}a_1 & a_2\\b_1 & b_2\end{vmatrix}}\right\} \Leftrightarrow$$

$$\left\{\frac{x}{\begin{vmatrix}a_2 & a_3\\b_2 & b_3\end{vmatrix}} = \frac{z}{\begin{vmatrix}a_1 & a_2\\b_1 & b_2\end{vmatrix}}\\ \frac{y}{\begin{vmatrix}a_3 & a_1\\b_3 & b_1\end{vmatrix}} = \frac{z}{\begin{vmatrix}a_3 & a_2\\b_1 & b_2\end{vmatrix}}\right\} \Leftrightarrow$$

$$\frac{x}{\begin{vmatrix}a_2 & a_3\\b_2 & b_3\end{vmatrix}} = \frac{y}{\begin{vmatrix}a_3 & a_1\\b_3 & b_1\end{vmatrix}} = \frac{z}{\begin{vmatrix}a_1 & a_2\\b_1 & b_2\end{vmatrix}}$$

and this completes the proof.

Example 4-11-1: Assuming that the system $\{a(y+z) = x, b(z+x) = y, c(x+y) = z\}$ admits a non zero solution ($x \neq 0, y \neq 0, z \neq 0$), show that $ab + bc + ca + 2abc = 1$.

Solution

$$\begin{cases}a(y+z) = x\\ b(z+x) = y\\ c(x+y) = z\end{cases} \Leftrightarrow \begin{cases}-x + ay + az = 0\\ bx - y + bz = 0\\ cx + cy - z = 0\end{cases} \quad (*)$$

and since this linear, homogeneous system admits a non trivial solution, the determinant of the coefficients must be zero, (see Th.4-11-1), i.e.

$$\begin{vmatrix}-1 & a & a\\ b & -1 & b\\ c & c & -1\end{vmatrix} = 0 \Leftrightarrow$$

$$(-1)\begin{vmatrix}-1 & b\\ c & -1\end{vmatrix} - a\begin{vmatrix}b & b\\ c & -1\end{vmatrix} + a\begin{vmatrix}b & -1\\ c & c\end{vmatrix} = 0 \Leftrightarrow$$

$$(-1)(1 - bc) - a(-b - bc) + a(bc + c) = 0 \Leftrightarrow ab + bc + ca + 2abc = 1$$

Example 4-11-2: Solve the system

$$\begin{cases}2x + y - z = 0\\ 3x - y + 4z = 0\\ 2x^3 - 3y^3 + 5z^3 = 1711\end{cases}$$

Solution

The first two equations constitute a linear, homogeneous system, 2 equations with 3 unknowns, and application of Theorem 4-11-2 yields,

$$\frac{x}{\begin{vmatrix} 1 & -1 \\ -1 & 4 \end{vmatrix}} = \frac{y}{\begin{vmatrix} -1 & 2 \\ 4 & 3 \end{vmatrix}} = \frac{z}{\begin{vmatrix} 2 & 1 \\ 3 & -1 \end{vmatrix}} = k \iff \quad (*)$$

$$\frac{x}{3} = \frac{y}{-11} = \frac{z}{-5} = k \implies \begin{cases} x = 3k \\ y = -11k \\ z = -5k \end{cases} \quad (**)$$

In equation (*) we have called k **the common value of the equal fractions**.

The third equation of the system, in view of equations (**) becomes,

$$2x^3 - 3y^3 + 5z^3 = 1711 \overset{(**)}{\implies} 2(3k)^3 - 3(-11k)^3 + 5(-5k)^3 = 1711 \iff$$

$$3422\, k^3 = 1711 \iff k^3 = \frac{1711}{3422} = \frac{1}{2} \iff k = \frac{1}{\sqrt[3]{2}}$$

and from equations (**) we obtain,

$$\left\{ x = \frac{3}{\sqrt[3]{2}}, \quad y = -\frac{11}{\sqrt[3]{2}}, \quad z = -\frac{5}{\sqrt[3]{2}} \right\}$$

Example 4-11-3: Given that the system
$\{ax + cy + bz = 0, \quad cx + by + az = 0, \quad bx + ay + cz = 0\}$ is satisfied by three numbers x, y, z **not all zeros**, show that $a^3 + b^3 + c^3 = 3abc$.

Solution

Since the given system admits **a non-trivial solution**, the determinant of its coefficients must be zero, (Theorem 4-11-1), i.e.

$$\begin{vmatrix} a & c & b \\ c & b & a \\ b & a & c \end{vmatrix} = 0 \iff a \begin{vmatrix} b & a \\ a & c \end{vmatrix} - c \begin{vmatrix} c & a \\ b & c \end{vmatrix} + b \begin{vmatrix} c & b \\ b & a \end{vmatrix} = 0 \iff$$

$$a(bc - a^2) - c(c^2 - ab) + b(ac - b^2) = 0 \iff$$

$$3abc = a^3 + b^3 + c^3$$

PROBLEMS

4-11-1) Show that the linear, homogeneous system $\{ax + by = 0, Ax + By = 0\}$ has the trivial solution as its only solution, if and only if $\begin{vmatrix} a & b \\ A & B \end{vmatrix} = 0$.

4-11-2) Show that if (X, Y, Z) is a non-trivial solution of the system $\{x = cy + bz, y = az + cx, z = bx + ay\}$ then the following equality holds true:

$$\frac{X^2}{1-a^2} = \frac{Y^2}{1-b^2} = \frac{Z^2}{1-c^2}$$

(Assume that the denominators are not zero).

Hint: Substitute $z = bx + ay$ in the first two equations, etc.

4-11-3) Solve the system

$$\begin{cases} \dfrac{x}{a} + \dfrac{y}{b} + \dfrac{z}{c} = \dfrac{1}{a} + \dfrac{1}{b} + \dfrac{1}{c} \\ \dfrac{x}{b} + \dfrac{y}{c} + \dfrac{z}{a} = \dfrac{1}{a} + \dfrac{1}{b} + \dfrac{1}{c} \\ \dfrac{x}{c} + \dfrac{y}{a} + \dfrac{z}{b} = \dfrac{1}{a} + \dfrac{1}{b} + \dfrac{1}{c} \end{cases}$$

(Ans: $x = 1, y = 1, z = 1$).

Hint: The first equation can be written as $\dfrac{x-1}{a} + \dfrac{y-1}{b} + \dfrac{z-1}{c} = 0$, etc...

4-11-4) If $\{ax + \lambda y + \mu z = 0, \lambda x + by + vz = 0, \mu x + vy + cz = 0\}$ show that:

1) $$\dfrac{x^2}{bc - v^2} = \dfrac{y^2}{ca - \mu^2} = \dfrac{z^2}{ab - \lambda^2}$$

2) $(bc - v^2)(ca - \mu^2)(ab - \lambda^2) = (\mu v - c\lambda)(\lambda \mu - av)(\lambda v - b\mu)$

4-11-5) Solve the system $\{7x - 2y + 3z = 0, \quad 3x + 2y - 6z = 0\}$.

(Ans: $x = 6k, y = 51k, z = 20k$, where k is any real number).

4-11-6) Solve the system $\{x + 3y + 2z = 0, \quad 2x - y + z = 0\}$.

4-11-7) Solve and investigate the system $\{(k^2 + 1)x + (k^2 - 1)y = 0, (k + 1)x + (k - 1)y = 0\}$, k is a real parameter.

(**Ans:** If $k(k - 1) \neq 0$ i.e. if $k \neq 0$ and $k \neq 1$, unique solution ($x = 0, y = 0$), if $k = 0$, infinite number of solutions of the form $x = y = \lambda$ (arbitrary), if $k = 1$, infinite number of solutions of the form $x = 0, y = \mu$, (arbitrary)).

4-11-8) Find the equation which should be satisfied by a and b so that every solution (x_0, y_0, z_0) of the system

$$\{x + y + z = 1, \quad ax + y + z = 0, \quad x + y + bz = 0\}$$

satisfies $y_0^2 = x_0 z_0$.

(**Ans:** $(1 - ab)^2 = (1 - a)(1 - b)$).

4-11-9) Find λ so that the system $\{x + 2y + 3z = 0, 4x + 5y + 6z = 0, 5x + 7y + \lambda z = 0\}$ has solutions other than the trivial one (0,0,0). What is the expression of these solutions?

(**Ans:** $\lambda = 9$, $x = -3k, y = 6k, z = -3k$ where k is any real number).

4-12) Special types of linear systems

Every linear system can be solved with the methods we have developed thus far, (Gauss's elimination method, substitution, Cramer's rule, etc). However, there are some types of systems which can be solved easily, by means of some **special techniques**. Hereunder we shall discuss some types of these systems.

1) Let us consider the system

$$\begin{cases} x + y = a \\ y + z = b \\ z + x = c \end{cases}, \quad a, b, c \text{ are known constants} \qquad (4-12-1)$$

Adding the three equations term wise we get

$$2(x + y + z) = a + b + c \Leftrightarrow x + y + z = \frac{a + b + c}{2} \qquad (*)$$

Subtracting the second equation of the system from (*) we obtain,

$$x = \frac{a+b+c}{2} - b = \frac{a+c-b}{2}$$

and similarly we obtain $y = \frac{b+a-c}{2}$ and $z = \frac{c+b-a}{2}$.

2) Let us consider the system

$$\begin{cases} \frac{2}{x} + \frac{3}{y} - \frac{1}{z} = 5 \\ \frac{1}{x} + \frac{1}{y} + \frac{1}{z} = 6 \\ \frac{2}{x} - \frac{2}{y} + \frac{3}{z} = 7 \end{cases} \qquad (4-12-2)$$

If we define, $X = \frac{1}{x}$, $Y = \frac{1}{y}$, $Z = \frac{1}{z}$, system (4-12-2) reduces to a linear system in X, Y and Z, i.e.

$$\{2X + 3Y - Z = 5, \quad X + Y + Z = 6, \quad 2X - 2Y + 3Z = 7\}$$

Solving the system we find $X = 1, Y = 2, Z = 3$ and the original unknowns are: $x = \frac{1}{X} = 1, y = \frac{1}{Y} = \frac{1}{2}, z = \frac{1}{Z} = \frac{1}{3}$.

3) Let us consider the system

$$\left\{\frac{x}{2} = \frac{y}{3} = \frac{z}{5}, \quad 3x - 4y + z = -2\right\} \qquad (4-12-3)$$

Let us call k the **common value of the equal fractions**, i.e. let

$$\frac{x}{2} = \frac{y}{3} = \frac{z}{5} = k \Rightarrow \begin{cases} x = 2k \\ y = 3k \\ z = 5k \end{cases} \qquad (**)$$

In (**) the **unknowns x, y, z are expressed in terms of k**, and substituting in the third equation in (4-12-3) we get an equation for k,

$$3 \cdot 2k - 4 \cdot 3k + 5k = -2 \Longleftrightarrow -k = -2 \Longleftrightarrow k = 2$$

The values of x, y, z are obtained from (**): $x = 4, y = 6, z = 10$.

Example 4-12-1: Solve the system $\left\{\frac{x-1}{3} = \frac{y+2}{4}, 2x + 3y = 7\right\}$.

Solution

From the first equation we have,

$$\frac{x-1}{3} = \frac{y+2}{4} = k \Leftrightarrow \begin{cases} \frac{x-1}{3} = k \\ \frac{y+2}{4} = k \end{cases} \Leftrightarrow \begin{cases} x = 3k+1 \\ y = 4k-2 \end{cases} \quad (*)$$

Inserting these expressions of x and y in the second equation we get,

$$2x + 3y = 7 \overset{(*)}{\Rightarrow} 2(3k+1) + 3(4k-2) = 7 \Leftrightarrow$$

$$6k + 2 + 12k - 6 = 7 \Leftrightarrow 18k = 11 \Leftrightarrow k = \frac{11}{18}$$

and from (*) we obtain, $\left\{x = \frac{33}{18} + 1, y = \frac{44}{18} - 2\right\}$, i.e. $\left\{x = \frac{51}{18}, y = \frac{8}{18}\right\}$.

Example 4-12-2: Solve the system $\{2x + y = 3xy, -x + 3y = 4xy\}$.

Solution

a) We note that an obvious solution of the system (the trivial solution) is $x = 0, y = 0$.

b) To find the **non-trivial solutions** ($x \neq 0$ and $y \neq 0$) we divide both equations of the system by xy, and the system is thus reduced to the following:

$$\left\{\frac{2}{y} + \frac{1}{x} = 3, \quad -\frac{1}{y} + \frac{3}{x} = 4\right\} \quad (*)$$

and if we set $X = \frac{1}{x}, Y = \frac{1}{y}$ system (*) assumes the form,

$$\{X + 2Y = 3 \quad 3X - Y = 4\} \Leftrightarrow X = \frac{11}{7}, Y = \frac{5}{7} \Leftrightarrow \begin{cases} x = 7/11 \\ y = 7/5 \end{cases}$$

The given system has two solutions, $(0,0)$ and $(7/11, 7/5)$.

Example 4-12-3: Solve the system

$$\{2x + 3(y+z) = 7, \quad y + 3(z+x) = 4, \quad 5z + 3(x+y) = 10\}$$

Solution

a) One possible method to solve the system is to simplify the three equations, as much as possible, and then apply, for example, Cramer's rule.

b) Here we present a different approach. If we set $S = x + y + z$, the three equations can be written as follows:

$$\begin{cases} 2x + 3(S - x) = 7 \\ y + 3(S - y) = 4 \\ 5z + 3(S - z) = 10 \end{cases} \Leftrightarrow \begin{cases} x = 3S - 7 \\ y = (3S - 4)/2 \\ z = (10 - 3S)/2 \end{cases} \quad (*)$$

Adding these three equations term wise, and recalling that $x + y + z = S$, we obtain the following equation for S,

$$S = 3S - 7 + \frac{3S - 4}{2} + \frac{10 - 3S}{2} \Leftrightarrow S = 2$$

Having found S, the first equation in system (*) yields $x = 3 \cdot 2 - 7 = -1$ and similarly, we find, $y = 1, z = 2$ (let the reader verify the calculations).

PROBLEMS

Solve the following systems:

4-12-1) $\{x + y = 6xy, 3x + 4y = 22xy\}$.

(**Ans:** $(x = 0, y = 0), (x = 1/4, y = 1/2)$).

4-12-2) $\{3x + 4y = 3xy, 8y - 3z = yz, 5z - 2x = 2xz\}$.

Hint: See Example 4-12-2.

4-12-3) $\left\{\frac{xy}{x+y} = \frac{12}{5}, \frac{yz}{y+z} = \frac{18}{5}, \frac{xz}{x+z} = \frac{36}{13}\right\}$, (**Ans:** $x = 4, y = 6, z = 9$).

4-12-4) $\{x + 3(y + z + w) = 32, y + 4(x + z + w) = 44, z + 5(x + y + w) = 58, w + 6(x + y + z) = 74\}$.

Hint: See Example 4-12-3.

4-12-5) $\{x + 2(y + z + w) = 19, 2y + 3(x + z + w) = 28, 3z + 4(x + y + w) = 37, w + 2(x + y + z) = 16\}$.

(**Ans:** $x = 1, y = 2, z = 3, w = 4$).

4-12-6) $\{x + y - z = 5, y + z - w = -3, z + w - t = -3, w + t - x = 6, t + x - y = 3\}$, (The unknowns are x, y, z, w, t).

(**Ans:** $x = 1, y = 2, z = -2, w = 3, t = 4$).

4-12-7) $\{xy + yz = axyz, \quad xy + xz = bxyx, \quad xz + yz = cxyz\}$.

(**Ans:** $x = 2/(a - b + c), y = 2/(-a + b + c), z = 2/(a + b - c)$, the trivial solution is (0,0,0)).

4-12-8) $\{y + z - x = xyz, z + x - y = \frac{xyz}{4}, x + y - z = \frac{xyz}{9}\}$.

Hint: Trivial solution (0,0,0), to find the non-trivial solutions ($xyz \neq 0$) divide through by xyz and solve the resulting system for xy, yz, zx, etc.

4-12-9) $\{\frac{1}{4x} = \frac{1}{2y} = \frac{3}{4z} = \frac{1}{w}, \quad x + y + z + w = 1\}$.

(**Ans:** $x = 1/10, y = 1/5, z = 3/10, w = 2/5$).

4-12-10) $\{\frac{1}{x} - \frac{1}{y} = \frac{1}{6}, \quad \frac{1}{y} - \frac{2}{z} = -\frac{1}{6}, \quad \frac{2}{z} + \frac{1}{x} = 1\}$.

4-12-11) $\{\frac{y+z}{yz} = \frac{5}{18}, \quad \frac{z+x}{zx} = \frac{13}{36}, \quad \frac{x+y}{xy} = \frac{5}{12}\}$, (**Ans:** $x = 4, y = 6, z = 9$).

4-12-12) $\{\frac{x+1}{2} = \frac{y-2}{3} = \frac{z+3}{4}, \quad 3x - 4y + 2z = 27\}$.

4-12-13) $\{\frac{1}{x} + \frac{1}{y} = \frac{1}{a}, \quad \frac{1}{y} + \frac{1}{z} = \frac{1}{b}, \quad \frac{1}{z} + \frac{1}{x} = \frac{1}{c}\}$.

(**Ans:** $x = \left(\frac{1}{2a} - \frac{1}{2b} + \frac{1}{2c}\right)^{-1}, y = \left(\frac{1}{2a} + \frac{1}{2b} - \frac{1}{2c}\right)^{-1}, z = \left(-\frac{1}{2a} + \frac{1}{2b} + \frac{1}{2c}\right)^{-1}$).

Hint: Adding term wise we find the sum $S = \frac{1}{x} + \frac{1}{y} + \frac{1}{z}$, etc.

4-12-14) $\left\{\dfrac{4}{2x+y-1} + \dfrac{3}{x+2y-3} = 2, \quad \dfrac{3}{2x+y-1} - \dfrac{2}{x+2y-3} = \dfrac{1}{12}\right\}.$

(Ans: $x = 4/3, y = 7/3$).

Hint: Set $X = 2x + y - 1, Y = x + 2y - 3$), find X, Y and then find x, y.

4-13) Word problems

One interesting application of linear systems is in the solution of word problems, when **two or more unknowns** are involved. The difficult part in solving word problems is to "translate" the words into equations. The following examples illustrate the method of approach.

Example 4-13-1: The sum of the digits of a two digit number is 13. If the digits are reversed, the number is decreased by 27. Find the number.

Solution

Let $a = xy$ the sought for two digit number. The sum of its two digits is 13, i.e.

$$x + y = 13 \quad (*)$$

The value of the number a is

$$a = 10x + y \quad (**)$$

(For example, let us consider the two digit number $a = 37$. In the decimal system, the value of this number is: $a = 3 \cdot 10 + 7 \cdot 1$).

According to the problem, if we reverse the digits, the number is decreased by 27, i.e. the number yx is 27 less than xy, i.e.

$$10y + x = 10x + y - 27 \quad (***)$$

Equations (*) and (***) constitute a linear systems for the digits x and y of the sought for number. We have,

$$\begin{cases} x + y = 13 \\ 10y + x = 10x + y - 27 \end{cases} \Leftrightarrow \begin{cases} x + y = 13 \\ 9x - 9y = 27 \end{cases} \Leftrightarrow \begin{cases} x = 8 \\ y = 5 \end{cases}$$

The sought for, two digit number, is $a = 85$.

Check: $8 + 5 = 13, 85 - 58 = 27$.

Note: Numeral Systems.

a) The number system we use in our daily transactions is based on **ten digits: 0,1,2,3,4,5,6,7,8,9. Position is important**, with the first position being **units**, the next on the left being **tens**, the next on the left being **hundreds**, etc. Every natural number can be represented with the aid of the aforementioned ten digits. We may use the symbol $xyz_{(10)}$ to represent the number having digits x, y and z in the **decimal numeral system**. We say that **the number 10 is the base of the decimal numeral system**. This means that,

$$xyz_{(10)} = x \cdot 10^2 + y \cdot 10 + z$$

where each of the digits x, y and z is one of the digits 0,1,2,3,4,5,6,7,8,9, z being the digit of units, y the digit of tens and x the digit of hundreds. For example,

$$378_{(10)} = 3 \cdot 10^2 + 7 \cdot 10 + 8 \cdot 1, \quad 409_{(10)} = 4 \cdot 10^2 + 0 \cdot 10 + 9 \cdot 1$$

Similarly, $xyzw_{(10)} = x \cdot 10^3 + y \cdot 10^2 + z \cdot 10 + w$, where now x is the digit of thousands, y is the digit of hundreds, z is the digit of tens and w is the digit of units. For example,

$$6943_{(10)} = 6 \cdot 10^3 + 9 \cdot 10^2 + 4 \cdot 10 + 3 \cdot 1$$

It has become customary, when working in the decimal numeral system, not to write the base 10, i.e. we write 6943 instead of writing $6943_{(10)}$.

b) It is possible to represent a given number in a numeral system other than the decimal one. Let b be a fixed positive integer, ($b \geq 2$). **The numeral system with base b makes use the digits $0, 1, 2, ..., (b-1)$**. Writing for instance, $xyz_{(b)}$, (where each one of the digits x, y, z is either 0, or 1, or 2, or...$(b-1)$), we mean the following:

$$xyz_{(b)} = x \cdot b^2 + y \cdot b + z$$

Similarly, $xyzw_{(b)} = x \cdot b^3 + y \cdot b^2 + z \cdot b + w$, etc.

The numeral system with base $b = 2$ is called the **binary numeral system**. The digits in the binary system are either 0 or 1. For example,

$$10110_{(2)} = 1 \cdot 2^4 + 0 \cdot 2^3 + 1 \cdot 2^2 + 1 \cdot 2^1 + 0 \cdot 1 = 16 + 4 + 2 = 22$$
$$= 22_{(10)}$$

and this shows that $10110_{(2)} = 22_{(10)}$.

The base 8 system ($b = 8$) is called the octal numeral system. In this system the digits used to represent numbers are 0,1,2,3,4,5,6,7. For example,

$$217_{(8)} = 2 \cdot 8^2 + 1 \cdot 8^1 + 7 \cdot 1 = 143 = 143_{(10)}$$

Example 4-13-2: Find a two digit number in the decimal system such that when its two digits are reversed we get the same number in the system with base 7.

Solution

Let $xy_{(10)}$ be the sought for number. Then, according to the problem we must have:

$$xy_{(10)} = yx_{(7)} \Leftrightarrow 10x + y = 7y + x \Leftrightarrow 9x = 6y \Leftrightarrow 3x = 2y \quad (*)$$

where each one of the integers x and y can take values from the set $\{1,2,3,4,5,6\}$. From (*) we get,

$$\frac{x}{2} = \frac{y}{3} = k \Leftrightarrow \begin{cases} x = 2k \\ y = 3k \end{cases}$$

For $k = 1$, we get $(x = 2, y = 3)$, for $k = 2$ we get $(x = 4, y = 6)$. Higher values of k are not allowed, since then y would be greater than 6. We see that there are two numbers satisfying the requirements of the problem, 23 and 46.

Example 4-13-3: A total of 200 seats for a concert are sold, producing a total income of $ 1065. If seats cost either $ 3.5 or $ 4.0 or $ 7.5 and the number of the most expensive seats exceeds the number of the cheapest seats by 20, how many seats of each kind were sold?

Solution

Let x, y, z be the number of seats of \$3.5, \$4.0, \$7.5 respectively. Then,

$$\begin{cases} x + y + z = 200 \\ 3.5x + 4y + 7.5z = 1065 \\ z = x + 20 \end{cases} \quad (*)$$

System (*) is a linear system in x, y and z. To solve this system we may use any one of the methods we have studied so far; we may, for example, substitute the third equation in the first two, the result being,

$$\begin{Bmatrix} 2x + y = 180 \\ 11x + 4y = 915 \end{Bmatrix} \Leftrightarrow \begin{Bmatrix} y = 180 - 2x \\ 11x + 4 \cdot (180 - 2x) = 915 \end{Bmatrix} \Leftrightarrow$$

$$\begin{Bmatrix} y = 180 - 2x \\ 11x - 8x = 915 - 720 \end{Bmatrix} \Leftrightarrow \begin{Bmatrix} y = 180 - 2x \\ x = 195/3 = 65 \end{Bmatrix} \Leftrightarrow$$

$$\{x = 65, \quad y = 180 - 130 = 50, \quad z = 65 + 20 = 85\}$$

PROBLEMS

4-13-1) The sum of the digits of a two digit number is 11. The difference between the number and that obtained by reversing the digits of the number is 45. Find the number.   (**Ans:** 83).

4-13-2) An airplane flying with a tail wind completes a trip of $2100 \, Km$ in $3 \, h$. Flying the reverse direction, the plane completes the same trip in $4 \, h$. What is the speed of the plain in still air and what is the speed of the air?

4-13-3) Verify that $25_{(7)} = 19_{(10)}, 1111_{(2)} = 15_{(10)}, 123_{(4)} = 27_{(10)}$.

Hint: See Example 4-13-2.

4-13-4) Consider the function $f(x) = ax^2 + bx + c$ and determine the constants a, b and c, so that $f(1) = 0, f(-1) = 6, f(2) = 3$.

4-13-5) In which numeral system the number 25 is doubled by interchanging its two digits?   (**Ans:** 8).

4-13-6) Consider a triangle with sides a, b and c and altitudes h_a, h_b and h_c respectively. If we know the two sides a and b and that $h_c = h_a + h_b$, find the third side c. Under what conditions does the problem have a solution?

(**Ans:** $c = ab/(a+b)$, $|a^2 - b^2| < ab < (a+b)^2$).

Hint: If A is the area of the triangle, then $2A = ah_a = bh_b = ch_c$. Notice that a, b, c should satisfy the "**triangle inequality**" $|a-b| < c < a+b$.

4-13-7) Find a three digit number which increases by 180 when the first two digits are interchanged, reduces by 36 when the second and third digits are interchanged and reduces by 198 when the first and the third digits are interchanged. (**Ans:** $240, 351, 462, 573, 684, 795$).

Hint: See Example 4-13-2.

4-13-8) Find a four digit number such that, $xyzw_{\langle 10 \rangle} = wzyx_{\langle 10 \rangle}$, $zw_{\langle 10 \rangle} - xy_{\langle 10 \rangle} = 36$, $x+y+z+w = 24$. (**Ans:** 4884).

CHAPTER 5: GRAPHICAL SOLUTION OF LINEAR SYSTEMS AND LINEAR INEQUALITIES

5-1) Graphical solution of a linear system of two equations in two unknowns

Let us, for definiteness, consider the following system

$$\{2x - 3y = 3, \qquad x + 2y = 5\} \qquad (*)$$

Of course, we may easily solve this system, (using for example, Cramer's rule or the method of substitution), the answer being $(x = 3, y = 1)$. However, we may also provide **a graphical solution** of the system, as described below:

Let us graph both equations, **on the same system xOy**, as show in Fig. 5-1.

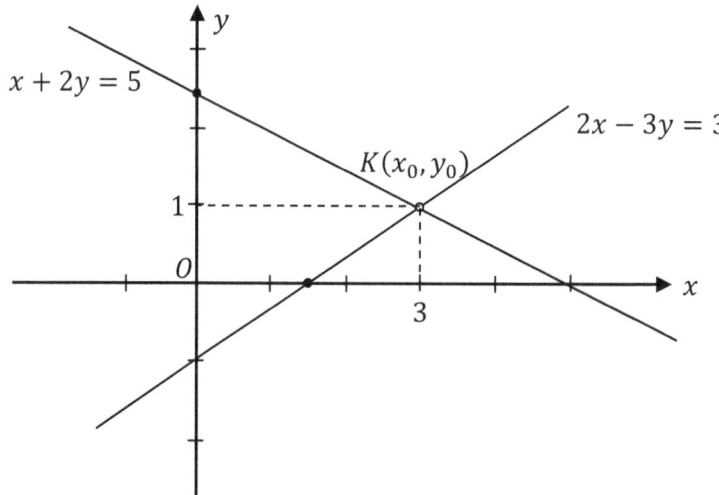

Fig. 5-1: Graphical solution of a linear system.

As we know (section 1-9), each linear equation represents a straight line, in the xOy plane. The two straight lines intersect at a point $K(x_0, y_0)$. This point belongs to the line $2x - 3y = 3$, and therefore its coordinates satisfy the equation of the line, i.e. $2x_0 - 3y_0 = 3$. The same point belongs also to the line $x + 2y = 5$ and hence $x_0 + 2y_0 = 5$. We thus see that **the coordinates (x_0, y_0) of the point of intersection K of the two lines is the solution of the**

system. In Fig. 5-1, we see that $x_0 = 3$ and $y_0 = 1$, which is, indeed, the solution of the system. This method provides **a graphical solution** of the system.

If the two lines intersect at **one point**, the system has **one solution**; if the two lines **are parallel**, the system has **no solution** (impossible system) while if the two lines **coincide**, then the system has **an infinite number of solutions** (infinite number of points).

PROBLEMS

5-1-1) Solve graphically the system $\{x - 2y = 5, 3x + y = 1\}$.

(Ans: $(x, y) = (1, -2)$).

5-1-2) Graph the straight lines $2x + y = 3, 4x + 2y = 4$ and show that the system is impossible.

5-2) Graphical solution of a linear inequality of the form $ax + by + c > 0$

Suppose that we want to solve the inequality $2x + 3y - 6 > 0$, i.e. to find **all** the pairs of numbers (x, y) which satisfy the given inequality. First of all, let us plot **the straight line** $2x + 3y - 6 = 0$, as in Fig. 5-2.

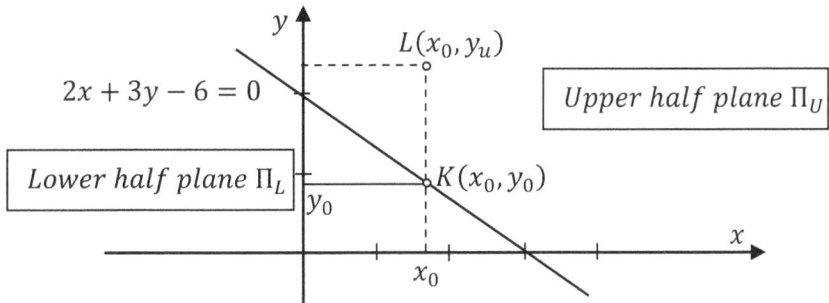

Fig. 5-2: Graphical solution of a linear inequality.

All the points on the line satisfy the equation $2x + 3y - 6 = 0$. This line divides the plane xOy in **two half planes**, let us call them the "**upper half plane** Π_U" relative to the line, and the "**lower half plane** Π_L". Let us take an arbitrary point $L(x_0, y_u)$ in the upper half plane, as in Fig. 5-2, and consider the quantity $(2x_0 + 3y_u - 6)$, which may be written as follows:

$$2x_0 + 3y_u - 6 = 2x_0 + 3y_u - 6 - 0 = 2x_0 + 3y_u - 6 - \underbrace{(2x_0 + 3y_0 - 6)}_{=0} \Rightarrow$$

$$2x_0 + 3y_u - 6 = 3\underbrace{(y_u - y_0)}_{Positive} > 0 \qquad (*)$$

Notice that since $L(x_0, y_u)$ is **an arbitrary point in the upper half plane**, $y_u > y_0$, and equation (*) shows that **all points (x, y) lying in the upper half plane**, satisfy the inequality $2x + 3y - 6 > 0$. Similarly, we may show that for all points (x, y) in the lower half plane $2x + 3y - 6 < 0$.

In general, any straight line $ax + by + c = 0$ divides the plane **in three parts**: the set of points (x, y) belonging in the line and satisfying the equation $ax + by + c = 0$, and the upper and the lower half planes relative to the line. **All the points in one half plane satisfy the inequality $ax + by + c > 0$, while all the points in the other half plane satisfy the inequality $ax + by + c < 0$.**

Let us consider the following illustrative examples.

Example 5-2-1: Solve the inequality $x + 2y - 4 < 0$.

Solution

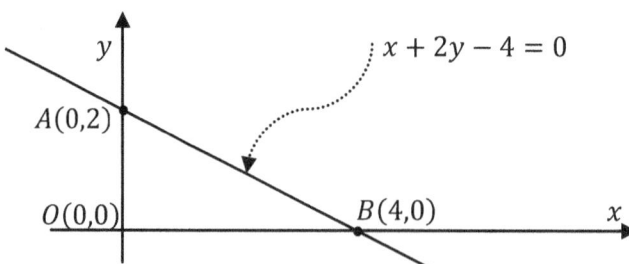

Fig. 5-3: Graphical solution of a linear inequality.

We know that the coordinates of all the points belonging in one of the two half planes (relative to the line) make the function $(x + 2y - 4)$ positive while the coordinates of all the points in the other half plane make the same function negative. The origin $O(0,0)$ belongs in **the lower half plane**, and since $0 + 2 \cdot 0 - 4 = -4 < 0$, we conclude that **all the points in the lower half plane** satisfy $x + 2y - 4 < 0$, i.e. the solution set of $x + 2y - 4 < 0$ is the totality of points belonging in the lower half plane, relative to the line $x + 2y - 4 = 0$. All the points in the upper half plane satisfy $x + 2y - 4 > 0$.

Example 5-2-2: Find the set of points in the xOy plane which satisfy both inequalities $x - y + 1 > 0$ and $x - 8y - 6 < 0$ simultaneously.

Solution

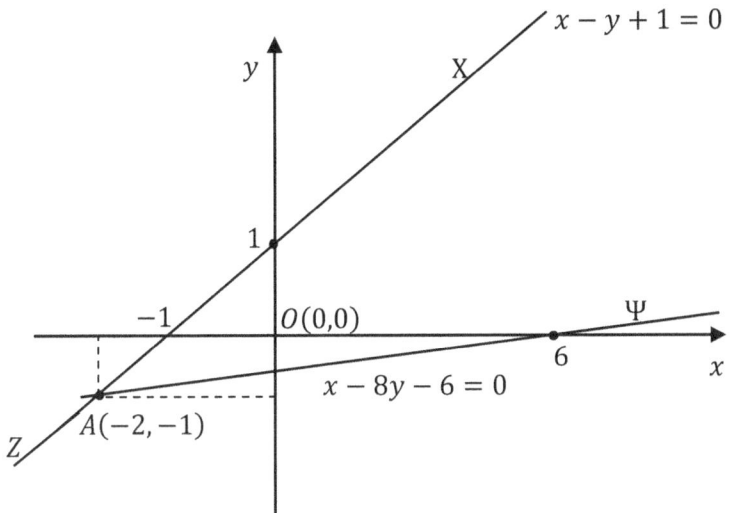

Fig. 5-4: Graphical solution of a system of linear inequalities.

The two lines $(x - y + 1 = 0)$ and $(x - 8y - 6 = 0)$ intersect at the point $A(-2, -1)$, (let the reader verify it).

The origin $(0,0)$ makes the function $x - y + 1$ positive $(0 - 0 + 1 = 1 > 0)$, and therefore the solution of $x - y + 1 > 0$ is **the lower half plane** (the one that contains the origin), relative to the line.

Similarly, the origin $(0,0)$ makes the function $x - 8y - 6$ negative $(0 - 0 - 6 = -6 < 0)$, and the solution of $x - 8y - 6 < 0$ is **the upper half plane**, relative to the corresponding line.

The set of points, satisfying **both equations simultaneously** is the **intersection** of the two solution sets, and as we see, **is the set of points in the interior of the angle XAΨ** in Fig. 5-4. The coordinates of any point, in the interior of the angle XAΨ, is a solution of the system of the two inequalities.

PROBLEMS

5-2-1) Find the set of points in the xOy plane which satisfy both inequalities $x - y + 1 > 0$ and $x - 8y - 6 > 0$ simultaneously, (see Fig. 5-4).

(**Ans:** The interior of the obtuse angle ΨAZ in Fig. 5-4).

5-2-2) Solve the inequality $(x + 2y - 4)(2x + 3y + 6) > 0$.

Hint: The product of two factors is positive if, either both factors are positive or both are negative, i.e. the given inequality is satisfied if either $(x + 2y - 4) > 0$ **and** $(2x + 3y + 6) > 0$, or $(x + 2y - 4) < 0$ **and** $(2x + 3y + 6) < 0$.

5-2-3) Consider the lines $(\varepsilon_1): 3x - 2y + 3 = 0$, $(\varepsilon_2): x + 3y - 10 = 0$ and $(\varepsilon_3): (2 - \lambda)x + (\lambda + 1)y + \lambda + 7 = 0$. **a)** Find $(\varepsilon_1) \cap (\varepsilon_2)$ (the point of intersection) and **b)** Determine λ so that (ε_3) passes from the point of intersection of (ε_1) and (ε_2), (**Ans:** $x = 1, y = 3, \lambda = -4$).

5-2-4) Solve the system of inequalities $x + y - 9 < 0$ and $x - y + 1 > 0$.

5-3) Equation of the circle

Let us consider a circle (c) having center at $K(a, b)$ and radius R, as in Fig. 5-5. If $M(x, y)$ is a point on the circle, then $MK = R$, (the radius).

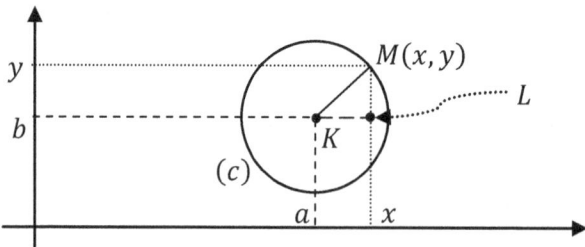

Fig. 5-5: Equation of a circle.

In the right triangle MLK, application of the Pythagorean theorem yields

$$KL^2 + LM^2 = KM^2 \Rightarrow$$

$$(x - a)^2 + (y - b)^2 = R^2 \qquad (5 - 3 - 1)$$

Equation (5-3-1) is the equation of a circle centered at (a, b) and having radius R. If the circle is centered at the origin, $(a = 0, b = 0)$, equation (5-3-1) assumes the form

$$x^2 + y^2 = R^2 \qquad (5-3-2)$$

The set of points (x, y) satisfying the inequality $(x - a)^2 + (y - b)^2 < R^2$ is the totality of the points **inside the circle**, while the set of points satisfying the inequality $(x - a)^2 + (y - b)^2 > R^2$ is the set of all the points **outside the circle**.

Using graphical methods we may solve systems of inequalities of the form $\{kx + \lambda y > \mu, (x - a)^2 + (y - b)^2 < R^2\}$.

Let us consider a few examples.

Example 5-3-1: Find the center and the radius of the circle $x^2 + y^2 - x + \frac{2y}{3} = \frac{17}{9}$.

Solution

We use the method of "**completion of the squares**" in order to, gradually, transform the given equation to the standard form (5-3-1).

$$x^2 + y^2 - x + \frac{2y}{3} = \frac{17}{9} \Leftrightarrow$$

$$\underbrace{x^2 - 2 \cdot \frac{1}{2} \cdot x + \left(\frac{1}{2}\right)^2}_{\left(x - \frac{1}{2}\right)^2} - \left(\frac{1}{2}\right)^2 + \underbrace{y^2 + 2 \cdot \frac{1}{3} \cdot y + \left(\frac{1}{3}\right)^2}_{\left(y + \frac{1}{3}\right)^2} - \left(\frac{1}{3}\right)^2 = \frac{17}{9} \Leftrightarrow$$

$$\left(x - \frac{1}{2}\right)^2 + \left(y + \frac{1}{3}\right)^2 = \frac{17}{9} + \left(\frac{1}{2}\right)^2 + \left(\frac{1}{3}\right)^2 = \frac{17}{9} + \frac{1}{4} + \frac{1}{9} = \frac{9}{4} \Leftrightarrow$$

$$\left(x - \frac{1}{2}\right)^2 + \left(y + \frac{1}{3}\right)^2 = \left(\frac{3}{2}\right)^2$$

This is the equation of a circle having center $K\left(\frac{1}{2}, -\frac{1}{3}\right)$ and radius $R = \frac{3}{2}$.

Example 5-3-2: Solve the inequality $x^2 + y^2 < 16$.

Solution

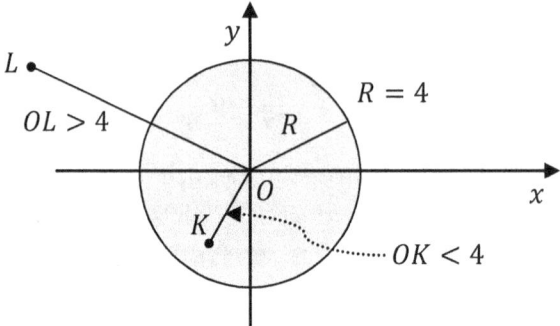

Fig. 5-6: Graphical solution of the inequality $x^2 + y^2 < 16$.

The equation $x^2 + y^2 = 16 = 4^2$ is the equation of a circle, centered at the origin $O(0,0)$ and having radius $R = 4$. All the points (x, y) in the interior of the circle, satisfy $x^2 + y^2 < 16$. All the points, outside the circle, satisfy the inequality $x^2 + y^2 > 16$.

Example 5-3-3: Solve the system of inequalities $2x + y > 3$ and $x^2 + y^2 < 4$.

Solution

We graph $2x + y = 3$ (a straight line) and $x^2 + y^2 = 4$ (a circle of radius $R = 2$) on the same plane xOy, as in Fig. 5-7.

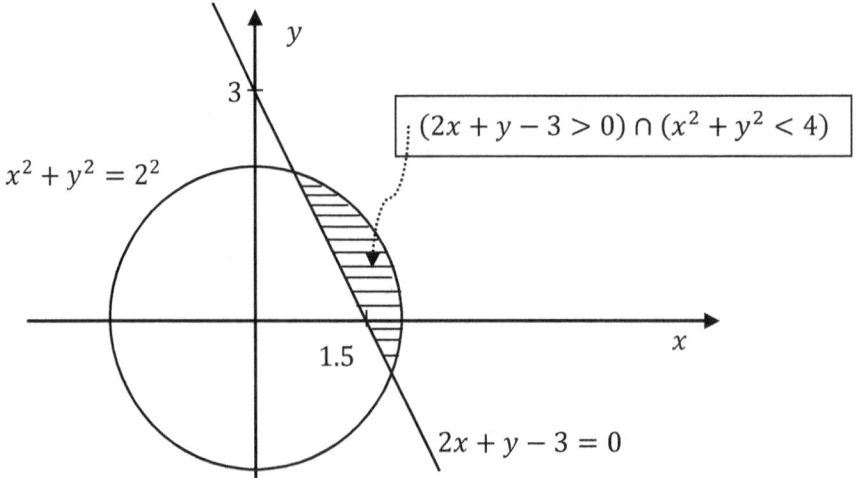

Fig. 5-7: Graphical solution of an inequality.

The solution of $2x + y - 3 > 0$ is the upper half plane, relative to the line, i.e. the half plane which does not contain the origin, while the solution of $x^2 + y^2 < 4$ is the interior of the circle of radius 2. From the graphical solution of the two inequalities, we see that the set of points which satisfy both equations **simultaneously** is the shaded area in Fig. 5-7.

If we call D the shaded area in Fig. 5-7, (the points which satisfy both equations simultaneously), we may write, using set notation,

$$D = (2x + y - 3 > 0) \cap (x^2 + y^2 < 4)$$

PROBLEMS

5-3-1) Find the center and the radius of the circle $x^2 + y^2 - 4x + 6y = -9$.

(**Ans:** Center $(2, -3)$, radius 2).

5-3-2) Solve graphically the system of inequalities $x + 2y > 5$ and $x^2 + y^2 < 9$.

5-3-3) Let D_1 be the solution set of the inequality $x^2 + y^2 \leq 1$, D_2 be the solution set of $x + 2y - 5 > 0$ and D_3 be the solution set of $x + 2y - 5 < 0$. Show that $D_1 \cap D_2 = \emptyset$ and $D_1 \cap D_3 = D_1$.

5-3-4) Solve graphically the system of inequalities $x + y < 1$ and $(x - 1)^2 + (y - 1)^2 < 1$.

5-3-5) If D is the solution set in Problem 5-3-4, find the area of D.

(**Ans:** $(\pi - 2)/4$).

CHAPTER 6: RATIONAL EQUATIONS AND INEQUALITIES

6-1) Rational algebraic fractions

Every function of the form (P/Q) where **P and Q are polynomials**, is called **a rational algebraic fraction**. The function P is the **numerator**, while the function Q is the **denominator**. For the values of the variable (or the variables) for which the denominator vanishes, **the rational fraction is not defined**. Therefore, **the domain of definition of the rational fraction (P/Q) does not include the values of the variable (or the variables) which make the denominator to be zero**.

For example the functions:

$$f(x) = \frac{x^3 + 3x^2 + 5x - 2}{x + 2}, \quad x + 2 \neq 0$$

$$g(x) = \frac{3x^2 + 10x - 7}{(x - 1)(x + 3)}, \quad (x - 1)(x + 3) \neq 0$$

$$u(x, y) = \frac{7x^3 - 15xy + 9y^5}{2x + 3y - 5}, \quad 2x + 3y - 5 \neq 0$$

$$w(x, y, z) = \frac{x^3 + y^3 - 2z^2x^5 + 8y^2z^3}{x^2 + 3y^2 - 6xyz^3}, \quad x^2 + 3y^2 - 6xyz^3 \neq 0$$

are all rational algebraic fractions.

A rational algebraic fraction can be simplified, provided that the numerator and the denominator share **a common factor**. In order to simplify a fraction, we factor the numerator and the denominator, and then cross out the common factors, if such factors exist.

Example 6-1-1: Simplify the rational fraction $f(x, y) = \frac{x^2+y^2-2xy}{x-y}$.

Solution

$$f(x, y) = \frac{x^2 + y^2 - 2xy}{x - y} = \frac{(x - y)^2}{x - y} = x - y, \quad (x - y \neq 0)$$

Example 6-1-2: Simplify the rational fraction $\frac{ab(x^2+y^2)+xy(a^2+b^2)}{ab(x^2-y^2)+xy(a^2-b^2)}$.

Solution

If we call N the numerator and D the denominator, we have:

$$N = ab(x^2 + y^2) + xy(a^2 + b^2) = abx^2 + aby^2 + xya^2 + xyb^2 \Rightarrow$$

$$N = ax(bx + ay) + by(ay + bx) = (bx + ay)(ax + by) \quad (*)$$

$$D = ab(x^2 - y^2) + xy(a^2 - b^2) = abx^2 - aby^2 + xya^2 - xyb^2 \Rightarrow$$

$$D = ax(bx + ay) - by(ay + bx) = (ax - by)(bx + ay) \quad (**)$$

By virtue of (*) and (**) we have,

$$\frac{N}{D} = \frac{ab(x^2 + y^2) + xy(a^2 + b^2)}{ab(x^2 - y^2) + xy(a^2 - b^2)} = \frac{(bx + ay)(ax + by)}{(ax - by)(bx + ay)} = \frac{ax + by}{ax - by}$$

PROBLEMS

6-1-1) Simplify the fraction $(x^3 - 1)/(x - 1)$, $x - 1 \neq 0$.

(**Ans:** $x^2 + x + 1$).

6-1-2) Simplify the fraction $(x^4 - y^4)/(x^2 + y^2)$, $x \neq 0, y \neq 0$.

6-1-3) Simplify the fraction $(a^4 - 1)/(a^2 - 1)$, (**Ans:** $a^2 + 1$).

6-2) Operations with rational algebraic fractions

The four operations with rational algebraic fractions (addition, subtraction, multiplication and division) are defined according to the same rules that apply for fractions with numerical values. For example,

$$\frac{P_1(x)}{Q(x)} \pm \frac{P_2(x)}{Q(x)} = \frac{P_1(x) \pm P_2(x)}{Q(x)}$$

$$\frac{P_1(x)}{Q_1(x)} \times \frac{P_2(x)}{Q_2(x)} = \frac{P_1(x) \cdot P_2(x)}{Q_1(x) \cdot Q_2(x)}$$

$$\frac{\frac{P_1(x)}{Q_1(x)}}{\frac{P_2(x)}{Q_2(x)}} = \frac{P_1(x) \cdot Q_2(x)}{P_2(x) \cdot Q_1(x)}$$

(The last formula shows how **a complex fraction is converted to a simple fraction**).

To add or subtract two algebraic fractions having **different denominators**, (unlike fractions), we have first of all, to convert them **to equivalent fractions** having **the same denominator**, (like fractions). Let us consider a few illustrative examples.

Example 6-2-1: Convert the following expression to a single rational fraction.

$$f(x,y) = \frac{x^2 + y^2}{x + y} + \frac{2xy}{x - y}$$

Solution

To add the two fractions, we have to convert them to like fractions, (to make them to have **the same denominator**). We multiply both terms of the first fraction by $(x - y)$ and both terms of the second fraction by $(x + y)$, (we know that **if we multiply both terms of a given fraction by the same number or expression, we get an equal fraction**). We have:

$$f(x,y) = \frac{x^2 + y^2}{x + y} + \frac{2xy}{x - y} = \frac{(x^2 + y^2)(x - y)}{(x + y)(x - y)} + \frac{(2xy)(x + y)}{(x - y)(x + y)} \Rightarrow$$

$$f(x,y) = \frac{x^3 + y^2 x - x^2 y - y^3}{x^2 - y^2} + \frac{2x^2 y + 2xy^2}{x^2 - y^2} \Rightarrow$$

$$f(x,y) = \frac{(x^3 + y^2 x - x^2 y - y^3) + (2x^2 y + 2xy^2)}{x^2 - y^2} \Rightarrow$$

$$f(x,y) = \frac{x^3 + x^2 y + 3xy^2 - y^3}{x^2 - y^2}$$

Example 6-2-2: Simplify the expression

$$F = \frac{y+z}{yz}(y^2+z^2-x^2) + \frac{z+x}{zx}(z^2+x^2-y^2) + \frac{x+y}{xy}(x^2+y^2-z^2)$$

Solution

Multiplying both terms of the first summand by x, both terms of the second by y and both terms of the third by z, we obtain,

$$F = \frac{x(y+z)}{xyz}(y^2+z^2-x^2) + \frac{y(z+x)}{yzx}(z^2+x^2-y^2)$$
$$+ \frac{z(x+y)}{zxy}(x^2+y^2-z^2) \qquad (*)$$

In equation (*) all the fractions have **the same denominator** (xyz), so all we have to do is to simplify the numerator, and then divide by the common denominator (xyz).

The numerator N is

$$N = x(y+z)(y^2+z^2-x^2) + y(z+x)(z^2+x^2-y^2)$$
$$+ z(x+y)(x^2+y^2-z^2)$$

Carrying out the multiplications and simplifying, we get (check it)

$$N = 2xyz(x+y+z)$$

and formula (*) yields,

$$F = \frac{2xyz(x+y+z)}{xyz} = 2(x+y+z)$$

Example 6-2-3: Show that

$$\left(\frac{1}{x-y} + \frac{1}{y-z} + \frac{1}{z-x}\right)^2 = \frac{1}{(x-y)^2} + \frac{1}{(y-z)^2} + \frac{1}{(z-x)^2}$$

for all values of x, y, z, which do not vanish the denominators.

Solution

Using a well known identity, (**the square of the sum of three numbers**), the left-hand side is written as

$$\left(\frac{1}{x-y}+\frac{1}{y-z}+\frac{1}{z-x}\right)^2$$

$$=\frac{1}{(x-y)^2}+\frac{1}{(y-z)^2}+\frac{1}{(z-x)^2}$$

$$+2\underbrace{\left\{\frac{1}{x-y}\cdot\frac{1}{y-z}+\frac{1}{y-z}\cdot\frac{1}{z-x}+\frac{1}{z-x}\cdot\frac{1}{x-y}\right\}}_{\text{Term A}} \quad (*)$$

Let us simplify the term A in equation (*):

$$A=\frac{1}{x-y}\cdot\frac{1}{y-z}+\frac{1}{y-z}\cdot\frac{1}{z-x}+\frac{1}{z-x}\cdot\frac{1}{x-y} \Rightarrow$$

$$A=\frac{1}{(x-y)(y-z)}+\frac{1}{(y-z)(z-x)}+\frac{1}{(z-x)(x-y)} \Rightarrow$$

$$A=\frac{z-x}{(z-x)(x-y)(y-z)}+\frac{x-y}{(x-y)(y-z)(z-x)}$$
$$+\frac{y-z}{(y-z)(z-x)(x-y)} \Rightarrow$$

$$A=\frac{z-x+x-y+y-z}{(x-y)(y-z)(z-x)}\equiv 0$$

and this completes the proof.

Example 6-2-4: Assuming that $abc=1$ simplify the expression

$$F=\frac{a}{ab+a+1}+\frac{b}{bc+b+1}+\frac{c}{ca+c+1}$$

Solution

Since $abc=1$, $ab=1/c$, and the first term becomes

$$\frac{a}{ab+a+1}=\frac{a}{\frac{1}{c}+a+1}=\frac{ac}{ca+c+1} \quad (*)$$

Similarly, $bc = 1/a$, $b = 1/ac$, and the second term becomes

$$\frac{b}{bc+b+1} = \frac{b}{\frac{1}{a}+\frac{1}{ac}+1} = \frac{abc}{ca+c+1} = \frac{1}{ca+c+1} \quad (**)$$

By virtue of (*) and (**) expression F assumes the form

$$F = \frac{ac}{ca+c+1} + \frac{1}{ca+c+1} + \frac{c}{ca+c+1} = \frac{ac+1+c}{ca+c+1} \equiv 1$$

Example 6-2-5: Convert the following complex fraction into a simple one.

$$F = \frac{x}{1+\dfrac{x}{1+\dfrac{x}{1-x}}}$$

Solution

We first simplify the denominator, which is the most complicated expression. We have,

$$1+\frac{x}{1+\dfrac{x}{1-x}} = 1+\frac{x}{\dfrac{(1-x)+x}{1-x}} = 1+\frac{x}{\dfrac{1}{1-x}} = 1+x(1-x) = 1+x-x^2$$

The fraction $F = x/(1+x-x^2)$.

PROBLEMS

6-2-1) Simplify the expression

$$\frac{x+y}{xy}\left(\frac{1}{x}-\frac{1}{y}\right) - \frac{y+z}{yz}\left(\frac{1}{z}-\frac{1}{y}\right)$$

(Ans: $\frac{1}{x^2} - \frac{1}{z^2}$).

6-2-2) Show that

$$\frac{\dfrac{a+b}{a-b} - \dfrac{a-b}{a+b}}{1+\dfrac{b^2}{a^2-b^2}} = \frac{4b}{a}$$

6-2-3) If $x + y + z = 0$, show that

$$\left(\frac{y-z}{x} + \frac{z-x}{y} + \frac{x-y}{z}\right)\left(\frac{x}{y-z} + \frac{y}{z-x} + \frac{z}{x-y}\right) = 9$$

6-2-4) Show that

$$\frac{a+b}{(b-c)(c-a)} + \frac{b+c}{(c-a)(a-b)} + \frac{c+a}{(a-b)(b-c)} = 0$$

6-2-5) Convert into a simple fraction the following complex fraction

$$A = \frac{\dfrac{1-y}{1-y+y^2} + \dfrac{1+y}{1+y+y^2}}{\dfrac{1+y}{1+y+y^2} - \dfrac{1-y}{1-y+y^2}}$$

(Ans: $1/y^3$).

6-3) Rational equations of the form $(P/Q) = 0$

Any equation which (perhaps after some simplifications) takes the form $\frac{P}{Q} = 0$, where P and Q are **polynomials in one or more variables**, is called a **rational equation**. Every solution of this equation, assigns the value zero to the fraction (P/Q), and this means that **this solution vanishes the numerator without vanishing the denominator**, i.e.

$$\frac{P}{Q} = 0 \Leftrightarrow \{P = 0 \quad and \quad Q \neq 0\} \qquad (6-3-1)$$

Example 6-3-1: Solve the equation $\frac{x-1}{x+2} = 0$.

Solution

$$\frac{x-1}{x+2} = 0 \Leftrightarrow x - 1 = 0 \Leftrightarrow x = 1$$

Since this solution does **not** vanish the denominator, it is a valid solution.

Example 6-3-2: Solve the equation $\frac{3-x}{4+x} + \frac{2x-1}{3x+2} = \frac{13}{24}$.

Solution

$$\frac{3-x}{4+x}+\frac{2x-1}{3x+2}=\frac{13}{24} \Leftrightarrow \frac{3-x}{4+x}+\frac{2x-1}{3x+2}-\frac{13}{24}=0 \Leftrightarrow$$

$$\frac{24(3x+2)(3-x)}{24(3x+2)(4+x)}+\frac{24(4+x)(2x-1)}{24(4+x)(3x+2)}-\frac{13(4+x)(3x+2)}{24(4+x)(3x+2)}=0 \Leftrightarrow$$

$$\frac{-9x^2+22x-8}{24(4+x)(3x+2)}=0 \qquad (*)$$

(Let the reader verify that the simplifications in the numerator leads to equation (*)).

$$\frac{-9x^2+22x-8}{24(4+x)(3x+2)}=0 \Leftrightarrow \begin{Bmatrix} -9x^2+22x-8=0 \\ \text{and} \\ 24(4+x)(3x+2)\neq 0 \end{Bmatrix} \qquad (**)$$

$$-9x^2+22x-8=0 \Leftrightarrow -9x^2+18x+4x-8=0 \Leftrightarrow$$

$$-9x(x-2)+4(x-2)=0 \Leftrightarrow (x-2)(4-9x) \Rightarrow \begin{Bmatrix} x=2 \\ \text{or} \\ x=4/9 \end{Bmatrix}$$

Since neither one of these two solutions vanishes the denominator, both are valid solutions. Let the reader verify, by direct substitution, that $x=2$ and $x=4/9$ satisfy the original equation.

Example 6-3-3: Solve and investigate the following equation, for the various values of the real numbers a and b, (parameters),

$$\frac{ax-1}{x-1}+\frac{b}{x+1}=\frac{a(x^2+1)}{x^2-1}$$

Solution

Obviously, we must have: $(x-1)(x+1)\neq 0$. Every solution of the original equation, must satisfy this condition.

$$\frac{ax-1}{x-1}+\frac{b}{x+1}=\frac{a(x^2+1)}{x^2-1}=\frac{a(x^2+1)}{(x-1)(x+1)} \Leftrightarrow$$

$$\frac{(ax-1)(x+1)}{(x-1)(x+1)}+\frac{b(x-1)}{(x+1)(x-1)}=\frac{a(x^2+1)}{(x-1)(x+1)} \Leftrightarrow$$

$$(ax - 1)(x + 1) + b(x - 1) = a(x^2 + 1) \Leftrightarrow$$

$$ax^2 - x + ax - 1 + bx - b = ax^2 + a \Leftrightarrow$$

$$(a + b - 1)x = a + b + 1 \qquad (*)$$

Case 1) If $(a + b - 1) \neq 0$, equation (*) has the unique solution $x = \frac{a+b+1}{a+b-1}$.

This solution will be valid, provided that it satisfies the condition $(x - 1)(x + 1) \neq 0$, i.e.

$$\left(\frac{a+b+1}{a+b-1} - 1\right)\left(\frac{a+b+1}{a+b-1} + 1\right) \neq 0 \Leftrightarrow \frac{2(a+b)}{a+b-1} \cdot \frac{2}{a+b-1} \neq 0 \Leftrightarrow$$

$$a + b \neq 0$$

Case2) Let us now assume that $(a + b - 1) = 0$, i.e. $a + b = 1$. Then equation (*) becomes $0 \cdot x = 2$, which is impossible.

In summary: If $(a + b)(a + b - 1) \neq 0$, equation (*) has the unique solution, $x = (a + b + 1)/(a + b - 1)$, while if $(a + b - 1) = 0$ the equation is impossible.

PROBLEMS

6-3-1) Solve the equation: $\frac{7x+8}{21} - \frac{x+4}{8x-11} = \frac{x}{3}$, (**Ans:** $x = 4$).

6-3-2) Solve the equation: $\frac{1}{x-2} + \frac{1}{x+2} + \frac{1}{x-3} + \frac{1}{x+3} = 0$.

6-3-3) Solve the equation: $\frac{x-2}{x-3} - \frac{x-3}{x-4} = \frac{x-5}{x-6} - \frac{x-6}{x-7}$, (**Ans:** $x = 5$).

6-3-4) Solve the equation: $\frac{2}{y-1} + \frac{3}{y+1} = \frac{14}{y-3} - \frac{9}{y-2}$.

6-3-5) Solve and investigate the equation: $\frac{a}{ax-1} + \frac{b}{bx-1} = \frac{a+b}{(a+b)x-1}$, ($a, b$ are real parameters).

(**Ans:** If $ab(a + b) \neq 0$ then $x = 0$ or $x = 2/(a + b)$, if $ab \neq 0$ and $(a + b) = 0$, then $x = 0$, while if $ab = 0$ the equation becomes an identity, (satisfied by all x)).

6-4) Rational inequalities of the form $(P/Q) > 0$ or $(P/Q) < 0$

Any inequality which (perhaps after some simplifications) assumes the form $\left(\frac{P}{Q}\right) > 0$ or $\left(\frac{P}{Q}\right) < 0$, where P and Q are **polynomials in one or more variables**, is called a **rational inequality**.

For example if we consider the inequality $\frac{2}{x+1} > \frac{1}{x+3}$, transfer the right side member in the left side and simplify, we obtain

$$\frac{2}{x+1} - \frac{1}{x+3} > 0 \Leftrightarrow \frac{2(x+3)-(x+1)}{(x+1)(x+3)} > 0 \Leftrightarrow \frac{x+5}{(x+1)(x+3)} > 0$$

and this is a rational inequality with $P = (x+5)$ and $Q = (x+1)(x+3)$.

Theorem 6-4-1: The rational inequality $(P/Q) > 0$ is equivalent to the inequality $P \cdot Q > 0$, and similarly, $(P/Q) < 0$ is equivalent to $P \cdot Q < 0$, i.e.

$$\begin{cases} \dfrac{P}{Q} > 0 \Leftrightarrow P \cdot Q > 0 \\ \\ \dfrac{P}{Q} < 0 \Leftrightarrow P \cdot Q < 0 \end{cases} \qquad (6-4-1)$$

Proof: Any value of x that makes the quotient (P/Q) positive, makes P and Q either both positive or both negative, i.e. makes the product $P \cdot Q$ positive. So, every solution of the inequality $(P/Q) > 0$ makes also the product $P \cdot Q$ positive, and vice versa. This shows that the inequalities $(P/Q) > 0$ and $P \cdot Q > 0$ have **the same solution set**, and are therefore **equivalent**. Similarly, we show the second formula in equation (6-4-1).

Of course, from the solution set of $(P/Q) > 0$, we must exclude the values of x, (if any), which vanish the denominator Q.

For the solution of inequalities $P \cdot Q > 0$ or $P \cdot Q < 0$, see section 3-4.

Example 6-4-1: Solve the inequality $\frac{x+1}{x-2} > \frac{x+3}{x-4}$.

Solution

$$\frac{x+1}{x-2} > \frac{x+3}{x-4} \Leftrightarrow \frac{x+1}{x-2} - \frac{x+3}{x-4} > 0 \Leftrightarrow$$

$$\frac{(x+1)(x-4) - (x+3)(x-2)}{(x-2)(x-4)} > 0 \Leftrightarrow \frac{-4x+2}{(x-2)(x-4)} > 0 \xRightarrow{(Th.\ 6-4-1)}$$

$$(-4x+2)(x-2)(x-4) > 0 \Leftrightarrow -4\left(x - \frac{1}{2}\right)(x-2)(x-4) > 0 \Leftrightarrow$$

$$\left(x - \frac{1}{2}\right)(x-2)(x-4) < 0 \qquad (*)$$

The original inequality is equivalent to inequality (*), which is **a factored inequality** and can be solved by the method developed in section 3-4.

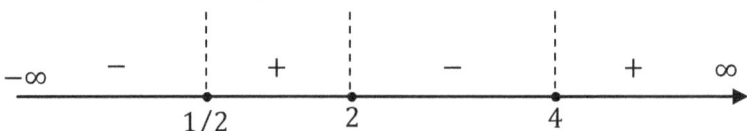

Fig. 6-1: Solution of the inequality in (*).

The solution set of the inequality in (*) is,

$$x \in \left(-\infty, \frac{1}{2}\right) \cup (2,4)$$

Example 6-4-2: Solve the inequality $\frac{(x+1)^2}{(x+3)^2} > \frac{x^2+1}{x^2+9}$.

Solution

$$\frac{(x+1)^2}{(x+3)^2} > \frac{x^2+1}{x^2+9} \Leftrightarrow \frac{(x+1)^2}{(x+3)^2} - \frac{x^2+1}{x^2+9} > 0 \Leftrightarrow$$

$$\frac{(x+1)^2(x^2+9) - (x+3)^2(x^2+1)}{(x+3)^3(x^2+9)} > 0 \Leftrightarrow$$

$$\frac{(x^2+2x+1)(x^2+9) - (x^2+6x+9)(x^2+1)}{(x+3)^3(x^2+9)} > 0 \Leftrightarrow$$

$$\frac{-4x^3 + 12x}{(x+3)^3(x^2+9)} > 0 \Leftrightarrow \frac{-4x(x^2-3)}{(x+3)^3(x^2+9)} > 0 \Leftrightarrow$$

$$\frac{x(x^2-3)}{(x+3)^3(x^2+9)} < 0 \xRightarrow{(Th.\ 6-4-1)}$$

$$x(x+\sqrt{3})(x-\sqrt{3})(x+3)^2(x^2+9) < 0 \qquad (*)$$

The original inequality is equivalent to the inequality in (*).

We notice that for any real number x, $x^2 + 9 > 0$ and $(x + 3)^2 > 0$ for any $x \in \mathbb{R} - \{-3\}$, (for $x = -3$, $(x + 3)^2 = 0$, but x is not allowed to take the value -3, since at this value of x the original inequality is not defined). So, in formula (*) the quantity $(x + 3)^2(x^2 + 9)$ is **strictly positive**, and hence inequality (*) is equivalent to the following

$$x(x+\sqrt{3})(x-\sqrt{3}) < 0 \qquad (**)$$

Solving this inequality using the method developed in section 3-4, we find

$$x \in (-\infty, -\sqrt{3}) \cup (0, \sqrt{3})$$

Example 6-4-3: Solve and investigate the inequality $a\frac{x-1}{x+1} > 1$, for the various values of the real parameter a.

Solution

$$a\frac{x-1}{x+1} > 1 \Leftrightarrow a\frac{x-1}{x+1} - 1 > 0 \Leftrightarrow \frac{a(x-1)-(x+1)}{x+1} > 0 \Leftrightarrow$$

$$\frac{(a-1)x-(a+1)}{x+1} > 0 \Leftrightarrow \{(a-1)x-(a+1)\}(x+1) > 0 \qquad (*)$$

a) If $a = 1$, the inequality in (*) becomes, $-2(x + 1) > 0$, i.e. $x < -1$.

b) If $a > 1$, i.e. $(a - 1) > 0$, the inequality in (*) becomes,

$$(a-1)\left(x - \frac{a+1}{a-1}\right)(x+1) > 0 \xRightarrow{(a-1)>0}$$

$$\left(x - \frac{a+1}{a-1}\right)(x+1) > 0 \Rightarrow x \in (-\infty, -1) \cup \left(\frac{a+1}{a-1}, \infty\right)$$

c) If $a = 0$, inequality (*) becomes: $-(x + 1)^2 > 0$, which is impossible.

d) If $a < 1$ $(a \neq 0)$, inequality (*) becomes

$$(a - 1)\left(x - \frac{a + 1}{a - 1}\right)(x + 1) > 0 \overset{(a-1)<0}{\Longleftrightarrow}$$

$$\left(x - \frac{a + 1}{a - 1}\right)(x + 1) < 0$$

and the numbers x which satisfy this inequality, lie between (-1) and $(a + 1)/(a - 1)$.

Example 6-4-4: Solve the inequality $\frac{x}{|x|-2} > \frac{3}{4}$.

Solution

$$\frac{x}{|x| - 2} > \frac{3}{4} \Leftrightarrow \frac{x}{|x| - 2} - \frac{3}{4} > 0 \Leftrightarrow \frac{4x - 3(|x| - 2)}{4(|x| - 2)} > 0 \Leftrightarrow$$

$$4\{4x - 3(|x| - 2)\}(|x| - 2) > 0 \Leftrightarrow \{4x - 3|x| + 6\}(|x| - 2) > 0 \quad (*)$$

Since the inequality in (*) contains the absolute value of x, we have to make certain assumptions about the range of x, in order to get rid of the absolute values.

a) Let us assume first that $x \geq 0$. In this case $|x| = x$, and equation (*) implies,

$$(4x - 3x + 6)(x - 2) > 0 \Leftrightarrow (x + 6)(x - 2) > 0 \quad (**)$$

The solution of the inequality in (**) which satisfies also the condition $x \geq 0$, is $x > 2$.

b) If $x < 0$, $|x| = -x$ and the inequality in (*) becomes

$$(4x + 3x + 6)(-x - 2) > 0 \Leftrightarrow (-1)(7x + 6)(x + 2) > 0 \Leftrightarrow$$

$$(7x + 6)(x + 2) < 0 \quad (***)$$

The solution of (***) which satisfies the condition $x < 0$ is $-2 < x < -\frac{6}{7}$.

In summary: The solution of the original inequality is

$$x \in \left(-2, -\frac{6}{7}\right) \cup (2, \infty)$$

PROBLEMS

6-4-1) Solve the inequality $\frac{2x-3}{7x+1} > 5$, (**Ans:** $-\frac{8}{33} < x < -\frac{1}{7}$).

6-4-2) Solve the inequality $\frac{3}{x+4} + \frac{4}{5} > \frac{1}{x-1}$.

6-4-3) Solve the inequality $\left|\frac{10x^2-3x-2}{(x-1)(x-2)}\right| < 1$, (**Ans:** $x \in \left(-\frac{2}{3}, 0\right) \cup \left(\frac{6}{11}, \frac{2}{3}\right)$).

Hint: The given inequality implies: $-1 < \frac{10x^2-3x-2}{(x-1)(x-2)} < 1$.

6-4-4) Solve the inequality $\frac{1}{x+1} + \frac{2}{x+2} < \frac{3}{x+3}$.

6-4-5) Solve the inequality $\frac{2-x}{x+2} > \frac{x}{1-x} + 1$, (**Ans:** $x \in (-2,0) \cup (1,4)$).

6-4-6) Solve the inequality $\frac{6+5x}{x^2+4x+5} > \frac{1}{2}$.

Hint: Notice that $\forall x \in \mathbb{R}, x^2 + 4x + 5 > 0$. Indeed, $x^2 + 4x + 5 = (x+2)^2 + 1 \geq 1 > 0$.

6-4-7) Find the values of x for which the following two inequalities are satisfied simultaneously

$$\left\{\frac{3x+5}{7-3x} < 3, \quad \frac{1}{(x-1)(x-2)} > \frac{1}{x(x+3)}\right\}$$

(**Ans:** $-3 < x < 0$, or $1 < x < \frac{4}{3}$, or $\frac{7}{3} < x < \infty$).

6-4-8) Solve the inequality $\frac{2}{3+|x|} < \frac{5}{2+|x-1|}$.

Hint: In order to get rid of the absolute values, consider the three intervals, $-\infty < x < 0$, $0 \leq x < 1$, $1 \leq x < \infty$.

6-4-9) Solve and investigate the inequality for the various values of the real parameter λ:

$$\frac{\lambda x}{\lambda+2} - \frac{x+1}{3} < \frac{2x-3}{4}$$

(Ans: If $(\lambda+2)(\lambda-10) > 0$, $x < -\frac{5}{2}\frac{\lambda+2}{\lambda-10}$, if $(\lambda+2)(\lambda-10) < 0$, then $x > -\frac{5}{2}\frac{\lambda+2}{\lambda-10}$, if $\lambda = 10$ impossible. Notice that $(\lambda+2) \neq 0$, since this term appears in the denominator in the left hand side of the inequality.

6-4-10) Solve and investigate the inequality for the various values of the real parameter k:

$$\frac{x}{5} > 3 + \frac{x}{k}$$

CHAPTER 7: IRRATIONAL EQUATIONS

The standard rules and laws governing the operations involving radicals and fractional exponents (**both positive and negative**) are developed in considerable depth and details in my book "**College Algebra, Vol. 1**". Even though the reader of this book is supposed to be familiar with the relevant material, we here under present a systematic summary of the main results.

7-1) Basic definitions and operations with radicals (a brief summary)

If n is an **odd** positive integer, the symbol $\sqrt[n]{x}$ is understood as **the only real number** whose n^{th} power is equal to x. In this case x could be either positive or negative.

If n is an **even** positive integer, the symbol $\sqrt[n]{x}$, where $x \geq 0$, is understood as **the only positive number** whose n^{th} power is equal to x.

For example, $\sqrt[3]{8} = 2$ (since $2^3 = 8$), $\sqrt[3]{-8} = -2$ (since $(-2)^3 = -8$), $\sqrt[5]{-64} = -2$ (since $(-2)^5 = -64$), etc. Also, $\sqrt{4} = +2$ (since $2^2 = 4$), $\sqrt[4]{81} = +3$ (since $3^4 = 4$), etc.

Notice that, even though $(-2)^2 = 4$ and $2^2 = 4$, the symbol $\sqrt{4}$ stands for **the positive square root** (by definition). So in general, we may write

$$\sqrt{x^2} = |x| = \begin{cases} x & if \ x \geq 0 \\ -x & if \ x < 0 \end{cases} \qquad (7-1-1)$$

Another important remark is that **the square root of a negative number does not exist (in the set of real numbers)**. For example, the symbol $\sqrt{-1}$ is not a real number. Indeed, if we call $y = \sqrt{-1}$, then according to the definition, we should have $y^2 = -1$, i.e. y should be a real number whose square is negative; but such real number does not exist, since the square of any real number is nonnegative, i.e. is zero if $y = 0$ or is strictly positive if $y \neq 0$.

In the symbol $\sqrt[n]{x}$, x is called the **radicand**, n is the **index** and the symbol $\sqrt{}$ is called the **radical**.

We may thus say that **if x is negative and the index is an even number, then the symbol $\sqrt[n]{x}$ is not defined in the set of real numbers**. However, as we shall see shortly, square roots of negative numbers are defined in a more general set, called the "**set of complex numbers**". The set of real numbers is a subset of the set of complex numbers.

The following rules of operation with radicals with positive radicands are proved easily, (n, m and k are **positive integers**):

$$\left\{\begin{array}{c} \sqrt[n]{ab} = \sqrt[n]{a}\,\sqrt[n]{b} \\ \sqrt[n]{\dfrac{a}{b}} = \dfrac{\sqrt[n]{a}}{\sqrt[n]{b}} \\ \sqrt[n]{\sqrt[m]{a}} = \sqrt[nm]{a} \\ \sqrt[n]{a^m} = \left(\sqrt[n]{a}\right)^m \\ p\sqrt[n]{a} = \sqrt[n]{p^n a}, \quad p \text{ is a positive number} \\ \sqrt[n]{a^m} = \sqrt[n\cdot k]{a^{m\cdot k}} \end{array}\right\} \qquad (7-1-2)$$

The proof of all formulas in (7-1-2) is based on the following proposition: **If the nth powers ($n = 2, 3, 4, ...$) of two positive numbers are equal, then these two numbers are equal**, i.e.

$$If\ A > 0\ and\ B > 0, then\ \ A^n = B^n \Leftrightarrow A = B \qquad (7-1-3)$$

Powers with fractional exponents: If m and n are two positive integers, and x is a positive real number, then the symbol $x^{m/n}$ represents the nth root of the power x^m, i.e.

$$x^{m/n} \stackrel{\text{def}}{=} \sqrt[n]{x^m} \qquad (7-1-4)$$

The symbol $x^{-(m/n)}$ represents the reciprocal of the power $x^{m/n}$, i.e.

$$x^{-(m/n)} \stackrel{\text{def}}{=} \frac{1}{x^{m/n}} \qquad (7-1-5)$$

For example,

$$x^{1/2} = \sqrt{x},\ x^{1/3} = \sqrt[3]{x},\ x^{3/7} = \sqrt[7]{x^3},\ x^{-(2/5)} = \frac{1}{x^{2/5}} = \frac{1}{\sqrt[5]{x^2}}$$

Making use of formulas (7-1-4) and (7-1-5) we see that we may, in general, define the symbol x^r, where r is any rational number, (recall that rational number is any number which can be expressed as a fraction (m/n), where m and n are integers, positive or negative, with $n \neq 0$). For example,

$$16^{-0.5} = 16^{-\frac{1}{2}} = \frac{1}{16^{1/2}} = \frac{1}{\sqrt{16}} = \frac{1}{4}$$

$$49^{0.5} = 49^{1/2} = \sqrt{49} = 7, \quad 8^{1/3} = \sqrt[3]{8} = 2$$

Remark: It is possible to define the power x^y, **where y is any real number, not just a rational one**. Functions of the form $f(x) = a^x$, where a is a positive number and x is a real variable, are known as "**exponential functions**".

It is not difficult to show that all the properties of powers with integer exponents, still hold for powers with fractional exponents, positive or negative. For example,

$$x^{\frac{m}{n}} \cdot x^{\frac{\kappa}{\lambda}} = x^{\frac{m}{n} + \frac{\kappa}{\lambda}} = x^{\frac{m\lambda + \kappa n}{n\lambda}}$$

$$\left(x^{\frac{m}{n}}\right)^{\frac{\kappa}{\lambda}} = x^{\frac{m}{n} \cdot \frac{\kappa}{\lambda}} = x^{\frac{m \cdot \kappa}{n \cdot \lambda}}$$

Example 7-1-1: Find $8^{1.2} \cdot 8^{0.7} \cdot 8^{0.3} \cdot 8^{0.8}$.

Solution

$$8^{1.2} \cdot 8^{0.7} \cdot 8^{0.3} \cdot 8^{0.8} = 8^{1.2+0.7+0.3+0.8} = 8^3 = 512$$

Example 7-1-2: Simplify the expression $\left(\frac{8}{27}\right)^{\frac{2}{3}} \cdot \left(\frac{64}{125}\right)^{-\frac{1}{3}}$.

Solution

$$\left(\frac{8}{27}\right)^{\frac{2}{3}} \cdot \left(\frac{64}{125}\right)^{-\frac{1}{3}} = \left(\frac{2^3}{3^3}\right)^{\frac{2}{3}} \cdot \left(\frac{4^3}{5^3}\right)^{-\frac{1}{3}} = \left(\left(\frac{2}{3}\right)^3\right)^{\frac{2}{3}} \cdot \left(\left(\frac{4}{5}\right)^3\right)^{-\frac{1}{3}}$$

$$= \left(\frac{2}{3}\right)^{3 \cdot \frac{2}{3}} \cdot \left(\frac{4}{5}\right)^{3 \cdot \left(-\frac{1}{3}\right)} = \left(\frac{2}{3}\right)^2 \cdot \left(\frac{4}{5}\right)^{-1} = \frac{\frac{4}{9}}{\frac{4}{5}} = \frac{4 \cdot 5}{4 \cdot 9} = \frac{5}{9}$$

Example 7-1-3: Find the number $((64^{0.75})^{0.5})^{0.5} \cdot \left(\frac{1}{32}\right)^{-\frac{1}{5}}$.

Solution

$$((64^{0.75})^{0.5})^{0.5} \cdot \left(\frac{1}{32}\right)^{-\frac{1}{5}} = 64^{(0.75)\cdot(0.5)\cdot(0.5)} \cdot \left(\frac{1}{2^5}\right)^{-\frac{1}{5}}$$

$$= 64^{\left(\frac{3}{4}\right)\cdot\left(\frac{1}{2}\right)\cdot\left(\frac{1}{2}\right)} \cdot (2^{-5})^{-\frac{1}{5}} = 64^{\frac{3}{16}} \cdot 2^{(-5)\cdot\left(-\frac{1}{5}\right)} = (8^2)^{\frac{3}{16}} \cdot 2$$

$$= 8^{2 \cdot \frac{3}{16}} \cdot 2 = 8^{\frac{3}{8}} \cdot 2 = 2\sqrt[8]{8^3}$$

Example 7-1-4: Assuming that $\frac{X}{x} = \frac{Y}{y} = \frac{Z}{z} = \frac{W}{w}$, where all the terms in the fractions are positive, show that

$$\sqrt{Xx} + \sqrt{Yy} + \sqrt{Zz} + \sqrt{Ww} = \sqrt{(X+Y+Z+W)(x+y+z+w)}$$

Solution

Let k be the common value of all the fractions, i.e.

$$\frac{X}{x} = \frac{Y}{y} = \frac{Z}{z} = \frac{W}{w} = k \Rightarrow \begin{cases} X = kx \\ Y = ky \\ Z = kz \\ W = kw \end{cases} \quad (*)$$

Then we have,

$$\sqrt{Xx} + \sqrt{Yy} + \sqrt{Zz} + \sqrt{Ww} = \sqrt{kx^2} + \sqrt{ky^2} + \sqrt{kz^2} + \sqrt{kw^2} \Rightarrow$$

$$\sqrt{Xx} + \sqrt{Yy} + \sqrt{Zz} + \sqrt{Ww} = \sqrt{k}\,x + \sqrt{k}\,y + \sqrt{k}\,z + \sqrt{k}\,w$$
$$= \sqrt{k}\,(x+y+z+w) \quad (**)$$

Also, the term $\sqrt{(X+Y+Z+W)(x+y+z+w)}$ becomes,

$$\sqrt{(X+Y+Z+W)(x+y+z+w)} = \sqrt{(kx+ky+kz+kw)(x+y+z+w)} \Rightarrow$$

$$\sqrt{(X+Y+Z+W)(x+y+z+w)} = \sqrt{k(x+y+z+w)^2}$$
$$= \sqrt{k}\,(x+y+z+w) \quad (***)$$

and the proof is completed.

PROBLEMS

7-1-1) Find the number $32^{-\frac{1}{5}} \cdot \left(\frac{1}{8}\right)^{\frac{1}{3}} \cdot \left(\frac{1}{4}\right)^{\frac{1}{2}}$, (Ans: 1/8).

7-1-2) If x and y are real numbers, show that

$$\left(x^{\frac{2}{3}} + y^{\frac{2}{3}} - (xy)^{\frac{1}{3}}\right)\left(x^{\frac{1}{3}} + y^{\frac{1}{3}}\right) = x + y$$

7-1-3) Simplify the expression $\left(\frac{8}{3}x^{-3}y^{-4}\right)^{-\frac{1}{6}} \cdot \left(\frac{2}{5}x^2 y^3\right)^{-\frac{1}{3}}$.

(Ans: $(8x/3)^{-\frac{1}{6}} (2y/5)^{-\frac{1}{3}}$).

7-1-4) Show that $(14 + 6 \cdot 5^{0.5})^{0.5} = 3 + 5^{0.5}$.

Hint: It suffices to show that the squares of both sides are equal, (see equation (7-1-3)).

7-1-5) Find the number $\sqrt{2} \cdot \sqrt{2 + \sqrt{2}} \cdot \sqrt{2 - \sqrt{2}}$, (Ans: 2).

7-1-6) Show that $\sqrt{3 - 2\sqrt{2}} = \sqrt{2} - 1$.

Hint: Since both numbers are positive, it suffices to show that their squares are equal, (see formula (7-1-3)).

7-1-7) If $\sqrt[3]{a} + \sqrt[3]{b} + \sqrt[3]{c} = 0$ show that $(a + b + c)^3 = 27abc$.

Hint: Consider **Cauchy's identity**

$$x^3 + y^3 + z^3 - 3xyz = \frac{1}{2}(x + y + z)\{(x - y)^2 + (y - z)^2 + (z - x)^2\}$$

and show that if $(x + y + z) = 0$ or $(x = y = z)$, then $x^3 + y^3 + z^3 = 3xyz$. In our problem, consider $x = \sqrt[3]{a}, y = \sqrt[3]{b}, z = \sqrt[3]{c}$.

7-1-8) If $x > 0$, $y > 0$ show that $\frac{x+y}{2} > \sqrt{xy} > \frac{2}{\frac{1}{x}+\frac{1}{y}}$.

7-1-9) Consider the number $X = \sqrt{2+\sqrt{3}} \cdot (\sqrt{2}-\sqrt{6})$. First, find X^2 and then show that $X = -2$, (Notice that X is a negative number, since $(\sqrt{2}-\sqrt{6}) < 0$ and $\sqrt{2+\sqrt{3}} > 0$).

7-2) Rationalization of the denominator

Irrational algebraic fractions are called the fractions that contain **at least one radical**, either in the numerator or in the denominator. For example, the following fractions are irrational:

$$\frac{2+\sqrt{3}}{\sqrt{5}-2}, \frac{7}{\sqrt{2}-1}, \frac{1}{\sqrt[5]{23}}, \frac{\sqrt{x}+\sqrt{y}}{\sqrt{x}-5\sqrt{y}+8}$$

If the denominator of a fraction contains a radical, we try to convert the fraction to **an equal fraction with rational denominator**. This process is known as "**the rationalization of the denominator**". We prefer to work with fractions with **rational denominators**, since in this case the operations and the calculations involved are easier. For example, consider the fraction $1/\sqrt{2}$, which means that we have to divide the number 1 by the number $\sqrt{2}$. This is not an easy task. However, since

$$\frac{1}{\sqrt{2}} = \frac{\sqrt{2}}{\sqrt{2} \cdot \sqrt{2}} = \frac{\sqrt{2}}{2}$$

The given fraction is equal to $\sqrt{2}/2$, and this division is much easier to be performed as compared to the division $1/\sqrt{2}$, (think about this).

The following forms of irrational fractions are often encountered in Algebra:

1) Fractions of the form $\frac{x}{y \pm \sqrt{z}}$ or $\frac{x}{\sqrt{y} \pm \sqrt{z}}$.

The fraction is converted to an equal fraction with rational denominator, if we multiply, both the numerator and the denominator, by the conjugate expression of the denominator. The "**conjugate expression**" of the denominator is the expression we obtain if we change the sign in front of one of the radicals. For example, let us rationalize the fraction $\frac{x}{y-\sqrt{z}}$. The conjugate expression of the denominator is $y + \sqrt{z}$, and we have,

$$\frac{x}{y-\sqrt{z}} = \frac{x(y+\sqrt{z})}{(y-\sqrt{z})(y+\sqrt{z})} = \frac{x(y+\sqrt{z})}{y^2-z}$$

2) Fractions of the form $\frac{x}{\sqrt{x}\pm\sqrt{y}\pm\sqrt{z}}$.

We multiply both numerator and denominator, by a conjugate expression of the denominator, and we thus reduce the number of radicals by one, and then apply again the same technique to reduce the number of radicals by one more, etc, until we obtain a fraction with no radicals in the denominator. For example, let us rationalize the fraction $x/(\sqrt{x}-\sqrt{y}+\sqrt{z})$. We have,

$$A \equiv \frac{x}{\sqrt{x}-\sqrt{y}+\sqrt{z}} = \frac{x(\sqrt{x}+\sqrt{y}+\sqrt{z})}{(\sqrt{x}-\sqrt{y}+\sqrt{z})(\sqrt{x}+\sqrt{y}+\sqrt{z})} \Rightarrow$$

$$A = \frac{x(\sqrt{x}+\sqrt{y}+\sqrt{z})}{(\sqrt{x}+\sqrt{z})^2 - y} = \frac{x(\sqrt{x}+\sqrt{y}+\sqrt{z})}{x+z-y+2\sqrt{xz}} \Rightarrow$$

$$A = \frac{x(\sqrt{x}+\sqrt{y}+\sqrt{z})(x+z-y-2\sqrt{xz})}{(x+z-y+2\sqrt{xz})(x+z-y-2\sqrt{xz})} \Rightarrow$$

$$A = \frac{x(\sqrt{x}+\sqrt{y}+\sqrt{z})(x+z-y-2\sqrt{xz})}{(x+z-y)^2 - 4xz}$$

3) Fractions of the form $\frac{x}{\sqrt[n]{y}}$.

The denominator becomes rational, if we multiply both numerator and denominator by $\sqrt[n]{y^{n-1}}$. Indeed,

$$\frac{x}{\sqrt[n]{y}} = \frac{x\sqrt[n]{y^{n-1}}}{\sqrt[n]{y}\sqrt[n]{y^{n-1}}} = \frac{x\sqrt[n]{y^{n-1}}}{\sqrt[n]{y \cdot y^{n-1}}} = \frac{x\sqrt[n]{y^{n-1}}}{\sqrt[n]{y^n}} = \frac{x\sqrt[n]{y^{n-1}}}{y}$$

Example 7-2-1: Simplify the expression $F = \frac{5}{\sqrt{3}-2} - \frac{2}{\sqrt{2}-3}$.

Solution

$$F = \frac{5}{\sqrt{3}-2} - \frac{2}{\sqrt{2}-3} = \frac{5(\sqrt{3}+2)}{(\sqrt{3}-2)(\sqrt{3}+2)} - \frac{2(\sqrt{2}+3)}{(\sqrt{2}-3)(\sqrt{2}+3)} \Rightarrow$$

$$F = \frac{5\sqrt{3} + 10}{3 - 4} - \frac{2\sqrt{2} + 6}{4 - 3} = -5\sqrt{3} - 10 - 2\sqrt{2} - 6 = -5\sqrt{3} - 2\sqrt{2} - 16$$

Example 7-2-2: Rationalize the fraction $A = \frac{1}{\sqrt{x}+\sqrt{y}-\sqrt{x+y}}$.

Solution

$$A = \frac{1}{\sqrt{x} + \sqrt{y} - \sqrt{x+y}} = \frac{\left(\sqrt{x} + \sqrt{y} + \sqrt{x+y}\right)}{\left(\sqrt{x} + \sqrt{y} - \sqrt{x+y}\right)\left(\sqrt{x} + \sqrt{y} + \sqrt{x+y}\right)} \Rightarrow$$

$$A = \frac{\left(\sqrt{x} + \sqrt{y} + \sqrt{x+y}\right)}{\left(\sqrt{x} + \sqrt{y}\right)^2 - \left(\sqrt{x+y}\right)^2} = \frac{\left(\sqrt{x} + \sqrt{y} + \sqrt{x+y}\right)}{x + y + 2\sqrt{xy} - x - y} \Rightarrow$$

$$A = \frac{\left(\sqrt{x} + \sqrt{y} + \sqrt{x+y}\right)}{2\sqrt{xy}} = \frac{\left(\sqrt{x} + \sqrt{y} + \sqrt{x+y}\right)\sqrt{xy}}{2\sqrt{xy}\sqrt{xy}} \Rightarrow$$

$$A = \frac{\left(\sqrt{x} + \sqrt{y} + \sqrt{x+y}\right)\sqrt{xy}}{2xy}$$

Example 7-2-3: Show that

$$\frac{2}{2 - \sqrt{3}} - \frac{2 + \sqrt{3}}{2} = \frac{3}{4}\left(\sqrt{3} + 1\right)^2$$

Solution

$$\frac{2}{2 - \sqrt{3}} - \frac{2 + \sqrt{3}}{2} = \frac{2(2 + \sqrt{3})}{(2 - \sqrt{3})(2 + \sqrt{3})} - \frac{2 + \sqrt{3}}{2} =$$

$$\frac{4 + 2\sqrt{3}}{4 - 3} - \frac{2 + \sqrt{3}}{2} = 4 + 2\sqrt{3} - \frac{2 + \sqrt{3}}{2} = \frac{8 + 4\sqrt{3} - 2 - \sqrt{3}}{2} =$$

$$\frac{6 + 3\sqrt{3}}{2} = \frac{12 + 6\sqrt{3}}{4} = \frac{3}{4}\left(4 + 2\sqrt{3}\right) = \frac{3}{4}\underbrace{\left(\left(\sqrt{3}\right)^2 + 1 + 2\sqrt{3}\right)}_{\left(\sqrt{3}+1\right)^2} = \frac{3}{4}\left(\sqrt{3} + 1\right)^2$$

PROBLEMS

7-2-1) Simplify the expression

$$F = \frac{4}{3\sqrt{3} - 2\sqrt{2}} + \frac{5}{2\sqrt{3} + \sqrt{2}}$$

(**Ans:** $(62\sqrt{3} - 3\sqrt{2})/38$).

7-2-2) Solve the equation $\frac{x-1}{x+1} = 2 - \sqrt{3}$.

7-2-3) Find the value of the expression

$$A = \sqrt{\frac{9}{\left(1 - \sqrt{3}\right)^2}} - \sqrt{\frac{9}{\left(1 + \sqrt{3}\right)^2}}$$

(**Ans:** $A = 3$).

Hint: Recall that $\sqrt{x^2} = |x|$.

7-2-4) Simplify the expression

$$F = \frac{2\sqrt{x-y}}{3\sqrt{x+y} - 2\sqrt{x-y}} + \frac{3\sqrt{x+y}}{3\sqrt{x+y} + 2\sqrt{x-y}}$$

7-2-5) Simplify the expression

$$A = \frac{1}{\sqrt{8} - \sqrt{6}} - \frac{3}{\sqrt{6} - \sqrt{3}} + \frac{4}{\sqrt{3} - \sqrt{8}}$$

(**Ans:** $A = -(3\sqrt{8} + 5\sqrt{6} + 18\sqrt{3})/10$).

7-3) Double radicals

Any expression of the form $\sqrt{A \pm \sqrt{B}}$, $A > 0$, $B > 0$, $A > \sqrt{B}$, where A and B are rational numbers or rational expressions, is called "**a double radical**". This double radical may be converted to the following form,

$$\sqrt{A \pm \sqrt{B}} = \sqrt{\frac{A+C}{2}} \pm \sqrt{\frac{A-C}{2}}, \text{ where } C = \sqrt{A^2 - B} \qquad (7-3-1)$$

The proof is not difficult. Indeed, let us consider the expressions $X = \sqrt{A + \sqrt{B}}$ and $Y = \sqrt{\frac{A+C}{2}} + \sqrt{\frac{A-C}{2}}$, where $C = \sqrt{A^2 - B}$. In order to show that

$X = Y$, it suffices to show that $X^2 = Y^2$, since both X and Y are positive numbers, (see formula (7-1-3)). We have:

$$X^2 = A + \sqrt{B} \qquad (*)$$

$$Y^2 = \left(\sqrt{\frac{A+C}{2}} + \sqrt{\frac{A-C}{2}}\right)^2 = \frac{A+C}{2} + \frac{A-C}{2} + 2 \cdot \sqrt{\frac{A+C}{2} \cdot \frac{A-C}{2}} \Rightarrow$$

$$Y^2 = A + 2 \cdot \frac{\sqrt{A^2 - C^2}}{2} \xrightarrow{(C=\sqrt{A^2-B})} Y^2 = A + \sqrt{B} \qquad (**)$$

From equations (*) and (**) we conclude that $X^2 = Y^2$ and hence $X = Y$, (since $X > 0, Y > 0$) and the proof is completed. Similar proof applies if we choose the negative sign in (7-3-1).

If the quantity $C = \sqrt{A^2 - B}$ is a rational number, (i.e. if $A^2 - B$ is a perfect square), then formula (7-3-1) converts a double radical into a sum or a difference of simple radicals.

Example 7-3-1: Convert the double radical $\sqrt{4 + \sqrt{15}}$ in a sum of simple radicals.

Solution

In this problem, $A = 4, B = 15, A^2 - B = 4^2 - 15 = 16 - 15 = 1$, $C = \sqrt{1} = 1$, and by virtue of formula (7-3-1) we have,

$$\sqrt{4 + \sqrt{15}} = \sqrt{\frac{4+1}{2}} + \sqrt{\frac{4-1}{2}} = \frac{\sqrt{5}}{\sqrt{2}} + \frac{\sqrt{3}}{\sqrt{2}} = \frac{\sqrt{5} + \sqrt{3}}{\sqrt{2}} = \frac{\sqrt{2}(\sqrt{5} + \sqrt{3})}{2}$$

Example 7-3-2: Simplify the expression $F = \sqrt{x + 2\sqrt{x-1}}, \ x > 2$.

Solution

$A = x, B = 4x - 4, A^2 - B = x^2 - 4x + 4 = (x - 2)^2, C = |x - 2| = x - 2$

$$F = \sqrt{\frac{x+x-2}{2}} + \sqrt{\frac{x-(x-2)}{2}} = \sqrt{x-1} + 1$$

Example 7-3-3: Express the double radical $\sqrt{9 + 4\sqrt{5}}$ as the sum of two simple radicals and then show that $\sqrt[3]{38 + 17\sqrt{5}} = \sqrt{9 + 4\sqrt{5}}$.

Solution

For the double radical $\sqrt{9 + 4\sqrt{5}}$, $A = 9, B = 16 \cdot 5 = 80, C = A^2 - B = 81 - 80 = 1$, and therefore

$$\sqrt{9 + 4\sqrt{5}} = \sqrt{\frac{9+1}{2}} + \sqrt{\frac{9-1}{2}} = \sqrt{5} + \sqrt{4} = 2 + \sqrt{5} \qquad (*)$$

In order to show that $\sqrt[3]{38 + 17\sqrt{5}} = 2 + \sqrt{5}$, it suffices to show that $\left(2 + \sqrt{5}\right)^3 = 38 + 17\sqrt{5}$. Indeed,

$$\left(2 + \sqrt{5}\right)^3 = 2^3 + 3 \cdot 2^2 \cdot \sqrt{5} + 3 \cdot 2 \cdot \left(\sqrt{5}\right)^2 + \left(\sqrt{5}\right)^3 =$$

$$8 + 12\sqrt{5} + 30 + 5\sqrt{5} = 38 + 17\sqrt{5}$$

and this completes the proof.

PROBLEMS

7-3-1) Convert to simple radicals the following double ones:

$$\sqrt{7 + 4\sqrt{3}}, \quad \sqrt{6 + \sqrt{35}}, \quad \sqrt{15 + 2\sqrt{56}}, \quad \sqrt{4 - \sqrt{15}}$$

(Ans: $2 + \sqrt{3}, \sqrt{7/2} + \sqrt{5/2}, \sqrt{7} + \sqrt{8}, \sqrt{5/2} - \sqrt{3/2}$).

7-3-2) Find the simplest expression possible of the number

$$P = \frac{\sqrt{26 + 15\sqrt{3}}}{5\sqrt{2} + \sqrt{38 - 5\sqrt{3}}}$$

(Ans: $P = \sqrt{3}/3$).

7-3-3) Find the simplest expression possible of the number

$$F = \sqrt{28 - 10\sqrt{3} - \frac{1}{\sqrt{7 - 4\sqrt{3}}}}$$

(Ans: $F = 3 - 2\sqrt{3}$).

7-3-4) If $x = \sqrt{2 + \sqrt{3}}, y = \sqrt{2 + x}, z = \sqrt{2 + y}$ and $w = \sqrt{2 - y}$, show that $xyzw = 1$.

7-3-5) If $a = \sqrt[n]{x + b^n}, b = \sqrt[n]{y + c^n}, c = \sqrt[n]{z + a^n}$ show that:

1) $x^3 + y^3 + z^3 = 3xyz$ and 2) $x^4 + y^4 + z^4 = 2(x^2y^2 + y^2z^2 + z^2x^2)$

Hint: Apply Cauchy's identity (see Problem 7-1-7).

7-3-6) If $a = \sqrt{11 + 2\sqrt{30}}, b = \sqrt{11 - 2\sqrt{30}}$, find the number $P = a^{-4} + b^{-4} + 5(ab)^2$.

7-3-7) a) Assuming that n is a positive integer, show that
$1/(n\sqrt{n+1} + (n+1)\sqrt{n}) = 1/\sqrt{n} - 1/\sqrt{n+1}$.

b) If $S(n) = \frac{1}{1\sqrt{2}+2\sqrt{1}} + \frac{1}{2\sqrt{3}+3\sqrt{2}} + \cdots + \frac{1}{n\sqrt{n+1}+(n+1)\sqrt{n}}$, show that $S(n) = 1 - \frac{1}{\sqrt{n+1}}$. What is $S(99)$?

(Ans: $S(99) = 9/10$).

7-3-8) If the positive numbers x, y, z satisfy the relation $xy + yz + zx = 1$, show that

$$x\sqrt{\frac{(y^2 + 1)(z^2 + 1)}{x^2 + 1}} + y\sqrt{\frac{(z^2 + 1)(x^2 + 1)}{y^2 + 1}} + z\sqrt{\frac{(x^2 + 1)(y^2 + 1)}{z^2 + 1}} = 2$$

7-3-9) If $x = \sqrt[3]{\sqrt{5} + 2} - \sqrt[3]{\sqrt{5} - 2}$, show that $x = 1$.

Hint: Apply the identity $(A - B)^3 = A^3 - B^3 - 3AB(A - B)$ with $A = \sqrt[3]{\sqrt{5} + 2}$ and $B = \sqrt[3]{\sqrt{5} - 2}$.

7-3-10) Assuming that $y > 0$, show that

$$\sqrt{x + 2y\sqrt{x - y^2}} + \sqrt{x - 2y\sqrt{x - y^2}} = \begin{cases} 2y & \text{if } y^2 \le x \le 2y^2 \\ 2\sqrt{x - y^2} & \text{if } x > 2y^2 \end{cases}$$

7-4) Binomials of the form $a + b\sqrt{c}$

Assuming that a, b, c are **rational numbers** and that $c > 0$ is not a perfect square, (i.e. $\sqrt{c}$ **is an irrational number**), the expression $a + b\sqrt{c}$ **is called a binomial of** $\sqrt{c}$. The binomial $a - b\sqrt{c}$ is the **conjugate binomial** of $a + b\sqrt{c}$.

As we easily verify, the square, the cube, the fourth power, etc. of $a + b\sqrt{c}$ is also a binomial of $\sqrt{c}$. For example,

$$\left(a + b\sqrt{c}\right)^2 = \underbrace{a^2 + b^2 c}_{A} + \underbrace{2ab}_{B} \sqrt{c} = A + B\sqrt{c}, (A, B \text{ rational})$$

$$\left(a + b\sqrt{c}\right)^3 = a^3 + 3a^2 b\sqrt{c} + 3a\left(b\sqrt{c}\right)^2 + \left(b\sqrt{c}\right)^3$$
$$= a^3 + 3a^2 b\sqrt{c} + 3ab^2 c + b^3 c\sqrt{c}$$
$$= \underbrace{a^3 + 3ab^2 c}_{A} + \underbrace{(3a^2 b + b^3 c)}_{B} \sqrt{c} = A + B\sqrt{c}$$

Also, if $P(x)$ is a polynomial in x, **with rational coefficients**, then the numbers $P(a + b\sqrt{c})$ and $P(a - b\sqrt{c})$ are conjugate binomials of $\sqrt{c}$. For example, let $P(x) = x^2 - 3x + 5$ and let us evaluate the value of $P(x)$ at $x = 1 + 2\sqrt{3}$. We have,

$$P(1 + 2\sqrt{3}) = (1 + 2\sqrt{3})^2 - 3(1 + 2\sqrt{3}) + 5 \Rightarrow$$

$$P(1 + 2\sqrt{3}) = 1 + 12 + 4\sqrt{3} - 3 - 6\sqrt{3} + 5 = 15 - 2\sqrt{3} \quad (*)$$

The value of $P(x)$ at $x = 1 - 2\sqrt{3}$ is

$$P(1 - 2\sqrt{3}) = (1 - 2\sqrt{3})^2 - 3(1 - 2\sqrt{3}) + 5 \Rightarrow$$

$$P(1 - 2\sqrt{3}) = 1 + 12 - 4\sqrt{3} - 3 + 6\sqrt{3} + 5 = 15 + 2\sqrt{3} \quad (**)$$

From (*) and (**) we see that $P(1 + 2\sqrt{3})$ and $P(1 - 2\sqrt{3})$ are conjugate binomials of $\sqrt{3}$.

Theorem 7-4-1: If r is a rational number and i is an irrational number, then the numbers $r + i$, $r - i$, ri, i/r are irrational numbers.

Proof: We prove the theorem by contradiction. For example, if we assume that $r + i$ is another rational r_1, then we would have

$$r + i = r_1 \Leftrightarrow i = r_1 - r$$

which cannot possibly be true, since the difference of two rational numbers is another rational and not an irrational number. We are therefore forced to conclude that $r + i$ is irrational. Similarly, we show that the product of a rational and an irrational number is irrational, etc.

Theorem 7-4-2: Assuming that a, b, c, d are rational numbers, with $b > 0, d > 0$ and not perfect squares, (i.e. $\sqrt{b}, \sqrt{d}$ are irrational numbers), then from the equality $a + \sqrt{b} = c + \sqrt{d}$ it follows that $a = c$ and $b = d$, i.e.

$$a + \sqrt{b} = c + \sqrt{d} \Leftrightarrow \begin{Bmatrix} a = c \\ b = d \end{Bmatrix} \qquad (7-4-1)$$

Proof: From $a + \sqrt{b} = c + \sqrt{d}$ we obtain,

$$\sqrt{b} = c - a + \sqrt{d} \Rightarrow b = \left(c - a + \sqrt{d}\right)^2 \Rightarrow$$

$$b = (c - a)^2 + d + 2(c - a)\sqrt{d} \Rightarrow 2(c - a)\sqrt{d} = b - (c - a)^2 \qquad (*)$$

If we assume that $(c - a) \neq 0$, then from (*) we would have $\sqrt{d} = \{b - (c - a)^2\}/\{2(c - a)\}$, i.e. we would have a rational number (right hand side) equals to an irrational number ($\sqrt{d}$), and since this cannot be true, we conclude that we must necessarily have $(c - a) = 0$, i.e. $a = c$, and then from $a + \sqrt{b} = c + \sqrt{d}$, $b = d$, and this completes the proof.

Remark: Similarly, under the conditions of Theorem 7-4-2, from $a - \sqrt{b} = c - \sqrt{d}$, it follows that $a = c$ and $b = d$.

Example 7-4-1: If $\sqrt[3]{a+\sqrt{b}} = x + \sqrt{y}$, where a, b, x, y rational numbers with b and y positive and not perfect squares, show that $\sqrt[3]{a-\sqrt{b}} = x - \sqrt{y}$.

Solution

Raising both sides of the given equality in the third power, we obtain:

$$\sqrt[3]{a+\sqrt{b}} = x + \sqrt{y} \Rightarrow \left(\sqrt[3]{a+\sqrt{b}}\right)^3 = \left(x+\sqrt{y}\right)^3 \Rightarrow$$

$$a + \sqrt{b} = x^3 + 3x^2\sqrt{y} + 3x\left(\sqrt{y}\right)^2 + \left(\sqrt{y}\right)^3 \Rightarrow$$

$$a + \sqrt{b} = \underbrace{x^3 + 3xy}_{Rational} + \underbrace{(3x^2 + y)}_{Rational}\sqrt{y} = x^3 + 3xy + \sqrt{(3x^2+y)^2 y} \xRightarrow{Th.\ 7-4-2}$$

$$\begin{cases} a = x^3 + 3xy \\ b = (3x^2 + y)^2 y \end{cases} \Rightarrow$$

$$a - \sqrt{b} = x^3 + 3xy - (3x^2+y)\sqrt{y} = x^3 - 3x^2\sqrt{y} + 3x\left(\sqrt{y}\right)^2 - \left(\sqrt{y}\right)^3 \Rightarrow$$

$$a - \sqrt{b} = \left(x - \sqrt{y}\right)^3 \Rightarrow \sqrt[3]{a-\sqrt{b}} = x - \sqrt{y}$$

Example 7-4-2: If x and y are positive integers, not perfect squares, show that the number $\sqrt{x} + \sqrt{y}$ is irrational. Also show that $\sqrt{x} - \sqrt{y}$ is irrational, provided that $x \neq y$.

Solution

a) If $x = y$, then $\sqrt{x} + \sqrt{y} = 2\sqrt{x}$, and since $\sqrt{x}$ is irrational, (by assumption x is not a perfect square), the number $2\sqrt{x}$ is likewise irrational, (see Theorem 7-4-1).

b) Let us now consider the case $x \neq y$. We shall prove that $\sqrt{x} + \sqrt{y}$ is irrational, by contradiction. Indeed, let us assume that $\sqrt{x} + \sqrt{y}$ is some positive rational number r, i.e.

$$\sqrt{x} + \sqrt{y} = r \Rightarrow \sqrt{x} = r - \sqrt{y} \Rightarrow \left(\sqrt{x}\right)^2 = \left(r - \sqrt{y}\right)^2 \Rightarrow$$

$$x = r^2 + y - 2r\sqrt{y} \Rightarrow \sqrt{y} = \frac{r^2 + y - x}{2r} \qquad (*)$$

However, equation (*) cannot possibly be true, since $\sqrt{y}$ is irrational (by assumption), while the right hand side is rational. We are therefore forced to conclude that $\sqrt{x} + \sqrt{y}$ is irrational, and this completes the proof.

Example 7-4-3: Assuming that the rational numbers a, b, c satisfy the equality $a + b\sqrt{2} = c\sqrt{3}$, show that $a = b = c = 0$.

Solution

Squaring both sides of the equality, we obtain,

$$a^2 + 2b^2 + 2\sqrt{2}\,ab = 3c^2 \Rightarrow 2\sqrt{2}\,ab = 3c^2 - a^2 - 2b^2 \qquad (*)$$

In equation (*), **ab must be zero**, since, otherwise, we were led to the contradiction $2\sqrt{2} = (3c^2 - a^2 - 2b^2)/(ab)$, (a rational equals an irrational). The product $ab = 0$, if either $a = 0$ or $b = 0$.

If $a = 0$, equation (*) implies,

$$0 = 3c^2 - 2b^2 \Rightarrow 2b^2 = 3c^2 \Rightarrow 4b^2 = 6c^2 \Rightarrow \pm 2b = c\sqrt{6} \Rightarrow c = 0$$

(otherwise $\sqrt{6} = \pm 2b/c$, which cannot be true, since $\sqrt{6}$ is irrational and $\pm 2b/c$ is rational). With $a = 0$ and $c = 0$, equation (*) implies that $b = 0$.

In the case where $b = 0$, equation (*) implies $a^2 = 3c^2$, i.e. $\pm a = \sqrt{3}\,c$, from which again, (similar reasoning as in the first case), $c = 0$ and $a = 0$.

We have thus shown that $a + b\sqrt{2} = c\sqrt{3}$ implies $a = b = c = 0$.

PROBLEMS

7-4-1) If $P(x) = x^3 - 2x^2 + 6x - 6$ find the numbers $P(3 - \sqrt{3})$ and $P(3 + \sqrt{3})$, (Ans: $42 - 24\sqrt{3},\ 42 + 24\sqrt{3}$).

7-4-2) If $f(x) = (x^2 - 2x + 5)/(x^3 - x^2 + 2)$ find the numbers $f(1 + \sqrt{3})$ and $f(1 - \sqrt{3})$.

7-4-3) Provided that a, b, c are rational numbers, show that the equality $a + b\sqrt{2} + c\sqrt[3]{2} = 0$ holds true if and only if $a = b = c = 0$.

Hint: $c\sqrt[3]{2} = -(a + b\sqrt{2})$, raise both sides to the third power, etc.

7-4-4) Show that $\sqrt[3]{2}$ cannot be expressed in the form $x + \sqrt{y}$, where x and y are rational, with $y > 0$ and not a perfect square.

7-4-5) If the binomial $(a + \sqrt{b})$, with a, b rational, $b > 0$ and not a perfect square, is a root of the polynomial $P(x) = x^2 + px + q$, with coefficients p, q rational numbers, then show that the conjugate binomial $(a - \sqrt{b})$ is also a root.

7-5) Irrational equations (equations with radicals)

If the unknown x, appears under a radical, the equation is called **"irrational" or equation with radicals**. For instance, the equations

$$\sqrt{x-1} = 5, \quad \sqrt{x+2} + \sqrt{x-7} = \sqrt{x+3}, \quad \sqrt[3]{x+3} + \sqrt[5]{x-2} = 7\sqrt{x+1}$$

are irrational equations. In order to solve an irrational equation, we try to eliminate the radicals, and thus convert the irrational equation to a rational equation. **Elimination of a radical can be achieved, in general, by raising both members of the equation to an appropriate power**. For example, to eliminate the radical in the equation $\sqrt{x-1} = 3$ we have to raise both sides of the equation to the second power, the result being $x - 1 = 3^2 = 9$, and hence $x = 10$. We check that $x = 10$ does indeed satisfy the original equation. However, at this point we have to be very cautious, since, **in general, raising an equation to a power, does not necessarily lead to an equivalent equation**. Recall that **two equations are equivalent if they have the same solution set**, i.e. when every solution of the first is a solution of the second and conversely, every solution of the second is a solution of the first.

The following two theorems are helpful in solving irrational equations.

Theorem 7-5-1: Let us consider the two equations

$$(1): P(x) = Q(x) \quad and \quad (2): P^2(x) = Q^2(x)$$

Then every solution of the first is also a solution of the second, but every solution of the second is either a solution of the first $P(x) = Q(x)$ or a solution of $P(x) = -Q(x)$.

Proof: Assuming that r is a solution of $P(x) = Q(x)$, $P(r) = Q(r)$ and then $P^2(r) = Q^2(r)$, which shows that r is a solution of $P^2(x) = Q^2(x)$.

Let us now assume that s is a solution of $P^2(x) = Q^2(x)$, i.e.

$$P^2(s) = Q^2(s) \Leftrightarrow P^2(s) - Q^2(s) = 0 \Leftrightarrow$$

$$\{P(s) - Q(s)\}\{P(s) + Q(s)\} = 0 \Leftrightarrow \begin{cases} P(s) - Q(s) = 0 \\ \text{or} \\ P(s) + Q(s) = 0 \end{cases} \Leftrightarrow \begin{cases} P(s) = Q(s) \\ \text{or} \\ P(s) = -Q(s) \end{cases}$$

and this shows that s is either a solution of $P(x) = Q(x)$ or a solution of $P(x) = -Q(x)$, and this completes the proof.

Remark: The equation obtained by squaring both sides of an equation, is not in general, equivalent to the original equation (since it contains the roots of $P(x) = -Q(x)$). A solution of $P^2(x) = Q^2(x)$ which is not a solution of the original equation $P(x) = Q(x)$, is called an **extraneous solution**. Thus, by squaring both sides of an equation will not eliminate any solutions, but it may introduce extraneous solutions.

Theorem 7-5-2: **Let us consider the two equations**

$$(1): \ P(x) = Q(x) \quad \text{and} \quad (2): \ P^3(x) = Q^3(x)$$

Then every solution of the first is a solution of the second, and conversely, every solution of the second is a solution of the first, (provided that this solution makes both sides of (1) real numbers).

Proof: The first part of the theorem is obvious. Let us now assume that s is a solution of $P^3(s) = Q^3(s)$. Since the real numbers $P(s)$ and $Q(s)$ have equal cubes, the numbers must be equal, and this completes the proof.

Remark: Theorem 7-5-1 holds true for the two equations

$$(1): \ P(x) = Q(x) \quad \text{and} \quad (2): \ P^{2n}(x) = Q^{2n}(x), \quad n = 1, 2, 3, \ldots$$

and Theorem 7-5-2 holds true for the two equations

(1): $P(x) = Q(x)$ and (2): $P^{2n+1}(x) = Q^{2n+1}(x)$, $n = 1,2,3,...$

In very general terms, **to solve an irrational equation we try to eliminate the radicals involved, by raising both sides of the equation to a suitable power. The thus obtained rational equation is then solved by the methods and techniques developed in Chapter 6. The final step is to check whether the roots of the rational equation satisfy the original (irrational) equation.**

The method of solution is illustrated by the following examples.

Example 7-5-1: Solve the equation $\sqrt{x-3} + 3 = x - 2$.

Solution

First we isolate the radical, by writing the given equation as $\sqrt{x-3} = x - 2 - 3 = x - 5$, and then square both sides:

$$\sqrt{x-3} = x - 5 \Rightarrow \left(\sqrt{x-3}\right)^2 = (x-5)^2 \Rightarrow x - 3 = x^2 - 10x + 25 \Rightarrow$$

$$x^2 - 10x + 25 - x + 3 = 0 \Rightarrow x^2 - 11x + 28 = 0 \Rightarrow$$

$$x^2 \underbrace{-7x - 4x}_{-11x} + 28 = 0 \Rightarrow x(x-7) - 4(x-7) = 0 \Rightarrow$$

$$(x-7)(x-4) = 0 \Rightarrow \{x = 7 \text{ or } x = 4\}$$

Check: We check whether the root $x = 7$ satisfies the original equation,

$$\sqrt{7-3} + 3 = 7 - 2, \text{ or } \sqrt{4} + 3 = 5, \text{ or } 2 + 3 = 5 \; (True)$$

Thus, **$x = 7$ is a valid solution.**

Next, we check whether the second root $x = 4$ satisfies the original equation,

$$\sqrt{4-3} + 3 = 4 - 2, \text{ or } \sqrt{1} + 3 = 2, \text{ or } 1 + 3 = 2 \; (False)$$

Thus, **$x = 4$ is not a valid solution.**

Conclusion: The only solution of the given equation is $x = 7$.

Remark: Notice that $x = 4$ is a solution of $\sqrt{x - 3} = -(x - 5)$, since $\sqrt{4 - 3} = -(4 - 5)$, $or\ 1 = -(-1)$, which is true, (see remark in Th. 7-5-1).

Alternative solution: Let us consider the equation $\sqrt{x - 3} = x - 5$. The radicand $(x - 3)$ must be greater than or equal to zero $(x - 3 \geq 0)$, (otherwise $\sqrt{x - 3}$ will not be real). Also, since $\sqrt{x - 3} \geq 0$, $x - 5 = \sqrt{x - 3} \geq 0$, i.e. $x \geq 5$. This shows that the allowed values of x are the ones satisfying the inequality $x \geq 5$, and hence the solution $x = 4$ must be rejected, leaving $x = 7$ as the only valid solution.

Example 7-5-2: Solve the equation $\sqrt{x - 2} + \sqrt{x + 5} = 7$.

Solution

$$\sqrt{x - 2} + \sqrt{x + 5} = 7 \Longrightarrow \sqrt{x - 2} = 7 - \sqrt{x + 5}$$

and squaring both sides results in the following equation

$$x - 2 = 49 + x + 5 - 14\sqrt{x + 5} \Longrightarrow x - 2 - 49 - x - 5 = -14\sqrt{x + 5} \Longrightarrow$$

$$-56 = -14\sqrt{x + 5} \Longrightarrow 4 = \sqrt{x + 5}$$

and squaring both sides once more, we obtain

$$16 = x + 5 \Longrightarrow x = 11$$

We have to check whether $x = 11$, is a root of the original (irrational) equation.

$$\sqrt{11 - 2} + \sqrt{11 + 5} = 7\ \ or\ \ \sqrt{9} + \sqrt{16} = 7\ \ or\ \ 3 + 4 = 7\ \ (True)$$

The solution $x = \mathbf{11}$ **is a valid solution.**

Example 7-5-3: Solve the equation

$$\frac{2 + x}{\sqrt{2} + \sqrt{2 + x}} + \frac{2 - x}{\sqrt{2} - \sqrt{2 - x}} = \sqrt{2}$$

Solution

First of all, the square roots will be real numbers, provided that $(2 + x) \geq 0$ and $(2 - x) \geq 0$, i.e. $-2 \leq x \leq 2$. We also notice that x is not allowed to take the value zero, since then we would have a zero in the denominator of the second fraction. The allowed values of x are thus

$$-2 \leq x \leq 2, \quad x \neq 0, \qquad (*)$$

Eliminating the denominators we get,

$$(2 + x)(\sqrt{2} - \sqrt{2 - x}) + (2 - x)(\sqrt{2} + \sqrt{2 + x})$$
$$= \sqrt{2}(\sqrt{2} + \sqrt{2 + x})(\sqrt{2} - \sqrt{2 - x})$$

which after some simplifications assumes the form,

$$x(\sqrt{2 - x} + \sqrt{2 + x}) = \sqrt{2}\left(2 + \sqrt{4 - x^2}\right) \qquad (**)$$

Squaring both sides of the equation in (**) yields,

$$x^2\left(2 - x + 2 + x + 2\sqrt{4 - x^2}\right) = 2\left(4 + 4 - x^2 + 4\sqrt{4 - x^2}\right) \Longrightarrow$$

$$x^2\left(4 + 2\sqrt{4 - x^2}\right) = 2\left(8 - x^2 + 4\sqrt{4 - x^2}\right)$$

and solving for the radical $\sqrt{4 - x^2}$ results in

$$(x^2 - 4)\sqrt{4 - x^2} = 8 - 3x^2$$

and squaring once more, (to eliminate the radical), we get

$$(x^2 - 4)^2(4 - x^2) = (8 - 3x^2)^2 \Longrightarrow$$

$$(x^4 + 16 - 8x^2)(4 - x^2) = 64 + 9x^4 - 48x^2 \Longrightarrow$$

$$4x^4 + 64 - 32x^2 - x^6 - 16x^2 + 8x^4 = 64 + 9x^4 - 48x^2 \Longrightarrow$$

$$x^6 - 3x^4 = 0 \Longrightarrow x^4(x^2 - 3) = 0 \Longrightarrow \begin{cases} x = 0 \\ x = -\sqrt{3} \\ x = \sqrt{3} \end{cases}$$

The solution $x = 0$ is rejected (see equation (*). Also, equation (**) implies that $x \geq 0$, and because of this restriction the second root $x = -\sqrt{3}$ is also rejected. By direct checking in the original equation, we see the third solution $x = \sqrt{3}$ does satisfy the equation, and therefore is the only valid solution.

If the irrational equation contains **a parameter**, it is possible, for some values of the parameter the equation to have a solution, while for some other values of the parameter, the equation to be impossible.

Finding the values of the parameter for which the equation has a solution, is called "**an investigation with respect to the parameter involved**". The method of approach is illustrated in the following example.

Example 7-5-4: Solve and investigate the equation $\sqrt{x-3} + \sqrt{x-7} = k$, for the various values of the real parameter k.

Solution

Since the left side of the equation is positive, the parameter k must likewise be positive, i.e. $\boldsymbol{k > 0}$. Squaring both sides of the equation yields,

$$\left(\sqrt{x-3} + \sqrt{x-7}\right)^2 = k^2 \Rightarrow x - 3 + x - 7 + 2\sqrt{(x-3)(x-7)} = k^2 \Rightarrow$$

$$2\sqrt{(x-3)(x-7)} = k^2 - 2x + 10 \qquad (*)$$

Since the left hand side in (*) is positive or zero, we must have

$$k^2 - 2x + 10 \geq 0 \qquad (**)$$

Squaring both sides of (*) yields,

$$4(x-3)(x-7) = (k^2 - 2x + 10)^2 \Rightarrow$$

$$4(x^2 - 10x + 21) = k^4 + 4x^2 + 100 - 4k^2 x + 20k^2 - 40x \Rightarrow$$

$$4k^2 x = k^4 + 20k^2 + 16 \xRightarrow{(k \neq 0)} x = \frac{k^4 + 20k^2 + 16}{4k^2} \qquad (***)$$

The solution in (***) must satisfy the restriction imposed in (**), i.e.

$$k^2 - 2\frac{k^4 + 20k^2 + 16}{4k^2} + 10 \geq 0 \Leftrightarrow \frac{4k^4 - 2k^4 - 40k^2 - 32 + 40k^2}{4k^2} \geq 0$$

or, since $4k^2 > 0$,

$$2k^4 - 32 \geq 0 \Leftrightarrow k^2 - 16 \geq 0 \Leftrightarrow$$

$$(k^2)^2 - 4^2 \geq 0 \Leftrightarrow (k^2 + 4)(k^2 - 4) \geq 0 \xLeftrightarrow{(k^2+4>0)}$$

$$k^2 - 4 = (k+2)(k-2) \geq 0 \Leftrightarrow \begin{Bmatrix} k \leq -2 \\ \text{or} \\ k \geq 2 \end{Bmatrix} \qquad (****)$$

The condition $k \leq -2$ is rejected, since $k > 0$. We are thus left with the second condition in (****), i.e. $k \geq 2$.

In summary: For any $k \geq 2$, the original equation has a solution, as given in (***), while for all other values of k the equation is impossible.

PROBLEMS

Solve the following irrational equations:

7-5-1) $\sqrt{2x + 12} = 4$, (**Ans:** $x = 2$).

7-5-2) $\sqrt{3 - x} = 2$.

7-5-3) $\sqrt{3x - 2} - \sqrt{x - 1} = \sqrt{2x - 3}$, (**Ans:** $x = 2$).

7-5-4) $\sqrt{4x^2 + 13} - 2x = 1$

7-5-5) $\sqrt[3]{x^2 - 4x + 6} = 3$, (**Ans:** $x = 7, -3$).

7-5-6) $\sqrt{x + 3} - \sqrt{x - 2} = 1$.

7-5-7) Solve the eq. $(4/\sqrt{10x - 4}) + \sqrt{10x - 4} = 5$, (**Ans:** 2, $\frac{1}{2}$).

7-5-8) Solve the equation $\sqrt{9x - 32} + (8/\sqrt{9x - 32}) = 3\sqrt{x}$.

7-5-9) Solve the equation

$$\sqrt{x} + \sqrt{x - \sqrt{1-x}} = 1$$

(Ans: $x = 16/25$).

7-5-10) Solve the equation $\sqrt{7x-5} + \sqrt{4x-1} = \sqrt{7x-4} + \sqrt{4x-2}$.

7-5-11) Solve the equation

$$\frac{x + \sqrt{3}}{\sqrt{x} + \sqrt{x + \sqrt{3}}} + \frac{x - \sqrt{3}}{\sqrt{x} - \sqrt{x - \sqrt{3}}} = \sqrt{x}$$

(Ans: $x = 2$).

7-5-12) Solve and investigate the equation for the various values of the real parameter k,

$$\sqrt{1 + x + x^2} + \sqrt{1 - x + x^2} = k$$

(Ans: $x = \pm(k/2)\sqrt{(k^2 - 4)/(k^2 - 1)}$, $k > 2$).

7-5-13) Solve and investigate the equation for the various values of the real parameters k and λ,

$$\frac{1 - kx}{1 + kx} \cdot \sqrt{\frac{1 + \lambda x}{1 - \lambda x}} = 1$$

(Ans: $x = 0$, or $x = \pm\sqrt{(2k - \lambda)/(k^2\lambda)}$ provided that $\frac{1}{2} \leq \frac{k}{\lambda} \leq 1$).

7-5-14) Solve the equation $\sqrt{x^2 + \frac{1}{x^2}} + \sqrt{x^2 - \frac{1}{x^2}} = \sqrt{3}\,x$.

(Ans: $x = \pm \sqrt[8]{4/3}$).

7-5-15) Solve the system:
$\{3\sqrt{x-1} - 4\sqrt{y+2} = -6, \quad 5\sqrt{x-1} + 2\sqrt{y+2} = 16\}$

(Ans: $x = 5, y = 7$).

Hint: Set $X = \sqrt{x - 1} \geq 0, Y = \sqrt{y - 2} \geq 0$.

7-5-16) Solve the system: $\left\{\begin{array}{l}\sqrt{4y-2}=\sqrt{x+2y-2}\\ \sqrt{x^2-2y}=x-1\end{array}\right\}$

(Ans: $x=1, y=1/2$**).**

7-5-17) Solve the equation $\sqrt{x}+\sqrt{x+2}=4/\sqrt{x+2}$, **(Ans:** $x=\frac{2}{3}$**).**

7-6) Equations of the form $\sqrt[3]{P(x)}+\sqrt[3]{Q(x)}+\sqrt[3]{R(x)}=0$

Recall that if the sum of three numbers is equal to zero, then the sum of their cubes equals three times their product, i.e.

$$\text{If } a+b+c=0 \quad \text{then} \quad a^3+b^3+c^3=3abc \qquad (7-6-1)$$

Equation (7-6-1) is a direct consequence of Cauchy's identity, (see Pr. 7-1-7). Conversely, if $a^3+b^3+c^3=3abc$ then $a+b+c=0$, provided that the three numbers a, b and c **are not equal to each other**.

By virtue of (7-6-1), equation $\sqrt[3]{P(x)}+\sqrt[3]{Q(x)}+\sqrt[3]{R(x)}=0$ implies $P(x)+Q(x)+R(x)=3\sqrt[3]{P(x)}\cdot\sqrt[3]{Q(x)}\cdot\sqrt[3]{R(x)}$ and this may, sometimes, help, to solve equations of the form $\sqrt[3]{P(x)}+\sqrt[3]{Q(x)}+\sqrt[3]{R(x)}=0$.

Example 7-6-1: Solve the equation $\sqrt[3]{x+1}+\sqrt[3]{x-1}=\sqrt[3]{5x}$.

Solution

$$\sqrt[3]{x+1}+\sqrt[3]{x-1}=\sqrt[3]{5x} \Leftrightarrow \sqrt[3]{x+1}+\sqrt[3]{x-1}-\sqrt[3]{5x}=0 \Leftrightarrow$$

$$\sqrt[3]{x+1}+\sqrt[3]{x-1}+\sqrt[3]{-5x}=0 \qquad (*)$$

Equation (*) implies that

$$(x+1)+(x-1)+(-5x)=3\sqrt[3]{x+1}\cdot\sqrt[3]{x-1}\cdot\sqrt[3]{-5x} \Leftrightarrow$$

$$-3x=3\sqrt[3]{x+1}\cdot\sqrt[3]{x-1}\cdot\sqrt[3]{-5x} \overset{(Th.\ 7-5-2)}{\Longleftrightarrow}$$

$$(-3x)^3=3^3\{(x+1)(x-1)(-5x)\} \Leftrightarrow x^3=(x^2-1)5x=5x^3-5x \Leftrightarrow$$

$$5x-4x^3=0 \Leftrightarrow x(5-4x^2)=0 \Rightarrow x=0 \quad or \quad x=\pm\sqrt{5}/2$$

Example 7-6-2: Solve the equation $\sqrt[3]{2-x} + \sqrt{x-1} = 1$.

Solution

The square root $\sqrt{x-1}$ will be a real number provided that $x \geq 1$.

$$\sqrt[3]{2-x} + \sqrt{x-1} = 1 \Leftrightarrow \sqrt[3]{2-x} = 1 - \sqrt{x-1} \Leftrightarrow$$

$$\sqrt[3]{2-x} = \frac{(1-\sqrt{x-1})(1+\sqrt{x-1})}{1+\sqrt{x-1}} = \frac{1-(x-1)}{1+\sqrt{x-1}} \Leftrightarrow$$

$$\sqrt[3]{2-x} = \frac{2-x}{1+\sqrt{x-1}} \qquad (*)$$

In order to eliminate the cubic root in (*) we may make the substitution $2-x = u^3$, and in terms of the unknown u, equation (*) becomes,

$$u = \frac{u^3}{1+\sqrt{1-u^3}} \Leftrightarrow u + u\sqrt{1-u^3} - u^3 = 0 \Leftrightarrow$$

$$u\left\{1 - u^2 + \sqrt{1-u^3}\right\} = 0 \Rightarrow \begin{cases} u = 0 \\ \text{or} \\ 1 - u^2 + \sqrt{1-u^3} = 0 \end{cases} \qquad (**)$$

From the second equation in (**) we get,

$$\sqrt{1-u^3} = u^2 - 1 \Rightarrow 1 - u^3 = (u^2-1)^2 \Rightarrow$$

$$1 - u^3 = u^4 - 2u^2 + 1 \Rightarrow u^4 + u^3 - 2u^2 = 0 \Rightarrow u^2(u^2+u-2) = 0 \Rightarrow$$

$$u^2(u-1)(u+2) = 0 \Rightarrow \{u = 0, 1, -2\} \qquad (***)$$

In summary, the solutions of the equation in (**), are $u = 0, 1, -2$, and since $x = 2 - u^3$, we find that the solutions of the original equation are $x = 2, 1, 10$. We can easily check that all solutions are valid solutions of the original equation.

PROBLEMS

7-6-1) Solve the equation $\sqrt[3]{x+1} - \sqrt[6]{x^2-1} = \sqrt[3]{x-1}$, (**Ans:** $x = \pm \frac{\sqrt{5}}{2}$).

7-6-2) Solve the equation $\sqrt[3]{3+x} + \sqrt[3]{3-x} = 2 - \sqrt[3]{2}$, **(Ans:** $x = \pm 5$**)**.

7-6-3) Solve and investigate the equation $\sqrt[3]{x-a} + \sqrt[3]{x-b} + \sqrt[3]{x-c} = 0$, where a, b, c are real parameters.

(Ans: If $a = b = c \stackrel{\text{def}}{=} A$ (common value), then $x = A$, otherwise
$$x = 2 \cdot \frac{(a+b+c)^3 - 27abc}{(a-b)^2 + (b-c)^2 + (c-a)^2}$$

Hint: Apply Cauchy's identity.

7-6-4) Using the result obtained in Pr. 7-6-3, solve the equation
$$\sqrt[3]{x-1} + \sqrt[3]{x+2} + \sqrt[3]{x-3} = 0$$

7-6-5) Using the result obtained in Pr. 7-6-3, solve the equation
$$\sqrt[3]{x^2-1} + \sqrt[3]{x^2+18} + \sqrt[3]{x^2-134} = 0$$

(Ans: $x = \pm 3$**)**.

Hint: Make the substitution $u = x^2$, find u and then x.

CHAPTER 8: COMPLEX NUMBERS

8-1) Introduction

The equation $x^2 - 1 = 0$ admits two real roots, $x = 1$ and $x = -1$. The equation $x^2 + 1 = 0$ **does not have any real roots**. This is obvious, since for any real x, $x^2 \geq 0$ and therefore $x^2 + 1 \geq 1 \neq 0$.

Equation $x^2 + 1 = 0$ is not the only one that does not have real roots (solutions). For example, the equations $x^2 + 25 = 0, x^2 + 2x + 10 = 0, x^6 + 20 = 0$, etc, do not have real roots.

Mathematicians' attempts to solve equations like $x^2 + 1 = 0$, led gradually to the invention of a new set of numbers, called **the set of Complex Numbers**. Within this set of complex numbers, the aforesaid equations do have solutions (not real solutions of course).

We shall temporarily define **a complex number as an ordered pair (a, b) of real numbers a and b**, subject to the following rules of operation:

Equality: $(x, y) = (a, b)$ if and only if $\{x = a \text{ and } y = b\}$ $\quad(8-1-1)$

Sum: $(a, b) + (c, d) = (a + c, b + d)$ $\quad(8-1-2)$

Difference: $(a, b) - (c, d) = (a - c, b - d)$ $\quad(8-1-3)$

Product: $(a, b) \cdot (c, d) = (ac - bd, ad + bc)$ $\quad(8-1-4)$

Multiplication of a real number x by the complex number (a, b):

$$x \cdot (a, b) = (xa, xb) \quad(8-1-5)$$

Quotient: The quotient of two complex numbers $(a, b) \div (c, d)$ is another complex number (x, y), which when multiplied by (c, d) yields (a, b).

Provided that $c^2 + d^2 \neq 0$, i.e. $c \neq 0$ and $d \neq 0$, the quotient (x, y) exists and is unique. Indeed, from the very definition of the quotient, we have:

$$(x, y) \cdot (c, d) = (a, b) \xRightarrow{(8-1-4)} (xc - yd, xd + yc) = (a, b) \xRightarrow{(8-1-1)}$$

$$\{xc - yd = a \quad \text{and} \quad xd + yc = b\}$$

and solving this system for x and y we obtain,

$$x = \frac{ac+bd}{c^2+d^2} \quad \text{and} \quad y = \frac{cb-ad}{c^2+d^2}$$

We have thus proved that

Quotient: $\dfrac{(a,b)}{(c,d)} = \left(\dfrac{ac+bd}{c^2+d^2}, \dfrac{cb-ad}{c^2+d^2}\right) \quad (c^2+d^2 \neq 0) \qquad (8-1-6)$

The zero complex number:

The complex number $(0,0)$ is called the zero complex number.

If (a,b) is any complex number, then $(a,b) + (0,0) = (a+0, b+0) = (a,b)$, i.e. **the zero complex number $(0,0)$ is the neutral element with respect to the addition.**

If we multiply any complex number (a,b) by the zero complex number we obtain (by using equation (8-1-4), $(a,b) \cdot (0,0) = (a \cdot 0 - b \cdot 0, a \cdot 0 + b \cdot 0) = (0,0)$. The following theorem states that the inverse statement is also true.

Theorem 8-1-1: **If the product of two complex numbers is zero, then at least one of them is the zero complex number.**

Proof: Assuming that $(a,b) \cdot (c,d) = (0,0)$ and that $(a,b) \neq (0,0)$, we shall show that $(c,d) = (0,0)$.

$$(a,b) \cdot (c,d) = (0,0) \Rightarrow (ac - bd, ad + bc) = (0,0) \Rightarrow$$

$$\begin{cases} ac - bd = 0 \\ ad + bc = 0 \end{cases}$$

Squaring and adding term wise the two equations results in the following (let the reader verify it),

$$(a^2+b^2) \cdot (c^2+d^2) = 0 \xrightarrow{(a^2+b^2 \neq 0)} c^2+d^2 = 0 \Rightarrow \begin{cases} c = 0 \\ \text{and} \\ d = 0 \end{cases} \Rightarrow$$

$$(c,d) = (0,0)$$

and this completes the proof.

The unit complex number:

The complex number $(1, 0)$ is called the unit complex number.

If (a, b) is any complex number, then

$$(a, b) \cdot (1, 0) = (a \cdot 1 - b \cdot 0, a \cdot 0 + b \cdot 1) = (a, b)$$

and this shows that $(1, 0)$ **is the neutral element with respect to the multiplication.**

The inverse of a complex number:

If $(a, b) \neq (0,0)$ then there exists another (**unique**) complex number (x, y) such that $(x, y) \cdot (a, b) = (1, 0)$. This complex number (x, y) is called the inverse of (a, b) with respect to the multiplication.

It is not difficult to show that (see Pr. 8-1-1),

$$(x, y) = \left(\frac{a}{a^2 + b^2}, \frac{-b}{a^2 + b^2}\right) \qquad (8-1-7)$$

Note: Usually we represent complex numbers by single letters, like z, w, etc. For example we may write, $z_1 = (a, b), z_2 = (c, d), w_1 = z_1 + z_2, w_2 = z_1 z_2$, etc.

In section 8-3, we shall develop a more convenient way of expressing complex numbers, in terms of the "**imaginary unit i**".

Example 8-1-1: If $z_1 = (1,1), z_2 = (2,3)$ find the complex number $w = z_1^2 + z_2$.

Solution

$$z_1^2 = z_1 \cdot z_1 = (1,1) \cdot (1,1) = (1 \cdot 1 - 1 \cdot 1, 1 \cdot 1 + 1 \cdot 1) = (0,2)$$

$$w = z_1^2 + z_2 = (0,2) + (2,3) = (0 + 2, 2 + 3) = (2,5)$$

Example 8-1-2: Given the complex numbers $z_1 = (2,4)$ and $z_2 = (3,1)$, find the complex number $w = \frac{z_1}{z_2} + \frac{z_2}{z_1}$.

Solution

Making use of equation (8-1-6) we have:

$$\frac{z_1}{z_2} = \frac{(2,4)}{(3,1)} = \left(\frac{6+4}{10}, \frac{12-2}{10}\right) = (1,1)$$

$$\frac{z_2}{z_1} = \frac{(3,1)}{(2,4)} = \left(\frac{6+4}{20}, \frac{2-12}{20}\right) = \left(\frac{1}{2}, -\frac{1}{2}\right)$$

$$w = \frac{z_1}{z_2} + \frac{z_2}{z_1} = (1,1) + \left(\frac{1}{2}, -\frac{1}{2}\right) = \left(1+\frac{1}{2}, 1-\frac{1}{2}\right) = \left(\frac{3}{2}, \frac{1}{2}\right)$$

Example 8-1-3: If $z = \left(\frac{1}{2}, \frac{\sqrt{3}}{2}\right)$ find z^3.

Solution

We first find the product $z \cdot z = z^2$, and then we find $z^3 = z^2 \cdot z$.

$$z^2 = z \cdot z = \left(\frac{1}{2}, \frac{\sqrt{3}}{2}\right) \cdot \left(\frac{1}{2}, \frac{\sqrt{3}}{2}\right) = \left(\frac{1}{2} \cdot \frac{1}{2} - \frac{\sqrt{3}}{2} \cdot \frac{\sqrt{3}}{2}, \frac{1}{2} \cdot \frac{\sqrt{3}}{2} + \frac{\sqrt{3}}{2} \cdot \frac{1}{2}\right)$$

$$= \left(-\frac{1}{2}, \frac{\sqrt{3}}{2}\right)$$

$$z^3 = z^2 \cdot z = \left(-\frac{1}{2}, \frac{\sqrt{3}}{2}\right) \cdot \left(\frac{1}{2}, \frac{\sqrt{3}}{2}\right) \Rightarrow$$

$$z^3 = \left(-\frac{1}{2} \cdot \frac{1}{2} - \frac{\sqrt{3}}{2} \cdot \frac{\sqrt{3}}{2}, -\frac{1}{2} \cdot \frac{\sqrt{3}}{2} + \frac{\sqrt{3}}{2} \cdot \frac{1}{2}\right) = (-1,0)$$

Example 8-1-4: If $z_1 = (\cos\theta, \sin\theta)$ and $z_2 = (\cos\phi, \sin\phi)$, show that $w = z_1 z_2 = (\cos(\theta+\phi), \sin(\theta+\phi))$.

Solution

$$w = z_1 z_2 = (\cos\theta, \sin\theta) \cdot (\cos\phi, \sin\phi) \Rightarrow$$

$$w = \left(\underbrace{\cos\theta \cdot \cos\phi - \sin\theta \sin\phi}_{\cos(\theta+\phi)}, \underbrace{\cos\theta \sin\phi + \cos\phi \sin\theta}_{\sin(\theta+\phi)}\right) \Rightarrow$$

$$w = (\cos(\theta + \phi), \sin(\theta + \phi))$$

PROBLEMS

8-1-1) Prove equation (8-1-7).

8-1-2) If $z = (-2,3)$, find $w = 2z^2 - 3z^3$.

8-1-3) If $z_1 = \left(\frac{1}{2}, \frac{\sqrt{3}}{2}\right)$ and $z_2 = \left(\frac{1}{2}, -\frac{\sqrt{3}}{2}\right)$, find $w_1 = \frac{1}{4}(z_1 + z_2)$ and $w_2 = z_1 z_2$, (**Ans:** $w_1 = \left(\frac{1}{4}, 0\right), w_2 = (1,0)$).

8-1-4) If $z_1 = (2,-3), z_2 = (4,3)$ and $z_3 = (-1,2)$ find $w = z_1^2 z_2 z_3^3$.

8-1-5) If $z_1 = (2,3)$ and $z_2 = (4,1)$, solve for z the equation $\frac{z}{z_1} + z_2 = (z_2 - z_1)^3$, (**Ans:** $z = (11, -94)$).

8-1-6) If $z_1 = (\cos \theta, \sin \theta)$ and $z_2 = (\cos \phi, \sin \phi)$, show that $w = \frac{z_1}{z_2} = (\cos(\theta - \phi), \sin(\theta - \phi))$.

8-1-7) If $z_1 = (\cos \theta, \sin \theta)$ and $z_2 = (\cos \theta, -\sin \theta)$, show that $z_1 z_2 = (1,0)$.

8-1-8) If $z = (\cos \theta, \sin \theta)$, show that $z^5 = (\cos 5\theta, \sin 5\theta)$.

8-1-9) Generalize Problem 8-1-8 to show that, if n is any positive integer and $z = (\cos \theta, \sin \theta)$ then $z^n = (\cos n\theta, \sin n\theta)$.

8-2) The fundamental laws in the Algebra of complex numbers

Let us, from now on, use the symbol $\mathbb{C}$ to represent the set of complex numbers. Writing $z \in \mathbb{C}$, means that z is a complex number (z belongs to the set of complex numbers).

The fundamental operations with complex numbers, as defined in sect. 8-1, i.e. addition, subtraction, multiplication and division, obey the following rules:

Let $z_1 \in \mathbb{C}, z_2 \in \mathbb{C}, z_3 \in \mathbb{C}$. Then,

1) $z_1 + z_2 \in \mathbb{C}$ and $z_1 \cdot z_2 \in \mathbb{C}$, (The closure property for addition and multiplication).

2) $z_1 + z_2 = z_2 + z_1$, (Commutative law of addition).

3) $z_1 + (z_2 + z_3) = (z_1 + z_2) + z_3$, (Associative law of addition).

4) $z_1 z_2 = z_2 z_1$, (Commutative law of multiplication).

5) $z_1(z_2 z_3) = (z_1 z_2)z_3$, (Associative law of multiplication).

6) $z_1(z_2 + z_3) = z_1 z_2 + z_1 z_3$, (Distributive law).

7) The complex number $(0,0)$ is the neutral element with respect to the addition, i.e. for any complex number z, $z + (0,0) = (0,0) + z = z$. For simplicity, **we write 0 for the zero complex number**, instead of the full notation $(0,0)$.

8) The complex number $(1,0)$ is the neutral element with respect to the multiplication, i.e. for any complex number z, $z \cdot (1,0) = (1,0) \cdot z = z$. For simplicity **we write 1 for the unit complex number**, instead of the full notation $(1,0)$.

9) For any complex number z, there exists another complex number, called **the inverse of z with respect to the addition and denoted as $(-z)$**, such that $z + (-z) = 0$. If $z = (a,b)$ then $(-z) = (-a,-b)$.

10) For any complex number $z \neq 0$, there exists another complex number, called **the inverse of z with respect to the multiplication and denoted as z^{-1}**, such that $zz^{-1} = z^{-1}z = 1$. If $z = (a,b)$ then $z^{-1} = \left(\frac{a}{a^2+b^2}, \frac{-b}{a^2+b^2}\right)$, (see equation (8-1-7).

Theorem 8-2-1: **The cancelation property holds true in the set of complex numbers, i.e. if $w \neq 0$ and $z_1 w = z_2 w$, then $z_1 = z_2$.**

Proof: The given equality implies that

$$z_1 w - z_2 w = 0 \iff (z_1 - z_2)w = 0$$

By virtue of theorem 8-1-1, at least one of the two factors in the product must be zero, and since $w \neq 0$, $z_1 - z_2 = 0$, i.e. $z_1 = z_2$.

PROBLEMS

8-2-1) If $z_1 = (1,2), z_2 = (2,3)$ and $z_3 = (4,5)$ verify the distributive law.

8-2-2) If $z_1 = (-2,5), z_2 = (3,1)$ verify that $(z_1 z_2)^2 = z_1^2 z_2^2$.

8-2-3) If $z_1 = \left(\cos\frac{\pi}{6}, \sin\frac{\pi}{6}\right), z_2 = \left(\cos\frac{\pi}{4}, \sin\frac{\pi}{4}\right)$, verify that $(z_1 z_2)^3 = z_1^3 z_2^3$.

8-2-4) If $x \in \mathbb{R}, y \in \mathbb{R}$ and $x^2 + y^2 = 0$, show that $x = 0$ and $y = 0$.

8-2-5) If $z_1 \in \mathbb{C}, z_2 \in \mathbb{C}$ does $z_1^2 + z_2^2 = 0$ necessarily implies that $z_1 = 0$ and $z_2 = 0$? Give an example where $z_1 \neq 0, z_2 \neq 0$ but $z_1^2 + z_2^2 = 0$.

(Ans: In general no, for example $z_1 = (1,0), z_2 = (0,1)$).

8-2-6) If $z_1 = (1,3), z_2 = (-2,-1), z_3 = (2,-4)$, verify that $\left(\frac{z_1-z_2}{z_1+z_3}\right)^2 = \frac{(z_1-z_2)^2}{(z_1+z_3)^2}$.

8-3) The imaginary unit $i = (0,1) = \sqrt{-1}$

In sect. 8-1 we defined **the unit complex number $1 = (1, 0)$**, which is the neutral element in the multiplication, i.e. $z \cdot 1 = 1 \cdot z = z$.

The complex number $(0, 1)$ is defined to be the imaginary unit and is denoted by the symbol i, i.e.

$$i \stackrel{\text{def}}{=} (0, 1) \qquad (8-3-1)$$

(Some authors use the symbol j instead of i, to represent the imaginary unit).

The symbol i for the imaginary unit, had a great impact and contributed immensely to the development of Algebra and Mathematics in general, during the past few centuries.

Let $z = (x, y)$ be any complex number. This number can be written in an equivalent form (in terms of the imaginary unit i), as follows:

$$z = (x, y) = (x, 0) + (0, y) = (x, 0) + (0,1)y = x + iy \qquad (8-3-2)$$

Equation (8-3-2) shows that **every complex number $z = (x, y)$ can be expressed equivalently as $z = x + iy$, where $i = (0, 1)$ is the imaginary unit.**

The real number x is called **the real part of z**, while the real number y is called **the imaginary part of z**, and we write

$$x = Re(z), \qquad y = Im(z) \qquad (8-3-3)$$

If $y = 0$, the complex number z coincides with the real number x, ($z = x + i \cdot 0 = x + 0 = x$). If $x = 0$, the complex number $z = iy$ is called "**a purely imaginary number**".

The set $\mathbb{R}$ of real numbers can be considered as a subset of the set $\mathbb{C}$ of complex numbers, i.e. $\mathbb{R} \in \mathbb{C}$. Any real number x can be written as

$$x = x + i \cdot 0 = (x, 0) \qquad (8-3-4)$$

In other words, **every complex number $(x, 0)$ coincides with the real number x.**

The imaginary unit i has the fundamental property

$$i^2 = -1 \qquad (8-3-5)$$

Indeed, $i^2 = (0,1) \cdot (0,1) = (0 \cdot 0 - 1 \cdot 1, 0 \cdot 1 + 1 \cdot 0) = (-1, 0) = -1$.

Equation (8-3-5) implies that $i = \sqrt{-1}$. Of course, $(-i)$ is another square root of (-1), since $(-i)^2 = (-i)(-i) = (0, -1)(0, -1) = (-1, 0) = -1$.

In summary, **the symbol $\sqrt{-1}$ can have two values, either i or $(-i)$.**

Starting with $i^2 = -1$, we may easily evaluate **higher powers of i**, for example,

$$\begin{cases} i^3 = i^2 \cdot i = -i \\ i^4 = i^2 \cdot i^2 = (-1) \cdot (-1) = 1 \\ i^5 = i^4 \cdot i = 1 \cdot i = i \\ \ldots \quad \ldots \quad \ldots \end{cases} \qquad (8-3-6)$$

In general, **if k is any positive integer, i^k can take on one of the four values $\{-1, 1, i, -i\}$**. Also, by definition $i^0 = 1$ and $i^{-k} = 1/i^k$, where k is any positive number.

In terms of the imaginary unit i, we may define square roots of negative real numbers, something that within the set of real numbers, is not possible. Indeed, if p is any positive number, $(-p) < 0$, and

$$\sqrt{-p} = i\sqrt{p}, \quad \text{where } p > 0 \qquad (8-3-7)$$

since $(ip)^2 = i^2 p^2 = (-1)p = -p$.

For example, $\sqrt{-16} = 4i$, $\sqrt{-3} = i\sqrt{3}$, etc. Of course, $\sqrt{-p} = -i\sqrt{p}$ as well, since again $\left(-i\sqrt{p}\right)^2 = -p$. However, **at this point we shall make the agreement that by the symbol $\sqrt{-p}$, where $p > 0$, we shall mean $i\sqrt{p}$, i.e. the square root $\sqrt{p}$ multiplied by $+i$**. Following this agreement, for example, $\sqrt{-64} = 8i$, while $-8i$ will be the $-\sqrt{-64}$, etc.

Finally, making use of the imaginary unit i, the operations with complex numbers are greatly simplified. **Algebraic operations are now performed, in the same way as between real numbers, applying commutative, associative and distributive laws, and in the final result replace $i^2, i^3, i^4, \ldots$ by $-1, -i, 1, \ldots$ respectively**.

Example 8-3-1: If $z_1 = 3 + 2i$, $z_2 = 2 - i$, find the complex numbers $z_1 + z_2$ and $z_1 z_2$.

Solution

$$z_1 + z_2 = (3 + 2i) + (2 - i) = (3 + 2) + i(2 - 1) = 5 + i$$

$$z_1 z_2 = (3 + 2i)(2 - i) = 3 \cdot 2 + (2i) \cdot 2 - 3 \cdot i - (2i) \cdot i = 6 + 4i - 3i + 2$$
$$= (6 + 2) + i(4 - 3) = 8 + i$$

Example 8-3-2: If $z = 2 + 3i$, find z^3.

Solution

$$z^3 = (2+3i)^3 = 2^3 + 3 \cdot 2^2 \cdot (3i) + 3 \cdot 2 \cdot (3i)^2 + (3i)^3$$
$$= 8 + 36i - 54 - 27i = -46 + i$$

Example 8-3-3: If $z_1 = 2 + i, z_2 = 1 + 3i$, find $w = \dfrac{z_1}{z_2}$.

Solution

$$w = \frac{z_1}{z_2} = \frac{2+i}{1+3i} = \frac{(2+i)(1-3i)}{(1+3i)(1-3i)} = \frac{2+i-6i+3}{1+3i-3i+9} = \frac{5-5i}{10} = \frac{5}{10} - i\frac{5}{10}$$

Remark: To find the quotient $w = \dfrac{2+i}{1+3i}$ we multiply both the numerator and the denominator by the complex number $(1 - 3i)$, which is called **the complex conjugate** of $(1 + 3i)$. We note that the product $(1 + 3i)(1 - 3i) = 1 + 3i - 3i + 9 = 10$, is a real number.

Complex conjugate numbers and their properties are studied in section 8-4.

Example 8-3-4: If $z_1 = \cos\theta + i\sin\theta$ and $z_2 = \cos\phi + i\sin\phi$, find the product $w = z_1 z_2$, (see also Example 8-1-4).

Solution

$$w = z_1 z_2 = (\cos\theta + i\sin\theta)(\cos\phi + i\sin\phi) \Rightarrow$$

$$w = \cos\theta\cos\phi + i\sin\theta\cos\phi + i\cos\theta\sin\phi + \underbrace{i^2}_{-1}\sin\theta\sin\phi \Rightarrow$$

$$w = (\cos\theta\cos\phi - \sin\theta\sin\phi) + i(\sin\theta\cos\phi + \cos\theta\sin\phi) \Rightarrow$$

$$w = \cos(\theta + \phi) + i\sin(\theta + \phi)$$

Example 8-3-5: If $a, b, c \in \mathbb{R} - \{0\}$ and $\dfrac{a}{3} = \dfrac{b}{2} = \dfrac{4}{c}$, show that $(a+b) + i(b-a) = \dfrac{20-4i}{c}$.

Solution

If we set $\dfrac{a}{3} = \dfrac{b}{2} = \dfrac{4}{c} = \lambda$, (i.e. λ is the common value of each one of the three equal fractions), then $a = 3\lambda, b = 2\lambda$ and $4 = c\lambda$.

The term $(a + b) + i(b - a)$ assumes the form (in terms of λ),

$$(a+b) + i(b-a) = (3\lambda + 2\lambda) + i(2\lambda - 3\lambda) = 5\lambda - i\lambda \xRightarrow{\left(\lambda = \frac{4}{c}\right)}$$

$$(a+b) + i(b-a) = 5 \cdot \frac{4}{c} - i\frac{4}{c} = \frac{20 - 4i}{c}$$

Example 8-3-6: If $(x + iy)^3 = i^5$, $x, y \in \mathbb{R}$, find x and y.

Solution

$$(x+iy)^3 = i^5 \Leftrightarrow x^3 + 3x^2(iy) + 3x(iy)^2 + (iy)^3 = i^4 \cdot i \Leftrightarrow$$

$$x^3 - 3xy^2 + i(3x^2y - y^3) = i \Leftrightarrow \begin{cases}(a): x^3 - 3xy^2 = 0\\ (b): 3x^2y - y^3 = 1\end{cases} \quad (*)$$

From equation (a) in (*) we have,

$$x(x^2 - 3y^2) = 0 \Leftrightarrow \{x = 0 \ \ or \ \ x = \pm\sqrt{3}y\} \quad (**)$$

1) If $x = 0$, equation (b) in (*) yields, $0 - y^3 = 1$, i.e. $y = -1$, and therefore one complex number that satisfies the original equation is $z_1 = 0 - i = -i$.

2) If $x = \pm\sqrt{3}y$, equation (b) in (*) yields, $3(\pm\sqrt{3}y)^2 y - y^3 = 1$, or $9y^3 - y^3 = 1$, i.e. $8y^3 = 1$ and finally, $y = \sqrt[3]{\frac{1}{8}} = \frac{1}{2}$. The corresponding values of x are $x = \pm\frac{\sqrt{3}}{2}$.

In summary, the complex numbers that satisfy the original equation are the following three:

$$z_1 = -i, \quad z_2 = \frac{\sqrt{3}}{2} + i\frac{1}{2}, \quad z_3 = -\frac{\sqrt{3}}{2} + i\frac{1}{2}$$

PROBLEMS

8-3-1) If $z_1 = 2 + 3i$, $z_2 = 2 - 3i$, $z_3 = 1 + i$, find the complex number $w = z_1 z_2 + z_3^2$, (**Ans:** $w = 13 + 2i$).

8-3-2) If $z_1 = \frac{1+i}{2+5i}$ and $z_2 = (1 + 2i)^3$, find $w = \frac{1}{z_1} + \frac{1}{z_2}$.

8-3-3) If $z_1 = 3 + 5i, z_2 = 2 - 4i$, find $w = (z_1/z_2)^2$, **(Ans:** $w = -\frac{36+77i}{50}$).

8-3-4) If $z_1 = 2 - i, z_2 = 3 + 2i, z_3 = \frac{1}{1+i}$, find $w = (z_1 + z_3)\left(z_2 - \frac{z_3}{z_1}\right)$.

8-3-5) Find two real numbers x and y, such that $(1 + i)x + (2 - i)(1 + iy) + 3 + 4i = 0$, **(Ans:** $x = -7, y = 2$).

8-3-6) If $z = 2 - 3i$ find z^3 and z^4.

8-3-7) Find the real numbers x and y, such that $(1 + i)x^2 + 2y^2 = 3 - i2xy + 2x^2 - iy^2$.

(Ans: $\{x = \sqrt{3}, y = -\sqrt{3}\}$ or $\{x = -\sqrt{3}, y = \sqrt{3}\}$).

8-3-8) Show that $(1 + i)^4 + 4 = 0$ and $(1 - i)^4 + 4 = 0$.

8-3-9) If $z = \cos\frac{\pi}{5} + i\sin\frac{\pi}{5}$, show that $z^5 + 1 = 0$.

Hint: See Problem 8-1-9.

8-3-10) If $z = a + ib$ and $w = \frac{z-i}{z+i}$, find $Re(w)$ and $Im(w^2)$.

8-3-11) Find all the complex numbers $z = x + iy$, such that $z^3 \in \mathbb{R}$ and $z^3 \geq 8$.

(Ans: $z = x + i \cdot 0, x \geq 2$, or $z = x(1 + i\sqrt{3}), x \leq -1$, or $z = x(1 - i\sqrt{3}), x \leq -1$).

8-3-12) Express in the form $a + ib$ the complex numbers

$$z = \frac{2}{1 + i\sqrt{3}} + 1 - i\sqrt{3}, \quad w = \frac{(1 + 2i)^3 - (1 + i)^2}{(3 + 2i)^2 - (3 + i)^3}$$

8-4) Complex conjugate numbers

Let $z = x + iy$ be a complex number. The complex number $x - iy$ is called **the complex conjugate of z**, and is denoted by $\bar{z}$.

$$z = x + iy, \quad \bar{z} = x - iy \qquad (8-4-1)$$

From equation (8-4-1) we find

$$Re(z) = x = \frac{z + \bar{z}}{2}, \quad Im(z) = y = \frac{z - \bar{z}}{2i} \qquad (8-4-2)$$

Also, the product of a complex number and its conjugate is

$$z \cdot \bar{z} = (x + iy) \cdot (x - iy) = x^2 + y^2 \qquad (8-4-3)$$

Using conjugate complex numbers, we may find easily the quotient of two complex numbers, **by multiplying both the numerator and the denominator by the complex conjugate of the denominator**, as shown below:

$$\frac{a + ib}{c + id} = \frac{(a + ib)(c - id)}{(c + id)(c - id)} = \frac{ac + bd}{c^2 + d^2} + i\frac{bc - ad}{c^2 + d^2} \qquad (8-4-4)$$

Notice that the result obtained in equation (8-4-4) is identical to that shown in equation (8-1-6).

Properties of conjugate complex numbers:

If $\bar{z} = x - iy$ is the complex conjugate of $z = x + iy$, the following properties are easily shown:

$$If\ z = \bar{z} \Leftrightarrow z = x\ is\ a\ Real\ number \qquad (8-4-5)$$

$$If\ z = -\bar{z} \Leftrightarrow z = iy\ is\ a\ purely\ imaginary\ number \qquad (8-4-6)$$

$$\overline{z_1 + z_2 + \cdots + z_n} = \bar{z}_1 + \bar{z}_2 + \cdots + \bar{z}_n \qquad (8-4-7)$$

$$\overline{z_1 - z_2} = \bar{z}_1 - \bar{z}_2 \qquad (8-4-8)$$

$$\overline{z_1 \cdot z_2 \cdots z_n} = \bar{z}_1 \cdot \bar{z}_2 \cdots \bar{z}_n \qquad (8-4-9)$$

$$\overline{\left(\frac{z_1}{z_2}\right)} = \frac{\bar{z}_1}{\bar{z}_2} \qquad (8-4-10)$$

$$If\ n\ is\ a\ positive\ integer\ then\ \overline{z^n} = (\bar{z})^n \qquad (8-4-11)$$

For example, to show (8-4-5), we assume that $z = a + ib$, then $\bar{z} = a - ib$, and $z = \bar{z}$ implies that $a + ib = a - ib$, i.e. $2ib = 0$, i.e. $b = 0$, and $z = a + 0$ is a real number. Similarly we may show equation (8-4-6).

To show equation (8-4-7), we assume that $z_1 = a + ib, z_2 = c + id$. Then,

$$\overline{z_1 + z_2} = \overline{(a + ib) + (c + id)} = \overline{(a + c) + i(b + d)} = a + c - i(b + d)) \Rightarrow$$

$$\overline{z_1 + z_2} = (a - ib) + (c - id) = \bar{z}_1 + \bar{z}_2$$

If we now consider three complex numbers z_1, z_2, z_3 we have,

$$\overline{z_1 + z_2 + z_3} = \overline{(z_1 + z_2) + z_3} = \overline{(z_1 + z_2)} + \overline{z_3} = \bar{z}_1 + \bar{z}_2 + \bar{z}_3$$

and the generalization to an arbitrary number of complex numbers is thus obvious.

To show equation (8-4-9), we again assume that $z_1 = a + ib, z_2 = c + id$. Then,

$$\overline{z_1 z_2} = \overline{(a + ib)(c + id)} = \overline{(ac - bd) + i(bc + ad)}$$
$$= (ac - bd) - i(bc + ad) \Rightarrow$$

$$\overline{z_1 z_2} = a(c - id) - ib(c - id) = (a - ib)(c - id) = \bar{z}_1 \bar{z}_2$$

The property is easily generalized to an arbitrary number of complex numbers.

In case where $z_1 = z_2 = \cdots = z_n \stackrel{\text{def}}{=} z$, then equation (8-4-9) implies,

$$\overline{z \cdot z \cdots z} = \bar{z} \cdot \bar{z} \cdots \bar{z} \Rightarrow \overline{(z^n)} = (\bar{z})^n$$

and this proves equation (8-4-11).

Finally, to show equation (8-4-10), we set $w = z_1/z_2$ which implies that

$$z_1 = w z_2 \Rightarrow \bar{z}_1 = \overline{w z_2} = \bar{w}\,\bar{z}_2 \Rightarrow \bar{w} = \frac{\bar{z}_1}{\bar{z}_2} \xrightarrow{\left(w = \frac{z_1}{z_2}\right)} \overline{\left(\frac{z_1}{z_2}\right)} = \frac{\bar{z}_1}{\bar{z}_2}$$

and this completes the proof.

Example 8-4-1: If $z = \frac{(3+2i)(1-2i)}{(1+i)(4-i)}$ find $\bar{z}$.

Solution

$$\bar{z} = \overline{\left(\frac{(3+2i)(1-2i)}{(1+i)(4-i)}\right)} = \frac{\overline{(3+2i)(1-2i)}}{\overline{(1+i)(4-i)}} = \frac{\overline{(3+2i)}\,\overline{(1-2i)}}{\overline{(1+i)}\,\overline{(4-i)}}$$
$$= \frac{(3-2i)(1+2i)}{(1-i)(4+i)}$$

Remark: Notice that the expression for $\bar{z}$ is obtained from the expression of z, if i (the imaginary unit) is replaced by $(-i)$.

Example 8-4-2: Find all the complex numbers z which satisfy the equation $z^2 + \bar{z} = 0$.

Solution

If we set $z = x + iy$, the given equation is equivalent to the following:

$$(x+iy)^2 + x - iy = 0 \Leftrightarrow x^2 - y^2 + x + i(2xy - y) = 0 \Leftrightarrow$$

$$\begin{Bmatrix} x^2 - y^2 + x = 0 \\ \text{and} \\ 2xy - y = 0 \end{Bmatrix} \quad (*)$$

From the second equation in (*) we obtain,

$$y(2x-1) = 0 \Longrightarrow \{y = 0 \quad or \quad 2x - 1 = 0\} \quad (**)$$

1) If $y = 0$, the first equation in (*) yields, $x^2 + x = 0$, i.e. $x(x+1) = 0$, i.e. $x = 0$ or $x = -1$. This shows that the complex numbers which satisfy both equations in (*) are $z_1 = 0 + i0 = 0$ and $z_2 = -1 + i0 = -1$.

2) From the second equation in (**) we find $x = \frac{1}{2}$, and the first equation in (*) yields,

$$\frac{1}{4} - y^2 + \frac{1}{2} = 0 \Leftrightarrow y^2 = \frac{3}{4} \Leftrightarrow y = \pm\frac{\sqrt{3}}{2}$$

and this shows that the complex numbers which satisfy both equations in (*) are $z_3 = \frac{1}{2} + i\frac{\sqrt{3}}{2}$ and $z_4 = \frac{1}{2} - i\frac{\sqrt{3}}{2}$.

In summary, the complex numbers which satisfy $z^2 + \bar{z} = 0$, are

$$z_1 = 0, z_2 = -1, z_3 = \frac{1}{2} + i\frac{\sqrt{3}}{2}, z_4 = \frac{1}{2} - i\frac{\sqrt{3}}{2}$$

Example 8-4-3: If z and w are any two complex numbers, show that the number $u = z\bar{w} + \bar{z}w$ is real.

Solution

It suffices to show that $\bar{u} = u$, (see property (8-4-1)).

$$\bar{u} = \overline{(z\bar{w} + \bar{z}w)} = \overline{(z\bar{w})} + \overline{(\bar{z}w)} = \bar{z}\,\overline{(\bar{w})} + \overline{(\bar{z})}\bar{w} = \bar{z}w + z\bar{w} = u$$

and this completes the proof, (notice that in general, $\overline{(\bar{z})} = z$, let the reader prove it).

PROBLEMS

8-4-1) If $z = \frac{1}{2}(-1 + i\sqrt{3})$, show that $z^3 = 1$ and $(\bar{z})^3 = 1$.

8-4-2) If $z = \frac{1}{2}(-1 + i\sqrt{3})$, show that $z^2 = \bar{z}$ and $(\bar{z})^2 = z$.

8-4-3) If $z^2 = (\bar{z})^2$, show that either z is real or z is a purely imaginary number.

8-4-4) If z, w are two complex numbers satisfying $20z^{36} + 17w^{27} + 3 = 0$, show that they will also satisfy, $20(\bar{z})^{36} + 17(\bar{w})^{27} + 3 = 0$ as well.

Hint: Use properties of conjugate complex numbers.

8-4-5) Show that the number $z = \left(1 + i\sqrt{3}\right)^7 + \left(1 - i\sqrt{3}\right)^7$ is real, while the number $w = \left(1 + i\sqrt{3}\right)^7 - \left(1 - i\sqrt{3}\right)^7$ is purely imaginary.

Hint: It suffices to show that $\bar{z} = z$ and $\bar{w} = -w$.

8-4-6) If $z_1\bar{z}_2 = 1$, show that $w_1 = \frac{1+z_1z_2}{z_1+z_2}$, $w_2 = i\frac{1-z_1z_2}{z_1+z_2}$, $w_3 = \frac{z_1-z_2}{z_1+z_2}$ are all real numbers.

Hint: It suffices to show $\bar{w}_1 = w_1, \bar{w}_2 = w_2, \bar{w}_3 = w_3$. Notice that from $z_1 \bar{z}_2 = 1$ it follows that $\bar{z}_1 z_2 = 1$, (prove it).

8-4-7) If z_1, z_2 are any two complex numbers and $w_1 = \frac{1+z_1 z_2}{z_1+z_2}, w_2 = i\frac{1-z_1 z_2}{z_1+z_2}, w_3 = \frac{z_1 - z_2}{z_1+z_2}$, show that $w_1^2 + w_2^2 + w_3^2 = 1$, (assume $z_1 + z_2 \neq 0$).

8-4-8) Express in the form $a + ib$ the following complex numbers:

$$(2+5i)^3, \quad \frac{2+3i}{4-7i} + \frac{(2+i)^2}{4+7i}, \quad \frac{(1-i)^4}{(1+i)^4},$$

$$\frac{3-\sqrt{-4}}{3+\sqrt{-4}}, \quad \frac{(2-3i)^3}{1+i^{23}}, \quad \frac{5+\sqrt{-9}}{3-\sqrt{-16}} + \frac{2+\sqrt{-1}}{7+\sqrt{-4}}$$

8-4-9) Express as a product of two complex factors the following algebraic expressions, $(x, y$ are real numbers).

$$x^2 + y^2, \quad (x+3)^2 + (y-1)^2$$

(Ans: $(x+iy)(x-iy), (x+3+i(y-1))(x+3-i(y-1))$).

8-4-10) Apply Lagrange's identity to the two triads of numbers (x, y, i) and (a, b, i), where x, y, a, b are real numbers, $i = \sqrt{-1}$, and show that the expression $(x^2 + y^2 - 1)(a^2 + b^2 - 1) + (x-a)^2 + (y-b)^2 \geq 0$, for all values of x, y, a, b.

8-4-11) Express the complex number $w = \frac{(3+i)^2}{2-i} \cdot \frac{1+i}{3-i}$ in the form $x + iy$.

(Ans: $w = -\frac{6}{5} + \frac{8}{5} i$).

8-4-12) Find the real numbers x and y satisfying the equation $\frac{y}{x-i} = \frac{4i+3y}{3x+y}$.

8-4-13) Find the complex numbers z satisfying the equation $3z\bar{z} + 2(z - \bar{z}) = 15 + 8i$, (Ans: $z_1 = 1 + 2i, z_2 = -1 + 2i$).

8-5) The absolute value (or modulus) of a complex number

If $z = x + iy$ is a complex number, **the absolute value or modulus of z, is a non negative real number, denoted by |z|,** and defined as

$$|z| = |x + iy| = +\sqrt{x^2 + y^2} \qquad (8-5-1)$$

For example, $|1 + 2i| = \sqrt{1^2 + 2^2} = \sqrt{5}$, $|3 - 4i| = \sqrt{3^2 + (-4)^2} = \sqrt{25} = 5$, $|i| = |0 + 1 \cdot i| = \sqrt{0^2 + 1^2} = 1$, etc.

The absolute value of a complex number is always a positive number, except when $z = 0$, in which case $|z| = 0$.

Notice that any complex number $z = x + iy$ and its conjugate $\bar{z} = x - iy$, both have the same absolute value, i.e. $|z| = |\bar{z}| = \sqrt{x^2 + y^2}$.

Theorem 8-5-1: **The absolute value of the product of two complex numbers is equal to the product of the absolute values of the two numbers,**

$$|z_1 z_2| = |z_1||z_2|$$

Proof: If $z_1 = x + iy, z_2 = a + ib$, then

$$|z_1 z_2| = |(x + iy)(a + ib)| = |(ax - by) + i(ay + bx)| \Rightarrow$$

$$|z_1 z_2| = \sqrt{(ax - by)^2 + (ay + bx)^2} \Rightarrow$$

$$|z_1 z_2| = \sqrt{a^2 x^2 + b^2 y^2 + a^2 y^2 + b^2 x^2} \Rightarrow$$

$$|z_1 z_2| = \sqrt{(x^2 + y^2)(a^2 + b^2)} = \sqrt{x^2 + y^2} \cdot \sqrt{a^2 + b^2} = |z_1||z_2|$$

and this completes the proof.

Corollary 1: The absolute value of the product of any number of complex numbers is equal to the product of the absolute values of the numbers, i.e.

$$|z_1 z_2 \cdots z_n| = |z_1||z_2| \cdots |z_n| \qquad (8-5-2)$$

Equation (8-5-2) is proved easily, for example,

$$|z_1 z_2 z_3| = |(z_1 z_2) z_3| = |(z_1 z_2)||z_3| = |z_1||z_2||z_3|$$

and similarly, we may generalize to any number of factors.

Corollary 2: If $z_1 = z_2 \cdots = z_n \stackrel{\text{def}}{=} z$, then (8-5-2) implies,

$$|z^n| = |z|^n \qquad (8-5-3)$$

For example, $|(1+i)^{10}| = |1+i|^{10} = (\sqrt{2})^{10} = 2^5 = 32$.

To appreciate this Theorem you may want to find first $(1+i)^{10}$ and then find its absolute value.

Theorem 8-5-2: **The absolute value of the quotient of two complex numbers is equal to the quotient of their absolute values, i.e.**

$$\left|\frac{z_1}{z_2}\right| = \frac{|z_1|}{|z_2|} \qquad (8-5-4)$$

Proof: If we set $w = z_1/z_2$, then $z_1 = wz_2$, and by virtue of Theorem 8-5-1,

$$|z_1| = |wz_2| = |w||z_2| \Rightarrow |w| = \frac{|z_1|}{|z_2|} \xRightarrow{\left(w=\frac{z_1}{z_2}\right)} \left|\frac{z_1}{z_2}\right| = \frac{|z_1|}{|z_2|}$$

Theorem 8-5-3: **For any complex number z, we have**

$$z \cdot \bar{z} = |z|^2 \qquad (8-5-5)$$

Proof: Obvious.

Theorem 8-5-4: **If z_1, z_2 are any two complex numbers, then**

$$|z_1 + z_2| \leq |z_1| + |z_2| \qquad (8-5-6)$$

Proof: if we set $z_1 = x + iy, z_2 = a + ib$, we have to prove that

$$|(x+iy) + (a+ib)| \leq |x+iy| + |a+ib| \Leftrightarrow$$

$$|(x+a) + i(y+b)| \leq |x+iy| + |a+ib| \Leftrightarrow$$

$$\sqrt{(x+a)^2 + (y+b)^2} \leq \sqrt{x^2+y^2} + \sqrt{a^2+b^2} \qquad (*)$$

If we show that equation (*) is true, then equation (8-5-6) shall be true as well.

Since both sides in (*) are **positive quantities**, it suffices to show that the square of the left member is less than or equal to the square of the right member, i.e. to show that

$$\left(\sqrt{(x+a)^2+(y+b)^2}\right)^2 \leq \left(\sqrt{x^2+y^2}+\sqrt{a^2+b^2}\right)^2 \Leftrightarrow$$

$$(x+a)^2+(y+b)^2 \leq (x^2+y^2)+(a^2+b^2)+2\sqrt{(x^2+y^2)(a^2+b^2)} \Leftrightarrow$$

$$ax+by \leq \sqrt{(x^2+y^2)(a^2+b^2)} \quad (**)$$

If $ax+by < 0$, then the inequality in (**) is satisfied (a negative number is less than a positive number).

To show (**) in the case where $ax+by \geq 0$, since both side are positive, it suffices to show that

$$(ax+by)^2 \leq \left(\sqrt{(x^2+y^2)(a^2+b^2)}\right)^2 \Leftrightarrow$$

$$a^2x^2+b^2y^2+2abxy \leq a^2x^2+a^2y^2+b^2x^2+b^2y^2 \Leftrightarrow$$

$$2abxy \leq a^2y^2+b^2x^2 \Leftrightarrow 0 \leq (ay-bx)^2$$

which is always true, and this completes the proof of the Theorem.

Corollary: If $z_1, z_2, \cdots, z_n$ are any complex numbers, then

$$|z_1+z_2+\cdots+z_n| \leq |z_1|+|z_2|+\cdots+|z_n| \qquad (8-5-7)$$

Theorem 8-5-5: **For any two complex numbers z_1 and z_2 the following inequalities hold true:**

$$|z_1|-|z_2| \leq ||z_1|-|z_2|| \leq |z_1 \pm z_2| \leq |z_1|+|z_2| \qquad (8-5-8)$$

Proof: Similar to the proof of Theorem 8-5-4.

Remark: Identical inequalities hold true for real numbers.

Example 8-5-1: Find the absolute value of $z = \dfrac{(1+i)^3(2-i)^2}{(3i)^2(1+2i)}$.

Solution

By virtue of the Theorems proved, we have:

$$|z| = \left|\frac{(1+i)^3(2-i)^2}{(3i)^2(1+2i)}\right| = \frac{|(1+i)^3(2-i)^2|}{|(3i)^2(1+2i)|} = \frac{|(1+i)^3||(2-i)^2|}{|(3i)^2||1+2i|} \Rightarrow$$

$$z = \frac{|1+i|^3|2-i|^2}{|3i|^2|1+2i|} = \frac{\left(\sqrt{2}\right)^3 \cdot \left(\sqrt{5}\right)^2}{9 \cdot \sqrt{5}} = \frac{2\sqrt{10}}{9}$$

Example 8-5-2: Find the absolute value of $z = \left(\frac{1-i}{\sqrt{3}}\right)^{20}\left(\frac{1-i\sqrt{3}}{2}\right)^{15}$.

Solution

$$|z| = \left|\left(\frac{1-i}{\sqrt{3}}\right)^{20}\left(\frac{1-i\sqrt{3}}{2}\right)^{15}\right| = \left|\left(\frac{1-i}{\sqrt{3}}\right)^{20}\right|\left|\left(\frac{1-i\sqrt{3}}{2}\right)^{15}\right| \Rightarrow$$

$$|z| = \frac{|1-i|^{20}}{\left(\sqrt{3}\right)^{20}} \cdot \frac{|1-i\sqrt{3}|^{15}}{2^{15}} = \frac{\left(\sqrt{2}\right)^{20}}{\left(\sqrt{3}\right)^{20}} \cdot \frac{2^{15}}{2^{15}} = \left(\sqrt{\frac{2}{3}}\right)^{20} = \left(\frac{2}{3}\right)^{10}$$

Example 8-5-3: If $z, w, u \in \mathbb{C}$ and $|z| = |w| = |u| = 1$, show that $|zw + wu + uz| = |z + w + u|$.

Solution

Since $|z|^2 = z \cdot \bar{z} = 1$, the complex conjugate $\bar{z} = 1/z$, and similarly $\bar{w} = 1/w$ and $\bar{u} = 1/u$ and therefore

$$|zw + wu + uz| = \left|\frac{1}{\bar{z}\,\bar{w}} + \frac{1}{\bar{w}\,\bar{u}} + \frac{1}{\bar{u}\,\bar{z}}\right| = \left|\frac{\bar{u} + \bar{z} + \bar{w}}{\bar{z}\,\bar{w}\,\bar{u}}\right| = \frac{|\bar{z} + \bar{w} + \bar{u}|}{|\bar{z}\,\bar{w}\,\bar{u}|}$$
$$= \frac{|\overline{z + w + u}|}{|\overline{z\,w\,u}|} = \frac{|z + w + u|}{|z\,w\,u|} = \frac{|z + w + u|}{|z||w||u|} = |z + w + u|$$

since $|z||w||u| = 1 \cdot 1 \cdot 1 = 1$.

Example 8-5-4: If $w \in \mathbb{C}$, show that $|w + 1| + |w + 2| \leq |w| + |w + 3|$.

Solution

Since both sides of the inequality are positive numbers, it suffices to show that

$$(|w + 1| + |w + 2|)^2 \leq (|w| + |w + 3|)^2 \Leftrightarrow$$

$|w+1|^2 + |w+2|^2 + 2|w+1||w+2| \leq |w|^2 + |w+3|^2 + 2|w||w+3| \Leftrightarrow$

$(w+1)\overline{(w+1)} + (w+2)\overline{(w+2)} + 2|w+1||w+2|$

$\leq (w)\overline{(w)} + (w+3)\overline{(w+3)} + 2|w||w+3| \xleftrightarrow{(8-4-7)(w\bar{w}=|w|^2)}$

$|(w+1)(w+2)| \leq |w(w+3)| + 2 \Leftrightarrow$

$|(w^2+3w) + 2| \leq |w^2+3w| + 2$

which is obviously true, by virtue of the inequality (8-5-6), with $z_1 = w^2 + 3w$ and $z_2 = 2$, and this completes the proof.

Example 8-5-5: If $z, w \in \mathbb{C}$ and a is a positive real number, show the inequality

$$|z+w|^2 \leq (1+a)|z|^2 + \left(a + \frac{1}{a}\right)|w|^2$$

Solution

In order to prove this inequality, we shall use the (easily proved) **permanent inequality**: If c, d, p, q are real numbers, then

$$(c^2 + d^2)(p^2 + q^2) \geq (cp + dq)^2$$

Application of this inequality with $c = \sqrt{a+1}\,|z|, d = \sqrt{\left(1+\frac{1}{a}\right)}\,|w|, p = \frac{1}{\sqrt{a+1}\,|z+w|}$ and $q = \frac{1}{\sqrt{\left(1+\frac{1}{a}\right)}\,|z+w|}$ yields:

$$\left\{(a+1)|z|^2 + \left(1+\frac{1}{a}\right)|w|^2\right\} \cdot \left\{\frac{1}{(a+1)|z+w|^2} + \frac{1}{\left(1+\frac{1}{a}\right)|z+w|^2}\right\}$$

$$\geq \left\{\frac{|z|}{|z+w|} + \frac{|w|}{|z+w|}\right\}^2 \Rightarrow$$

$$\left\{(a+1)|z|^2 + \left(1+\frac{1}{a}\right)|w|^2\right\} \cdot \left\{\frac{1}{(a+1)|z+w|^2} + \frac{a}{(a+1)|z+w|^2}\right\}$$

$$\geq \left\{\frac{|z|+|w|}{|z+w|}\right\}^2 \geq 1, (since\ |z|+|w| \geq |z+w|) \Rightarrow$$

$$\left\{(a+1)|z|^2 + \left(1+\frac{1}{a}\right)|w|^2\right\} \cdot \left\{\frac{(a+1)}{(a+1)|z+w|^2}\right\} \geq 1 \Longrightarrow$$

$$(a+1)|z|^2 + \left(1+\frac{1}{a}\right)|w|^2 \geq |z+w|^2$$

and the proof is completed.

PROBLEMS

8-5-1) Find the absolute value of the complex number $z = \frac{(2+i)(1+2i)}{(1+3i)(3-2i)}$.

(Ans: $|z| = 5/\sqrt{130}$).

8-5-2) If $z = 1 + 3i$ and $w = (2+5i)/(1-i)$ find: $|zw|, |z/w|, |z+w^2|, \left|\frac{1}{z}-\frac{1}{w}\right|$ and $\left|2iz+\frac{5}{w}\right|$.

8-5-3) Find the absolute value of $z = \left(\frac{2+3i}{3-2i}\right)^{100}$, **(Ans:** $|z| = 1$).

8-5-4) If $z = 2+7i, w = 1+3i, u = -4-3i$, find $|z\bar{w}^2(u+z)|$.

8-5-5) If a complex number z satisfies $|z|^2 = |z^2+1|$, show that $Re(z^2) = -\frac{1}{2}$.

Hint: $|z|^2 = |z^2+1| \Longrightarrow |z^2| = |z^2+1| \Longrightarrow (|z^2|)^2 = |z^2+1|^2$ and then apply equation (8-5-5).

8-5-6) If θ is a positive real number, m is a real number greater than 1, and z is a complex number satisfying $|(m^2-1)z + \theta m^2| = m \cdot |(m^2-1)z + \theta|$, show that $|z| = \frac{m\theta}{m^2-1}$.

Hint: Use equation (8-5-5).

8-5-7) If $z \neq 0$ and $zw = |z|^2$, show that $w = \bar{z}$.

8-5-8) Find the absolute value of $z = \left(\frac{4i(3+3i)(\sqrt{3}+i)}{2+3i}\right)^4$.

8-5-9) If $z = 2 + 3i, w = 3 + 5i$, find $|z|, |w|, |z + w|, ||z| - |w||$, and verify the inequality $||z| - |w|| \leq |z + w| \leq |z| + |w|$.

8-5-10) If $z = x + iy$, show that $|x| + |y| \leq \sqrt{2}\,|z|$, $(x, y \in \mathbb{R})$.

8-5-11) If $w_1 w_2 \neq 0$ and $|w_1|^2 + |w_2|^2 = 1$, show that $\left|\frac{z_1}{w_1}\right|^2 + \left|\frac{z_2}{w_2}\right|^2 \geq |z_1 + z_2|^2$.

Hint: Apply the permanent inequality in Example 8-5-5, with $p = |w_1|, q = |w_2|, c = |z_1/w_1|, d = |z_2/w_2|$

8-5-12) If $z^2 + w^2 = 0$, $(z, w \in \mathbb{C})$, show that $|z + w| = |z - w|$.

8-5-13) If $z^2 + w^2 + u^2 = zw + wu + uz$, show that $|z - w| = |w - u| = |u - z|$.

Hint: $z^2 + w^2 + u^2 - zw - wu - uz = \frac{1}{2}\{(z-w)^2 + (w-u)^2 + (u-z)^2\}$. If we set $a = z - w, b = w - u, c = u - z$, the given condition implies $a^2 + b^2 + c^2 = 0$ and $a + b + c = 0$, and it suffices to show that $|a| = |b| = |c|$.

8-5-14) If $z^2 + |z|^2 = 0$, show that $Re(z) = 0$, and conversely, $(z \neq 0)$.

8-5-15) If $z^2 - |z|^2 = 0$, show that $Im(z) = 0$, and conversely, $(z \neq 0)$.

8-5-16) If $|z + 9| = 3\,|z + 1|$, show that $|z| = 3$.

Hint: The given equality implies $|z + 9|^2 = 9|z + 1|^2$, i.e. $(z + 9)\overline{(z + 9)} = 9(z + 1)\overline{(z + 1)}$, etc.

8-5-17) Find the complex number w such that $|w - i| = |w - 1| = |w - 2|$.

(Ans: $w = \frac{3}{2}(1 + i)$).

8-5-18) If $|z| = |w| = |u| = 1$ and $z + w + u = 1$, $(z, w, u \in \mathbb{C})$, show that $\frac{1}{z} + \frac{1}{w} + \frac{1}{u} = 1$.

8-5-19) If $|z| = 1, |w| = 2, |u| = 3$ and $zw + wu + uz = zwu$, show that $\bar{z} + \frac{\bar{w}}{4} + \frac{\bar{u}}{9} = 1$.

8-6) The trigonometric or polar form of a complex number

Any complex number $z = (x + iy) \neq 0$ can be written as

$$z = x + iy = \sqrt{x^2 + y^2}\left(\frac{x}{\sqrt{x^2 + y^2}} + i\frac{y}{\sqrt{x^2 + y^2}}\right)$$

If we set,

$$\frac{x}{\sqrt{x^2 + y^2}} = \cos\theta, \qquad \frac{y}{\sqrt{x^2 + y^2}} = \sin\theta, \qquad r \stackrel{\text{def}}{=} |z| = \sqrt{x^2 + y^2}$$

the complex number z can be expressed in the following form, known as **the trigonometric or the polar form**,

$$z = r(\cos\theta + i\sin\theta) \qquad (8-6-1)$$

The expression $z = x + iy$ is called **the Rectangular or the Cartesian form** of the complex number z.

The angle θ, having cosine equal to $\left(x/\sqrt{x^2 + y^2}\right)$ and sine equal to $\left(y/\sqrt{x^2 + y^2}\right)$, is called **the argument of z**.

It is obvious that if θ is an argument of z, then $(\theta + k \cdot 2\pi)$, where k is any integer, will also be an argument, and this means that **a complex number $z \neq 0$, has an infinite number of arguments, differing one from the other by an integral multiple of 2π**. To denote that θ is an argument of z, we write $\theta = \arg(z)$. Usually we take the argument of z to lie in the interval $(-\pi, \pi]$, i.e. $-\pi < \theta \leq \pi$. In this case, θ is called **the principal argument of z**, and we write $\theta = Arg(z)$. So, by definition,

$$\theta = Arg(z) \Leftrightarrow -\pi < \theta \leq \pi \qquad (8-6-2)$$

Remark: If two complex numbers z and w are equal, then they have equal absolute values and arguments either equal or differing by an integral multiple of 2π, i.e.

$$\text{If } r_1(\cos\theta + i\sin\theta) = r_2(\cos\phi + i\sin\phi) \Leftrightarrow \begin{Bmatrix} r_1 = r_2 \\ \theta = \phi + k \cdot 2\pi \end{Bmatrix} \qquad (8-6-3)$$

where $k = 0, \pm 1, \pm 2, \pm 3, \ldots$.

As an example, the number $z = 1 + i = \sqrt{2}\left(\frac{1}{\sqrt{2}} + i\frac{1}{\sqrt{2}}\right) = \sqrt{2}\left(\cos\frac{\pi}{4} + i\sin\frac{\pi}{4}\right)$. **The principal argument of z is** $\theta = Arg(z) = \frac{\pi}{4}$. However, in general, $\frac{\pi}{4} + k \cdot 2\pi$ is also an argument of z.

Similarly, the principal argument of $z = 1 - i$, is $\theta = Arg(z) = -\frac{\pi}{4}$, the principal argument of $z = 1 + i\sqrt{3}$ is $\theta = Arg(z) = \frac{\pi}{3}$, etc.

Theorem 8-6-1: If $z_1 = r_1(\cos\theta_1 + i\sin\theta_1)$ and $z_2 = r_2(\cos\theta_2 + i\sin\theta_2)$, then the product $z_1 z_2$ is equal to

$$z_1 z_2 = r_1 r_2(\cos(\theta_1 + \theta_2) + i\sin(\theta_1 + \theta_2)) \qquad (8-6-4)$$

Proof:

$$z_1 z_2 = r_1 r_2 (\cos\theta_1 + i\sin\theta_1)(\cos\theta_2 + i\sin\theta_2) \Rightarrow$$

$$z_1 z_2 = r_1 r_2 \left\{ \underbrace{\cos\theta_1 \cos\theta_2 - \sin\theta_1 \sin\theta_2}_{\cos(\theta_1+\theta_2)} + i\left(\underbrace{\sin\theta_1 \cos\theta_2 + \sin\theta_2 \cos\theta_1}_{\sin(\theta_1+\theta_2)} \right) \right\}$$
$$= r_1 r_2 (\cos(\theta_1 + \theta_2) + i\sin(\theta_1 + \theta_2))$$

This theorem shows that, **in order to multiply two complex numbers expressed in trigonometric form, we multiply the absolute values and add the arguments.**

Corollary: If $z_1 = r_1(\cos\theta_1 + i\sin\theta_1), z_2 = r_2(\cos\theta_2 + i\sin\theta_2), \cdots, z_n = r_n(\cos\theta_n + i\sin\theta_n)$, then

$$z_1 z_2 \cdots z_n = r_1 r_2 \cdots r_n (\cos(\theta_1 + \theta_2 + \cdots \theta_n) + i\sin(\theta_1 + \theta_2 + \cdots \theta_n))$$

Theorem 8-6-2: If $z_1 = r_1(\cos\theta_1 + i\sin\theta_1)$ and $z_2 = r_2(\cos\theta_2 + i\sin\theta_2)$, then the quotient z_1/z_2 is equal to

$$\frac{z_1}{z_2} = \frac{r_1}{r_2}(\cos(\theta_1 - \theta_2) + i\sin(\theta_1 - \theta_2)) \qquad (8-6-5)$$

Proof:

$$\frac{z_1}{z_2} = \frac{r_1(\cos\theta_1 + i\sin\theta_1)}{r_2(\cos\theta_2 + i\sin\theta_2)} = \frac{r_1}{r_2} \cdot \frac{(\cos\theta_1 + i\sin\theta_1)(\cos\theta_2 - i\sin\theta_2)}{(\cos\theta_2 + i\sin\theta_2)(\cos\theta_2 - i\sin\theta_2)} \Rightarrow$$

$$\frac{z_1}{z_2} = \frac{r_1}{r_2} \cdot \frac{\cos(\theta_1 - \theta_2) + i\sin(\theta_1 - \theta_2)}{(\cos\theta_2)^2 + (\sin\theta_2)^2} \Rightarrow$$

$$\frac{z_1}{z_2} = \frac{r_1}{r_2}(\cos(\theta_1 - \theta_2) + i\sin(\theta_1 - \theta_2))$$

Theorem 8-6-3 (De Moivre's Theorem): If $z = r(\cos\theta + i\sin\theta)$ and n is a positive integer, then

$$z^n = (r(\cos\theta + i\sin\theta))^n = r^n(\cos(n\theta) + i\sin(n\theta)) \quad (8-6-6)$$

Proof: The proof follows immediately from the corollary of Theorem 8-6-1, if we take $z_1 = z_2 = \cdots = z_n = r(\cos\theta + i\sin\theta)$.

In Example 8-6-7 we shall show that De Moivre's Theorem **holds true for negative integers as well**. This means that **formula (8-6-6) holds true for all integer values of n, positive or negative**).

One of the most important applications of De Moivre's Theorem is to express $\cos(n\theta)$ and $\sin(n\theta)$ in terms of $\cos\theta$ and $\sin\theta$, (see Example 8-6-8).

Example 8-6-1: Express in polar form the number $z = 1 + i\sqrt{3}$.

Solution

$$r = |z| = \sqrt{1^2 + \left(\sqrt{3}\right)^2} = \sqrt{1+3} = 2$$

The principal value of the argument is

$$\theta = Arg(z) = \frac{\pi}{3}$$

(since $\left(\frac{\pi}{3}\right)$ lies in the interval $(-\pi, \pi]$, $\cos\frac{\pi}{3} = \frac{1}{2}$, $\sin\frac{\pi}{3} = \frac{\sqrt{3}}{2}$). The polar form of z is

$$z = 1 + i\sqrt{3} = 2\left(\cos\frac{\pi}{3} + i\sin\frac{\pi}{3}\right)$$

Example 8-6-2: Find the polar form of $z = 1 - i$.

Solution

Working as in Example 8-6-1, we find
$$z = \sqrt{2}\left(\cos\left(-\frac{\pi}{4}\right) + i\sin\left(-\frac{\pi}{4}\right)\right) = \sqrt{2}\left(\cos\frac{\pi}{4} - i\sin\frac{\pi}{4}\right)$$

Example 8-6-3: Find the absolute value and the argument of the complex number $z = (-3i)(2 + 2i)(-\sqrt{3} - i)/(3 - 3i)$.

Solution

$$|z| = \left|\frac{(-3i)(2+2i)(-\sqrt{3}-i)}{3-3i}\right| = \frac{|-3i| \cdot |2+2i| \cdot |-\sqrt{3}-i|}{|3-3i|} \Rightarrow$$

$$|z| = \frac{3 \cdot \sqrt{2^2 + 2^2} \cdot \sqrt{(-\sqrt{3})^2 + (-1)^2}}{\sqrt{3^2 + (-3)^2}} = \frac{3 \cdot 2\sqrt{2} \cdot \sqrt{4}}{3\sqrt{2}} = 4$$

The numerator of z is $w \overset{\text{def}}{=} (-3i)(2+2i)(-\sqrt{3}-i) = 3i(2+2i)(\sqrt{3}+i)$, and its argument is (see Theorem 8-6-1),

$$\arg w = \arg(3i) + \arg(2+2i) + \arg(\sqrt{3}+i) = \frac{\pi}{2} + \frac{\pi}{4} + \frac{\pi}{6} = \frac{11\pi}{12} \quad (*)$$

The argument of the denominator of z is

$$\arg(3 - 3i) = -\frac{\pi}{4} \quad (**)$$

The argument of z is (see Theorem 8-6-2),

$$\phi = \arg z = \frac{11\pi}{12} - \left(-\frac{\pi}{4}\right) = \frac{11\pi}{12} + \frac{\pi}{4} = \frac{14\pi}{12} = \frac{7\pi}{6} \quad (***)$$

However, **this is not the principal value of the argument**, since the principal value of the argument lies in the interval $(-\pi, \pi]$, while the argument of z in

(***) is outside of this interval. To find the principal argument $\Theta = Arg(z)$ of z we think as follows:

Any two values of the argument of a complex number z differ by an integral multiple of (2π), i.e. $\phi - \Theta = k \cdot 2\pi$, where $k = 0, \pm 1, \pm 2, \cdots$, in other words $\Theta = \phi - 2k\pi$, and since $-\pi < \Theta \leq \pi$, we have,

$$-\pi < \frac{7\pi}{6} - 2k\pi \leq \pi \Leftrightarrow -1 < \frac{7}{6} - 2k \leq 1 \Leftrightarrow \frac{13}{12} > k \geq \frac{1}{12}$$

and the only integer value of k that satisfies this double inequality is $k = 1$, and finally, $\Theta = \phi - 2\pi = \frac{7\pi}{6} - 2\pi = -\frac{5\pi}{6}$.

Example 8-6-4: If z is a complex number and θ is any angle, show that $Re(|z| - z(\cos\theta + i\sin\theta)) \geq 0$.

Solution

Let $z = |z|(\cos\phi + i\sin\phi)$ be the polar form of the complex number z. Then,

$$|z| - z(\cos\theta + i\sin\theta) = |z| - |z|(\cos\phi + i\sin\phi)(\cos\theta + i\sin\theta)$$
$$= |z| - |z|(\cos(\theta + \phi) + i\sin(\theta + \phi))$$
$$= |z|\{1 - \cos(\theta + \phi) - i\sin(\theta + \phi)\} \Rightarrow$$

$$Re(|z| - z(\cos\theta + i\sin\theta)) = |z|(1 - \cos(\theta + \phi)) \geq 0$$

since $\cos(\theta + \phi) \leq 1$ and $|z| > 0$.

Example 8-6-5: If $z = (1 + 5i)/(2 - i)$ and $w = (1 - 6i)/(2 + i)$, express $(z + w)$ in polar form.

Solution

$$z + w = \frac{1 + 5i}{2 - i} + \frac{1 - 6i}{2 + i} = \frac{(1 + 5i)(2 + i) + (1 - 6i)(2 - i)}{(2 - i)(2 + i)} = \frac{-7 - 2i}{5}$$

The absolute value $|z + w| = \frac{1}{5}\sqrt{(-7)^2 + (-2)^2} = \frac{\sqrt{53}}{5}$.

The principal value of $(z + w)$ is $\Theta = Arg(z + w) = -180° + Arctan\left(\frac{-2}{-7}\right) \cong -164°$, (see Example 8-6-4), and the polar form of the complex number z is, $z = \frac{\sqrt{53}}{5}(\cos 164° - i \sin 164°)$.

Example 8-6-6: Consider the complex numbers $z = 2 + i, w = 3 + i$. Find the product $u = zw$ and then show that: $\tan^{-1}\left(\frac{1}{2}\right) + \tan^{-1}\left(\frac{1}{3}\right) = \frac{\pi}{4}$.

Solution

$$u = zw = (2+i)(3+i) = 5 + 5i = 5\sqrt{2}\left(\cos\frac{\pi}{4} + i\sin\frac{\pi}{4}\right) \quad (*)$$

The polar form of the numbers z and w are:

$$z = 2 + i = \sqrt{3}(\cos\theta + i\sin\theta), \text{ where } \theta = \tan^{-1}\left(\frac{1}{2}\right) \quad (**)$$

$$w = 3 + i = 2(\cos\phi + i\sin\phi), \text{ where } \phi = \tan^{-1}\left(\frac{1}{3}\right) \quad (***)$$

Application of Theorem 8-6-1, shows that $\theta + \phi = \frac{\pi}{4}$, i.e.

$$\tan^{-1}\left(\frac{1}{2}\right) + \tan^{-1}\left(\frac{1}{3}\right) = \frac{\pi}{4}$$

and this completes the proof.

Example 8-6-7: Show that De Moivre's Theorem holds true for negative values of n, as well.

Solution

Let $n = -k$, where k is **a positive integer**. Then,

$$\{r(\cos\theta + i\sin\theta)\}^{-k} = \frac{1}{\{r(\cos\theta + i\sin\theta)^k\}} = \frac{1}{r^k(\cos k\theta + i\sin k\theta)} =$$

$$r^{-k}\frac{\cos k\theta - i\sin k\theta}{(\cos k\theta + i\sin k\theta)(\cos k\theta - i\sin k\theta)} = r^{-k}\frac{\cos k\theta - i\sin k\theta}{\sqrt{(\cos k\theta)^2 + (\sin k\theta)^2}} =$$

$$r^{-k}(\cos k\theta - i\sin k\theta) = r^{-k}\{\cos(-k\theta) + i\sin(-k\theta)\}$$

and this shows that De Moivre's Theorem is valid for negative integers as well.

Example 8-6-8: Express $\cos 3\theta$ and $\sin 3\theta$ in terms of $\cos \theta$ and $\sin \theta$.

Solution

Using De Moivre's Theorem with $r = 1$ and $n = 3$, we have:

$$(\cos \theta + i \sin \theta)^3 = \cos 3\theta + i \sin 3\theta$$

or expanding the left side term according to the well known identity,

$$(\cos \theta)^3 + 3(\cos \theta)^2(i \sin \theta) + 3 \cos \theta \, (i \sin \theta)^2 + (i \sin \theta)^3$$
$$= \cos 3\theta + i \sin 3\theta \Leftrightarrow$$

$$(\cos \theta)^3 - 3 \cos \theta \, (\sin \theta)^2 + i(3(\cos \theta)^2 \sin \theta - (\sin \theta)^3)$$
$$= \cos 3\theta + i \sin 3\theta$$

and **equating real and imaginary parts** we get,

$$\begin{cases} \cos 3\theta = (\cos \theta)^3 - 3 \cos \theta \, (\sin \theta)^2 \\ \sin 3\theta = 3(\cos \theta)^2 \sin \theta - (\sin \theta)^3 \end{cases}$$

Expressing $(\sin \theta)^2 = 1 - (\cos \theta)^2$ (in the first equation) and $(\cos \theta)^2 = 1 - (\sin \theta)^2$ (in the second equation), we find the known formulas

$$\begin{cases} \cos 3\theta = 4(\cos \theta)^3 - 3 \cos \theta \\ \sin 3\theta = 3 \sin \theta - 4(\sin \theta)^3 \end{cases}$$

Example 8-6-9: If $z = 1 + i$, express z^{100} in the form $a + ib$, (Cartesian form).

Solution

a) One possible way to solve the problem would be to find $z^2 = z \cdot z$, then find $z^3 = z^2 \cdot z$, etc. It is evident that this approach of solving the problem would require an enormous amount of calculations and computing time.

b) An easy method to solve the problem is to use De Moivre's Theorem, as shown below:

$$z = 1 + i = \sqrt{2}\left(\cos\frac{\pi}{4} + i\sin\frac{\pi}{4}\right) \Rightarrow z^{100} = \left\{\sqrt{2}\left(\cos\frac{\pi}{4} + i\sin\frac{\pi}{4}\right)\right\}^{100} \Rightarrow$$

$$z^{100} = (\sqrt{2})^{100}\left(\cos\left(100 \cdot \frac{\pi}{4}\right) + i\sin\left(100 \cdot \frac{\pi}{4}\right)\right) \Rightarrow$$

$$z^{100} = 2^{50}(\cos(25\pi) + i\sin(25\pi)) \qquad (*)$$

However, $25\pi = \pi + 12 \cdot (2\pi)$, and therefore $\cos(25\pi) = \cos\pi = -1$, $\sin 25\pi = \sin\pi = 0$, and equation (*) implies,

$$z^{100} = 2^{50}(-1 + i \cdot 0) = -2^{50}$$

Example 8-6-10: Show that
$$z = (1+i)^n + (1-i)^n = 2^{(n+2)/2}\cos\left(\frac{n\pi}{4}\right), n = 0, 1, 2, \cdots.$$

Solution

$$(1+i)^n = \left(\sqrt{2}\left(\cos\frac{\pi}{4} + i\sin\frac{\pi}{4}\right)\right)^n = (\sqrt{2})^n\left(\cos\left(\frac{n\pi}{4}\right) + i\sin\left(\frac{n\pi}{4}\right)\right)$$

$$(1-i)^n = \left(\sqrt{2}\left(\cos\frac{\pi}{4} - i\sin\frac{\pi}{4}\right)\right)^n = (\sqrt{2})^n\left(\cos\left(\frac{n\pi}{4}\right) - i\sin\left(\frac{n\pi}{4}\right)\right)$$

and adding term wise results in

$$(1+i)^n + (1-i)^n = 2(\sqrt{2})^n\cos\left(\frac{n\pi}{4}\right) = 2^{(n+2)/2}\cos\left(\frac{n\pi}{4}\right)$$

PROBLEMS

8-6-1) Express $z = -1 + i$ in polar form, **(Ans:** $z = \sqrt{2}\left(\cos\frac{3\pi}{4} + i\sin\frac{3\pi}{4}\right)$).

8-6-2) Express $z = -1 - i\sqrt{3}$ in polar form.

8-6-3) Express $z = \sqrt{3} - i$ in polar form, **(Ans:** $z = 2\left(\cos\frac{\pi}{6} - i\sin\frac{\pi}{6}\right)$).

8-6-4) If $z = 4\left(\cos\frac{\pi}{3} + i\sin\frac{\pi}{3}\right)$ and $w = 2\left(\cos\frac{\pi}{4} + i\sin\frac{\pi}{4}\right)$, find the Cartesian form of the numbers zw and z/w.

8-6-5) Find the principal argument of the complex number $z = (-1+i)(\sqrt{3}-i)$, (**Ans:** $Arg(z) = 7\pi/12$).

8-6-6) Show that the principal argument of $w = (-1+i)/(\sqrt{3}-i)$ is $\frac{11\pi}{12}$.

8-6-7) If $x, y \in \mathbb{R}$, show that: $\tan^{-1}\left(\frac{1}{x+y}\right) + \tan^{-1}\left(\frac{y}{x^2+xy+1}\right) = \tan^{-1}\left(\frac{1}{x}\right)$.

Hint: Consider the numbers $z = x + iy$ and $w = (x^2 + xy + 1) + iy$, and work as in Example 8-6-6.

8-6-8) Express $z = (\sqrt{2} - 2i)/(1+i)$ and $w = (1 - \sqrt{3}\,i)/(1 + \sqrt{3}\,i)$ in polar form.

8-6-9) If z is a complex number ($z \neq 1$), show that

$$S = 1 + z + z^2 + z^3 + \cdots + z^k = \frac{1 - z^{k+1}}{1 - z}$$

Hint: Show that the difference $S - z \cdot S = 1 - z^{k+1}$.

8-6-10) If k is any positive number, show that:

$$S_1 = 1 + \cos\theta + \cos 2\theta + \cdots + \cos k\theta = \frac{1}{2} + \frac{\sin\left\{\left(k + \frac{1}{2}\right)\theta\right\}}{2\sin\left(\frac{\theta}{2}\right)}$$

$$S_2 = \sin\theta + \sin 2\theta + \cdots + \sin k\theta = \frac{1}{2}\cot\left(\frac{\theta}{2}\right) - \frac{\cos\left\{\left(k + \frac{1}{2}\right)\theta\right\}}{2\sin\left(\frac{\theta}{2}\right)}$$

Hint: Use the result found in Pr. 8-6-9, with $z = \cos\theta + i\sin\theta$, apply De Moivre's Theorem, etc.

8-6-11) If $|z| = |w| = 1$, show that the number $(z+w)^n/(z^n + w^n)$ is a real number, for any positive integer n.

8-6-12) Show that $(1+i)^n - (1-i)^n = i\, 2^{(n+2)/2} \sin\left(\frac{n\pi}{4}\right)$.

8-6-13) Express $z = \left(\frac{1+i\sqrt{3}}{2}\right)^{77}$ in the form $a + ib$, (**Ans:** $\frac{1-i\sqrt{3}}{2}$).

8-6-14) Express $z = \left(\sqrt{3} + i\right)^{-57}$ in the form $a + ib$.

8-6-15) Using De Moivre's Theorem show:

$$\sin 5\theta = 16(\sin\theta)^5 - 20(\sin\theta)^3 + 5\sin\theta$$

$$\cos 5\theta = 16(\cos\theta)^5 - 20(\cos\theta)^3 + 5\cos\theta$$

8-6-16) Find the Cartesian form of $z = \left(1 + i\sqrt{3}\right)^{167}$.

8-6-17) Show that

$$\tan 8x = \frac{8\tan x - 56(\tan x)^3 + 56(\tan x)^5 - 8(\tan x)^7}{1 - 28(\tan x)^2 + 70(\tan x)^4 - 28(\tan x)^6 + (\tan x)^8}$$

Hint: Use De Moivre's Theorem to express $\sin 8x$ and $\cos 8x$ in terms of $\sin x$ and $\cos x$, and then find $\tan 8x = \sin 8x / \cos 8x$.

8-6-18) Express $z = \left(1 + i\sqrt{2}\right)^{15} / \left(\sqrt{2} - i\right)^{10}$ in the form $a + ib$.

8-6-19) Use De Moivre's Theorem to simplify the expression $w = \left(1 + i\sqrt{3}\right)^5 + \left(1 - i\sqrt{3}\right)^5$, **(Ans:** $w = 32$**)**.

8-6-20) If $z = \cos\phi + i\sin\phi$, $w = \cos\theta + i\sin\theta$ and m and n are positive integers, show that $\frac{z^m}{w^n} + \frac{w^n}{z^m} = 2\cos(m\phi - n\theta)$.

8-6-21) If $n = 1, 2, 3, \cdots$, show that

$$\{(1+i)\cos\theta + (1-i)\sin\theta\}^{2n} = 2^n\left\{\cos\left(\frac{n\pi}{2} - 2n\theta\right) + i\sin\left(\frac{n\pi}{2} - 2n\theta\right)\right\}$$

8-7) Roots of complex numbers

A complex number w is called the $n^{\underline{th}}$ root of another complex number z, if $w^n = z$, (n is a positive integer). In this case we write $w = \sqrt[n]{z}$.

For example, since $(1+i)^2 = 2i$, the complex number $(1+i)$ is the square root of $(2i)$. Of course, the number $(-1-i)$ is another square root of $(2i)$, since again, $(-1-i)^2 = 2i$.

It turns out that every positive number z has exactly $n, n^{\underline{th}}$ roots. This important result is stated in the following Theorem.

Theorem 8-7-1: **If n is a positive integer, then any complex number $z = r(\cos\theta + i\sin\theta) \neq 0$, has exactly n distinct (i.e. different from each other) $n^{\underline{th}}$ roots, given by the formula**

$$w_k = \sqrt[n]{r}\left\{\cos\left(\frac{\theta}{n} + k\cdot\frac{2\pi}{n}\right) + i\sin\left(\frac{\theta}{n} + k\cdot\frac{2\pi}{n}\right)\right\} \qquad (8-7-1)$$

where $k = 0, 1, 2, \cdots, (n-1)$.

Proof: The proof follows directly from De Moivre's Theorem, since

$$w_k^n = r\{\cos(\theta + 2k\pi) + i\sin(\theta + 2k\pi)\} = r(\cos\theta + i\sin\theta) = z$$

and this is true for every $k = 0, 1, 2, \cdots, (n-1)$. The n values $w_0, w_1, \cdots, w_{n-1}$ are all distinct (why?). We note that if $k = n$, $w_n = w_0$, and similarly, $w_{n+1} = w_1, w_{n+2} = w_2,\ldots$, i.e. we do not get different $n^{\underline{th}}$ roots of z.

The only distinct $n^{\underline{th}}$ roots of z are the ones given by (8-7-1) when k runs over the values $0, 1, 2, \cdots, (n-1)$.

Notice that all the $n^{\underline{th}}$ roots of z, have the same absolute value ($\sqrt[n]{r}$), but they have different arguments.

Therefore, the symbol $\sqrt[n]{z} = z^{1/n}$ can take on n different values, i.e. **is an n − valued function of z.**

Similarly, **if m is any positive integer, the symbol $z^{\frac{m}{n}} = \left(z^{\frac{1}{n}}\right)^m$ represents n different values, $w_0^m, w_1^m, w_2^m, \cdots, w_{n-1}^m$.**

In order to find the n distinct roots of a given number $z = a + ib$, we first convert the number in polar form and then apply formula (8-7-1).

The $n^{\underline{th}}$ roots of unity:

If n is a positive integer, the roots of the equation $z^n - 1 = 0$, are called **the $n^{\underline{th}}$ roots of unity.** Starting with $z^n - 1 = 0$, or equivalently, $z^n =$

$1(\cos 0 + i \sin 0)$, the $n^{\underline{th}}$ roots of unity are obtained from (8-7-1) with $r = 1$, i.e.

$$z_k = \cos\left(k\frac{2\pi}{n}\right) + i \sin\left(k\frac{2\pi}{n}\right), \quad k = 0,1,2,\cdots,(n-1) \quad (8-7-2)$$

If n is odd, $z_0 = 1$ is the only real root, all other roots are complex. For example $z^3 - 1 = 0$ has only one real root, $z = 1$, the other two roots are complex.

If n is even, there exist two real roots, $z_0 = 1$ and $z_{(n/2)} = -1$, all other are complex roots. For example, $z^4 - 1 = 0$ has two real roots ($z = 1, z = -1$) and two complex roots.

The equation $z^n - 1 = 0, n = 1,2,3,\cdots$, is called **a cyclotomic equation of degree n**. The roots of a cyclotomic equation, i.e. the $n^{\underline{th}}$ roots of unity, in the complex plane, **divide the unit circle into n equal arcs, starting from the point $z_0 = 1 + i0 = (1,0)$. In other words, the roots $\{z_0, z_1, z_2, \cdots, z_{n-2}, z_{n-1}\}$ of the cyclotomic equation $z^n - 1 = 0$, are the vertices of a normal polygon with n sides, inscribed in the unit circle**, (see also section 8-8).

Example 8-7-1: Find the cubic roots of $z = 1 + i\sqrt{3}$.

Solution

$$z = 1 + i\sqrt{3} = 2\left(\cos\frac{\pi}{3} + i \sin\frac{\pi}{3}\right)$$

$$w_k = \sqrt[3]{2}\left\{\cos\left(\frac{\pi}{9} + k\frac{2\pi}{3}\right) + i \sin\left(\frac{\pi}{9} + k\frac{2\pi}{3}\right)\right\}, \quad k = 0,1,2$$

$$k = 0: \quad w_0 = \sqrt[3]{2}\left(\cos\frac{\pi}{9} + i \sin\frac{\pi}{9}\right)$$

$$k = 1: \quad w_1 = \sqrt[3]{2}\left(\cos\left(\frac{\pi}{9} + \frac{2\pi}{3}\right) + i \sin\left(\frac{\pi}{9} + \frac{2\pi}{3}\right)\right) = \sqrt[3]{2}\left(\cos\frac{7\pi}{9} + i \sin\frac{7\pi}{9}\right)$$

$$k = 2: \quad w_2 = \sqrt[3]{2}\left(\cos\left(\frac{\pi}{9} + \frac{4\pi}{3}\right) + i \sin\left(\frac{\pi}{9} + \frac{4\pi}{3}\right)\right)$$
$$= \sqrt[3]{2}\left(\cos\frac{13\pi}{9} + i \sin\frac{13\pi}{9}\right)$$

Example 8-7-2: Find the sixth roots of the number (-64).

Solution

$$z = -64 = 64(\cos \pi + i \sin \pi)$$

Recall that $r = |z|$ is always a positive number.

The sixth roots of (-64) are given by formula (8-7-1) with $n = 6$, i.e.

$$w_k = \sqrt[6]{64}\left(\cos\left(\frac{\pi}{6} + k\frac{2\pi}{6}\right) + i \sin\left(\frac{\pi}{6} + k\frac{2\pi}{6}\right)\right), \quad k = 0,1,2,3,4,5$$

$$k = 0: w_0 = 2\left(\cos\frac{\pi}{6} + i \sin\frac{\pi}{6}\right) = 2\left(\frac{\sqrt{3}}{2} + i\frac{1}{2}\right) = \sqrt{3} + i$$

$$k = 1: w_1 = 2\left(\cos\frac{3\pi}{6} + i \sin\frac{3\pi}{6}\right) = 0 + 2i$$

$$k = 2: w_2 = 2\left(\cos\frac{5\pi}{6} + i \sin\frac{5\pi}{6}\right) = -\sqrt{3} + i$$

$$k = 3: w_3 = 2\left(\cos\frac{7\pi}{6} + i \sin\frac{7\pi}{6}\right) = -\sqrt{3} - i$$

$$k = 4: w_4 = 2\left(\cos\frac{9\pi}{6} + i \sin\frac{9\pi}{6}\right) = 0 - 2i$$

$$k = 5: w_5 = 2\left(\cos\frac{11\pi}{6} + i \sin\frac{11\pi}{6}\right) = \sqrt{3} - i$$

Example 8-7-3: Find the square root of the number $z = \frac{(2+3i)(1-i)}{(3+i)^2}$.

Solution

The given number is easily simplified to the standard Cartesian form $= \frac{46}{100} - i\frac{22}{100}$, or to the equivalent polar form,

$$z = \frac{\sqrt{26}}{10}(\cos\theta + i \sin\theta), \quad \theta = \tan^{-1}\left(\frac{-22}{46}\right) \cong -0.446 \, rad$$

The square root of z is

$$w_k = \sqrt{\left(\frac{\sqrt{26}}{10}\right)} \left(\cos\left(\frac{\theta}{2} + k\frac{2\pi}{2}\right) + i\sin\left(\frac{\theta}{2} + k\frac{2\pi}{2}\right)\right), k = 0,1$$

$$k = 0: w_0 = \sqrt{\left(\frac{\sqrt{26}}{10}\right)} \left(\cos\left(\frac{\theta}{2}\right) + i\sin\left(\frac{\theta}{2}\right)\right)$$

$$= 0.71(\cos(-0.223) + i\sin(-0.223)) = 0.692 - 0.157i$$

$$k = 1: w_1 = \sqrt{\left(\frac{\sqrt{26}}{10}\right)} \left(\cos\left(\frac{\theta}{2} + \frac{2\pi}{2}\right) + i\sin\left(\frac{\theta}{2} + \frac{2\pi}{2}\right)\right)$$

$$= 0.71(\cos(2.917) + i\sin(2.917)) = -0.692 + 0.157i$$

Example 8-7-4: In this example, we present a different method to find the square root of a given complex number $z = a + ib$, i.e. to find a complex number $w = x + iy$, such that $w^2 = z$.

Solution

$$w^2 = z \Rightarrow (x + iy)^2 = a + ib \Rightarrow x^2 - y^2 + i2xy = a + ib \Rightarrow$$

$$\begin{cases} x^2 - y^2 = a \\ 2xy = b \end{cases} \tag{*}$$

Squaring and adding term by term the two equations in (*) results in

$$(x^2 + y^2)^2 = a^2 + b^2 \Rightarrow x^2 + y^2 = \sqrt{a^2 + b^2} \tag{**}$$

Adding this equation with the first equation in (*) results in

$$2x^2 = a + \sqrt{a^2 + b^2} \Rightarrow x = \pm \frac{1}{\sqrt{2}} \sqrt{a + \sqrt{a^2 + b^2}} \tag{***}$$

and then from the second equation in (*) we obtain y,

$$y = \frac{b}{2x} = \pm \frac{b}{\sqrt{2}\sqrt{a + \sqrt{a^2 + b^2}}} \tag{****}$$

In summary, the complex number $z = a + ib$ has two square roots w_0 and w_1, given by the formulas:

$$w_0 = \frac{1}{\sqrt{2}}\sqrt{a+\sqrt{a^2+b^2}} + i\,\frac{b}{\sqrt{2}\sqrt{a+\sqrt{a^2+b^2}}}$$

$$w_1 = -w_0 = -\left(\frac{1}{\sqrt{2}}\sqrt{a+\sqrt{a^2+b^2}} + i\,\frac{b}{\sqrt{2}\sqrt{a+\sqrt{a^2+b^2}}}\right)$$

Let the reader find the square roots of the number $z = 0.46 - 0.22i$ in the preceding example, by means of the formulas found here, and verify that that the results obtained are identical to the ones found in Example 8-7-3.

Example 8-7-5: Solve the equation $(1+x)^5 = (1-x)^5$.

Solution

We first note that $x \neq \pm 1$, therefore the given equation is equivalent to the equation $(1-x)^5/(1+x)^5 = 1$, i.e. the number $w = (1-x)/(1+x)$ is the fifth root of the number $z = 1 \cdot (\cos 0 + i \sin 0)$, i.e.

$$w_k = \sqrt[5]{1}\left(\cos\left(k\frac{2\pi}{5}\right) + i\sin\left(k\frac{2\pi}{5}\right)\right), \quad k = 0,1,2,3,4 \qquad (*)$$

or equivalently,

$$\frac{1-x_k}{1+x_k} = \cos\left(k\frac{2\pi}{5}\right) + i\sin\left(k\frac{2\pi}{5}\right) \Leftrightarrow x_k = \frac{1-w_k}{1+w_k} \overset{(*)}{\Leftrightarrow}$$

$$x_k = \frac{1-\cos\left(k\frac{2\pi}{5}\right) - i\sin\left(k\frac{2\pi}{5}\right)}{1+\cos\left(k\frac{2\pi}{5}\right) + i\sin\left(k\frac{2\pi}{5}\right)} \quad k = 0,1,2,3,4 \qquad (**)$$

Making use of the trigonometric identities

$$1 - \cos x = 2\left(\sin\frac{x}{2}\right)^2, \quad 1 + \cos x = 2\left(\cos\frac{x}{2}\right)^2, \quad \sin x = 2\sin\frac{x}{2}\cos\frac{x}{2}$$

equation (*) yields,

$$x_k = \frac{2\left(\sin\frac{k\pi}{5}\right)^2 - i\,2\sin\frac{k\pi}{5}\cos\frac{k\pi}{5}}{2\left(\cos\frac{k\pi}{5}\right)^2 + i\,2\sin\frac{k\pi}{5}\cos\frac{k\pi}{5}} = \frac{2\sin\frac{k\pi}{5}\left(\sin\frac{k\pi}{5} - i\cos\frac{k\pi}{5}\right)}{2\cos\frac{k\pi}{5}\left(\cos\frac{k\pi}{5} + i\sin\frac{k\pi}{5}\right)} \Rightarrow$$

$$x_k = \tan\frac{k\pi}{5} \cdot \frac{\sin\frac{k\pi}{5} - i\cos\frac{k\pi}{5}}{\cos\frac{k\pi}{5} + i\sin\frac{k\pi}{5}} \qquad (***)$$

We notice that

$$\frac{\sin\frac{k\pi}{5} - i\cos\frac{k\pi}{5}}{\cos\frac{k\pi}{5} + i\sin\frac{k\pi}{5}} = \frac{-i\left(\cos\frac{k\pi}{5} + i\sin\frac{k\pi}{5}\right)}{\cos\frac{k\pi}{5} + i\sin\frac{k\pi}{5}} = -i$$

and equation (***) yields,

$$x_k = -i \tan\frac{k\pi}{5} \qquad k = 0, 1, 2, 3, 4$$

and these are the sought for solutions of the original equation.

PROBLEMS

8-7-1) Show that the two square roots of the imaginary unit i, are $\pm(1+i)/\sqrt{2}$, **a)** Using the polar form of i and applying formula (8-7-1) and **b)** Using the method developed in Example 8-7-4.

8-7-2) Show that the three cubic roots of 1 are: $1, (-1 \pm i\sqrt{3})/2$.

8-7-3) Find the sixth roots of 64.

(**Ans:** $2, -2, 1 + i\sqrt{3}, 1 - i\sqrt{3}, -1 + i\sqrt{3}, -1 - i\sqrt{3}$).

8-7-4) Find all the distinct values of $(1+i)^{3/4}$ and $(1 - i\sqrt{3})^{5/4}$.

Hint: $(1+i)^{3/4} = \left(\sqrt[4]{1+i}\right)^3$, etc.

8-7-5) Solve the equation $(x+1)^6 + (x-1)^6 = 0$.

(**Ans:** $x_k = -i \cot\left((2k+1)\frac{\pi}{12}\right)$, $k = 0,1,2,3,4,5$).

Hint: See Example 8-7-5.

8-7-6) Find the seventh roots of the number $z = (1+i)(4-3i)/(2+i)^2$.

Hint: Express z in the form $a + ib$.

8-7-7) Solve the equation: $\left(\dfrac{1+ix}{1-ix}\right)^3 = \dfrac{1+i\sqrt{3}}{1-i\sqrt{3}}$.

(**Ans:** $\tan\dfrac{\pi}{9}, \tan\dfrac{4\pi}{9}, \tan\dfrac{7\pi}{9}$).

8-7-8) Find: $\sqrt[6]{-1}, \sqrt[7]{1}, \sqrt[3]{-3+3i}$.

8-7-9) Solve the equation: $z^3 + 1 = i(z^3 - 1)$.

(**Ans:** $i, \dfrac{\sqrt{3}-i}{2}, \dfrac{-\sqrt{3}-i}{2}$).

8-7-10) Find all the distinct values of $\sqrt[4]{(1+i\sqrt{3})^3}$ and $\sqrt[4]{2\sqrt{3}+2i}$.

8-7-11) Solve the equation $(z-1)^6 = 1$.

(**Ans:** $0, 2, \dfrac{3\pm i\sqrt{3}}{2}, \dfrac{1\pm i\sqrt{3}}{2}$).

8-7-12) Solve the equation $\left(\dfrac{z+1}{z-i}\right)^2 = -i$.

8-7-13) Solve the equation $(z+1)^n = z^n$, $n = 2, 3, 4, \ldots$

(**Ans:** $z_k = -\dfrac{1}{2} - \dfrac{1}{2}i\cot\left(k\cdot\dfrac{\pi}{n}\right)$, $k = 1, 2, \cdots, (n-1)$).

8-7-14) If w is any one of the nth root of 1, other than 1, (i.e. $w \neq 1$), show that $1 + w + w^2 + \cdots + w^{n-1} = 0$.

8-8) The complex plane (or the Argand's diagram)

As we know, the ordered pair (a, b) is associated with **a unique** point M in the plane xOy, as shown in Fig. 8-1, (see Sect. 1-7). We write $M(a, b)$ to denote that the point M represents the ordered pair (a, b).

The number a is called **the abscissa of M**, while b is called **the ordinate of M**. Note that the coordinates a and b are signed numbers, their signs depending on the quadrant in which M lies. The origin O of the system corresponds to the pair $(0,0)$.

We accept that the point $M(a, b)$ represents the complex number $z = a + ib$, and we write $M = M(z) = M(a, b)$. In that sense, it is convenient to call the plane, **the complex plane or the Argand diagram**.

Every point on the complex plane represents a complex number, and conversely, every complex number is associated with a point on the complex plane. **The $x'x$ axis is called the real axis and the $y'y$ axis is called the imaginary axis**.

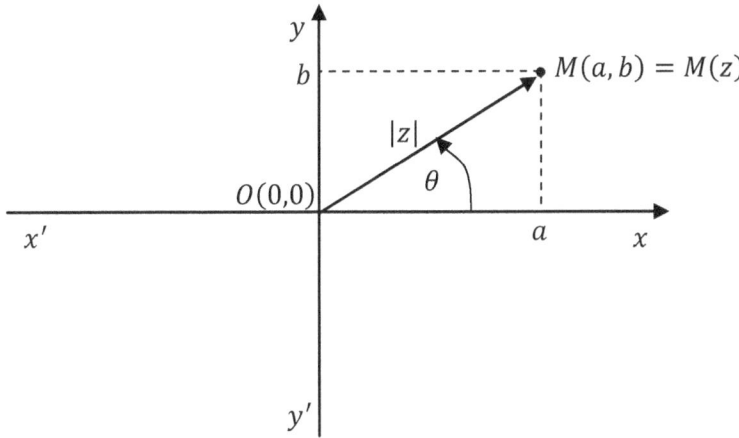

Fig. 8-1: The Complex plane (Argand's diagram).

Points on the real axis $(x'x)$ represent real numbers, while points on the imaginary axis $(y'y)$ represent pure imaginary numbers. **Note that the point $(0, 1)$ corresponds to the imaginary unit $i = (0, 1)$.**

Let $M = M(a, b) = M(z)$ be a point on the complex plane, corresponding to the complex number $z = a + ib$. We may as well consider that **the vector $\overrightarrow{OM}$ also represents the complex number $z = a + ib$**, as shown in Fig. 8-1.

1) The length of the vector $|\overrightarrow{OM}|$ is equal to the absolute value of z, i.e.

$$OM = |\overrightarrow{OM}| = \sqrt{a^2 + b^2} = |z| \qquad (8-8-1)$$

2) The angle between the positive x axis and the vector $\overrightarrow{OM}$ represents the argument of z, i.e.

$$\theta = Arg(z), \quad \text{since} \quad \begin{cases} \cos\theta = \dfrac{a}{\sqrt{a^2+b^2}} \\ \sin\theta = \dfrac{b}{\sqrt{a^2+b^2}} \end{cases} \quad (8-8-2)$$

Since complex numbers are represented by vectors, **addition or subtraction of complex numbers is equivalent to the addition or subtraction of vectors.**

Multiplication of a complex number by the imaginary unit i, geometrically means rotation of the vector $\overrightarrow{OM}$ representing the vector by $\left(+\dfrac{\pi}{2}\right)$ about the origin O. This is so since the argument of the complex number (iz) is $\left(\theta + \dfrac{\pi}{2}\right)$, while the absolute value of (iz) is the same as the absolute value of z.

Conjugate complex numbers are represented by vectors symmetric with respect to the real axis, while opposite complex numbers are represented by vectors symmetric with respect to the origin.

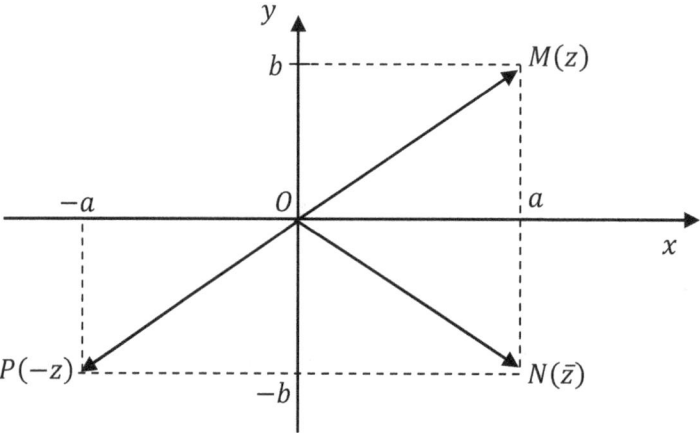

Fig. 8-2: Geometrical representation of z, $\bar{z}$ and $(-z)$.

Example 8-8-1: What is the locus of points z, whose magnitude is $|z| = 2$?

Solution

The locus of points on the complex plane with $|z| = 2$, is a circle, centered at the origin and having radius $R = 2$. In general, $|z - z_0| = R > 0$ represents a circle, center at z_0 and radius R. This is so since $(z - z_0)$ represents a vector having its origin at z_0 and its terminus at z.

Example 8-8-2: If $A(z) = r(\cos\theta + i\sin\theta)$ and $B(w) = \rho(\cos\phi + i\sin\phi)$, show that the distance AB from z to w, is given by the formula $AB = \sqrt{r^2 + \rho^2 - 2r\rho\cos(\theta - \phi)}$.

Solution

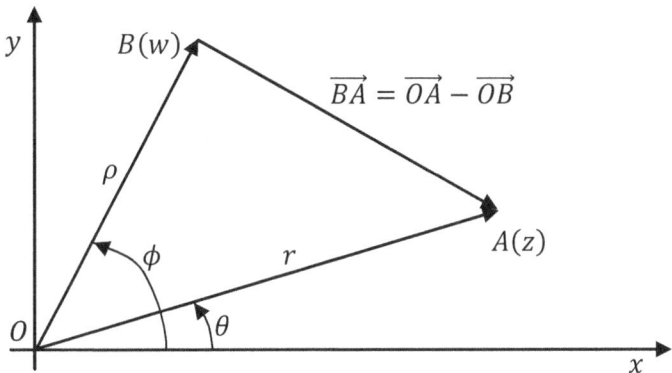

Fig. 8-3: Distance between two complex numbers.

The distance from A to B is $AB = |\overrightarrow{AB}| = |z - w|$.

$$AB^2 = |z - w|^2 = (z - w)\overline{(z - w)} = (z - w)(\bar{z} - \bar{w}) \Rightarrow$$

$$AB^2 = z\bar{z} + w\bar{w} - z\bar{w} - \bar{z}w = |z|^2 + |w|^2 - z\bar{w} - \overline{z(\bar{w})} \Rightarrow$$

$$AB^2 = r^2 + \rho^2 - 2Re(z\bar{w}) \qquad (*)$$

Since $(z\bar{w} + \overline{z(\bar{w})}) = Re(z\bar{w})$, (let the reader prove it).

$$z\bar{w} = \big(r(\cos\theta + i\sin\theta)\big)\big(\rho(\cos\phi - i\sin\phi)\big) \Rightarrow$$

$$z\bar{w} = (r\rho)(\cos(\theta - \phi) + i\sin(\theta - \phi)) \Rightarrow Re(z\bar{w}) = r\rho\cos(\theta - \phi)$$

and substituting in $(*)$ we find

$$AB^2 = r^2 + \rho^2 - r\rho\cos(\theta - \phi) \Rightarrow AB = \sqrt{r^2 + \rho^2 - 2r\rho\cos(\theta - \phi)}.$$

PROBLEMS

8-8-1) On the complex plane plot the complex numbers $z_1 = 2 + i, z_2 = 3 - i, z_3 = -1 - 3i, z_4 = -2 + 4i$.

8-8-2) On the complex plane plot $z = 3(\cos 60° + i \sin 60°), w = 2(\cos 45° - i \sin 45°), z \cdot w, z/w, w/z$.

8-8-3) Assuming that $z = x + iy$ find and sketch the set of points satisfying
a) $(z + \bar{z}) > 4$, **b)** $|z - 2 - i| \leq |z + 1 + 2i|$ and **c)** $|z| > |z - 2|$.

(**Ans: a)** $x > 2$, **b)** $x + y \geq 0$, **c)** $x > 1$).

8-8-4) Describe and graph the locus represented by each of the following: **a)** $|z + 1 - i\sqrt{3}| = 1$, **b)** $z(\bar{z} + 1) = 5$, **c)** $Re(z^2) = 2$, **d)** $Im(z^2) = 1$.

8-8-5) If $|z - w| = |z| + |w|$, show that (z/w) is a negative real number, (assume $zw \neq 0$).

8-8-6) Consider the complex numbers $A(1 + 3i), B(-1 + i), C(-\sqrt{3} - i)$ and find the lengths of the sides of the triangle ABC.

Hint: See Example 8-8-2.

8-8-7) Show that the set of points z on the Argand diagram that satisfy the relation $|(z - 1)/(z + i)| = \lambda$, where λ is a positive constant $\neq 1$, is a circle.

8-8-8) In Problem 8-8-7, find the center and the radius of the circle, assuming $\lambda = 2$.

8-8-9) In Problem 8-8-7, if $\lambda = 1$, show that the set of points is a straight line.

8-8-10) Find the set of points z, such that the number $(z - i/z + i)$ is purely imaginary, (**Ans:** The unit circle, $|z| = 1$).

CHAPTER 9: QUADRATIC EQUATIONS

Any equation (which perhaps after some simplifications) takes the form $ax^2 + bx + c = 0$, is called **a second order equation or a quadratic equation**. We shall assume that the coefficients a, b, c of the equation are real numbers, while any number r, (**real or complex**) that satisfies the equation, (i.e. such that $ar^2 + br + c = 0$) is called **a solution or a root of the equation**.

9-1) The solutions of a quadratic equation

In order to solve a quadratic equation we transform the equation as follows:

$$ax^2 + bx + c = 0 \Leftrightarrow a\left\{x^2 + \frac{b}{a}x + \frac{c}{a}\right\} = 0 \Leftrightarrow$$

$$a\left\{\underbrace{x^2 + 2\frac{b}{2a}x + \frac{b^2}{4a^2}}_{\left(x+\frac{b}{2a}\right)^2} - \frac{b^2}{4a^2} + \frac{c}{a}\right\} = 0 \Leftrightarrow$$

$$a\left\{\left(x + \frac{b}{2a}\right)^2 - \frac{b^2 - 4ac}{4a^2}\right\} = 0 \qquad (*)$$

We consider the following three cases:

1) The quantity $b^2 - 4ac$ is positive: Then the number $\frac{b^2-4ac}{4a^2}$ is positive, and as a matter of fact, is the square of the real number $\frac{\sqrt{b^2-4ac}}{2a}$, and equation (*) yields,

$$a\left\{\left(x + \frac{b}{2a}\right)^2 - \left(\frac{\sqrt{b^2 - 4ac}}{2a}\right)^2\right\} = 0 \Leftrightarrow$$

$$a\left\{x + \frac{b}{2a} + \frac{\sqrt{b^2 - 4ac}}{2a}\right\}\left\{x + \frac{b}{2a} - \frac{\sqrt{b^2 - 4ac}}{2a}\right\} = 0 \overset{(a\neq 0)}{\Longleftrightarrow}$$

$$x + \frac{b}{2a} + \frac{\sqrt{b^2 - 4ac}}{2a} = 0 \Leftrightarrow x = \frac{-b - \sqrt{b^2 - 4ac}}{2a}, \quad or \qquad (**)$$

$$x + \frac{b}{2a} - \frac{\sqrt{b^2 - 4ac}}{2a} = 0 \Leftrightarrow x = \frac{-b + \sqrt{b^2 - 4ac}}{2a} \qquad (***)$$

So, in case where the quantity $b^2 - 4ac > 0$, the quadratic equation $ax^2 + bx + c = 0$ has **two real and distinct roots** (solutions) given by formulas (**) and (***).

2) The quantity $b^2 - 4ac$ is equal to zero: Then the term $\frac{b^2-4ac}{4a^2}$ in (*) is zero, and

$$x = -\frac{b}{2a} \qquad (****)$$

So, in case $b^2 - 4ac = 0$, the quadratic equation has two real and equal roots. As we say, in this case, **the quadratic equation has a double, real root**.

3) The quantity $b^2 - 4ac$ is negative: Then equation (*) assumes the form

$$a\left\{\left(x + \frac{b}{2a}\right)^2 + \frac{4ac - b^2}{4a^2}\right\} = 0$$

and since both terms inside the braces are positive, the equation vanishes for no real value of x. The quadratic equation **has no real roots**. However, the equation **has complex roots**, since $\sqrt{b^2 - 4ac} = i\sqrt{4ac - b^2}$, and equation (*) in this case yields **two complex conjugate roots** given by the formulas

$$x = \frac{-b + i\sqrt{4ac - b^2}}{2a} \quad \text{or} \quad x = \frac{-b - i\sqrt{4ac - b^2}}{2a} \qquad (*****)$$

Remark: The quantity $\Delta \stackrel{\text{def}}{=} b^2 - 4ac$, which plays an important role in the solution of a quadratic equation, is called **the discriminant of the equation**.

If $\Delta > 0$, the equation has two real and distinct roots, if $\Delta = 0$, the equation has a double real root while if $\Delta < 0$, the equation has two complex conjugate roots. In **all cases** the solutions are given by the formula

$$r_{1,2} = \frac{-b \pm \sqrt{\Delta}}{2a} = \frac{-b \pm \sqrt{b^2 - 4ac}}{2a} \qquad (9-1-1)$$

(The first root r_1 is associated with the + sign in the square root, and the second root r_2 is associated with the − sign in the square root).

The sum and the product of the roots of a quadratic equation:

From formula (9-1-1) we obtain,

$$r_1 + r_2 = -\frac{b}{a} \quad \text{and} \quad r_1 r_2 = \frac{c}{a} \qquad (9-1-2)$$

Indeed,

$$r_1 + r_2 = \frac{-b + \sqrt{b^2 - 4ac}}{2a} + \frac{-b - \sqrt{b^2 - 4ac}}{2a} = \frac{-2b}{2a} = -\frac{b}{a}$$

$$r_1 r_2 = \left(\frac{-b + \sqrt{b^2 - 4ac}}{2a}\right)\left(\frac{-b - \sqrt{b^2 - 4ac}}{2a}\right) = \frac{(-b)^2 - \left(\sqrt{b^2 - 4ac}\right)^2}{4a^2} \Rightarrow$$

$$r_1 r_2 = \frac{b^2 - (b^2 - 4ac)}{4a^2} = \frac{4ac}{4a^2} = \frac{c}{a}$$

Remarks:

1) If the sum S and the product P of two (unknown) numbers are known, then these two numbers are the roots of the equation $x^2 - Sx + P = 0$, and are therefore found from equation (9-1-1).

Indeed, let a and b, be two numbers having sum S and product P, i.e.

$$\begin{cases} a + b = S \\ ab = P \end{cases} \Rightarrow a + \frac{P}{a} = S \Rightarrow a^2 - Sa + P = 0$$

and this shows that a is a root of the equation $x^2 - Sx + P = 0$. The other number b is the other root of the same equation.

2) The formation of an equation having roots the numbers r_1 and r_2.

Given two numbers r_1 and r_2 (real or complex), we may form a quadratic equation having roots these two given numbers. Indeed, the equation $(x - r_1)(x - r_2) = 0$ is a quadratic equation having roots r_1 and r_2, which eventually assumes the form

$$x^2 - (r_1 + r_2)x + r_1 r_2 = 0 \qquad (9-1-3)$$

Example 9-1-1: Solve the following equations: **1)** $x^2 - x - 6 = 0$, **2)** $x^2 + 12x + 36 = 0$, **3)** $x^2 - x + 1 = 0$.

Solution

1) $x^2 - x - 6 = 0$:

$$\Delta = (-1)^2 - 4 \cdot 1 \cdot (-6) = 1 + 24 = 25$$

$$r_{1,2} = \frac{-(-1) \pm \sqrt{25}}{2 \cdot 1} = \frac{1 \pm 5}{2} \Rightarrow \begin{cases} r_1 = 3 \\ r_2 = -2 \end{cases}$$

2) $x^2 + 12x + 36 = 0$:

$$\Delta = 12^2 - 4 \cdot 36 = 144 - 144 = 0$$

$$r_1 = r_2 = -\frac{12}{2 \cdot 1} = -6 \quad (Double\ root)$$

3) $x^2 - x + 1 = 0$:

$$\Delta = (-1)^2 - 4 \cdot 1 = -3$$

$$r_{1,2} = \frac{-(-1) \pm \sqrt{-3}}{2 \cdot 1} = \frac{1 \pm i\sqrt{3}}{2} \Rightarrow \begin{cases} r_1 = \dfrac{1 + i\sqrt{3}}{2} \\ r_2 = \dfrac{1 - i\sqrt{3}}{2} \end{cases} (complex\ conjugate)$$

Example 9-1-2: Find two numbers whose sum and product are $S = 5$ and $P = 7$, respectively.

Solution

The sought for numbers are the roots of the equation

$$x^2 - Sx + P = 0 \Leftrightarrow x^2 - 5x + 7 = 0 \Rightarrow$$

$$r_{1,2} = \frac{-(-5) \pm \sqrt{25 - 28}}{2} = \frac{5 \pm i\sqrt{3}}{2}$$

Example 9-1-3: Solve the equation $2x + \dfrac{18}{x+3} = 9$.

Solution

$$2x + \frac{18}{x+3} = 9 \Leftrightarrow 2x(x+3) + 18 = 9(x+3) \Leftrightarrow$$

$$2x^2 + 6x + 18 - 9x - 27 = 0 \Leftrightarrow 2x^2 - 3x - 9 = 0 \Rightarrow$$

$$x_{1,2} = \frac{-(-3) \pm \sqrt{(-3)^2 - 4 \cdot 2 \cdot (-9)}}{2 \cdot 2} = \frac{3 \pm \sqrt{81}}{4} = \frac{3 \pm 9}{4} \Rightarrow$$

$$x_1 = 3, \quad x_2 = -\frac{6}{4} = -\frac{3}{2}$$

Example 9-1-4: Determine the real values of the parameter k for which the two roots of the equation $kx^2 + 2kx + 9 = x^2 - 2x$ are equal.

Solution

$$kx^2 + 2kx + 9 = x^2 - 2x \Leftrightarrow (k-1)x^2 + 2(k+1)x + 9 = 0 \quad (*)$$

This equation (quadratic in x) will have **a double root** if and only if its discriminant is zero, i.e.

$$\Delta = 0 \Leftrightarrow 4(k+1)^2 - 4(k-1) \cdot 9 = 0 \Leftrightarrow 4(k^2 + 2k + 1 - 9k + 9) = 0 \Leftrightarrow$$

$$k^2 - 7k + 10 = 0 \Rightarrow k_{1,2} = \frac{7 \pm \sqrt{49 - 40}}{2} = \frac{7 \pm 3}{2} = 5 \text{ or } 2$$

Example 9-1-5: If a and b are the roots of $x^2 + x - 1 = 0$, show that $a^2 = b + 2$ and $b^2 = a + 2$, without finding the roots.

Solution

Since the number a is a root of $x^2 + x - 1 = 0$, it must satisfy the equation, i.e. $a^2 + a - 1 = 0$, or $a^2 = 1 - a$, (*)

From equation (9-1-2) we have $a + b = -1$, i.e. $a = -1 - b$, and substituting in (*) yields, $a^2 = 1 - (-1 - b) = b + 2$. Similarly, we show that $b^2 = a + 2$.

Example 9-1-6: If the real coefficients of the quadratic equation $ax^2 + bx + c = 0$ satisfy the inequality $|a| > |a + b + c|$, then if the equation has real roots, show that at least one of the roots lies between 0 and 2.

Solution

$$|a| > |a + b + c| \Leftrightarrow 1 > \frac{|a + b + c|}{|a|} = \left|\frac{a + b + c}{a}\right| = \left|1 + \frac{b}{a} + \frac{c}{a}\right| \overset{(9-1-2)}{\Longrightarrow}$$

$$1 > |1 - r_1 - r_2 + r_1 r_2| \Rightarrow |1 - r_1 - r_2(1 - r_1)| < 1 \Rightarrow$$

$$|(1 - r_1)(1 - r_2)| < 1 \Rightarrow |1 - r_1||1 - r_2| < 1 \qquad (*)$$

If the product of two positive numbers is less than 1, then **at least one** of the numbers must be smaller than 1. So equation (*) implies that at least one of the two factors must be smaller than 1, let us for example assume that the first factor is smaller than 1, i.e.

$$|1 - r_1| < 1 \Leftrightarrow -1 < 1 - r_1 < 1 \Leftrightarrow -2 < -r_1 < 0 \Leftrightarrow 2 > r_1 > 0$$

and this completes the proof.

Example 9-1-7: If (a_1, b_1) and (a_2, b_2) are two roots of the equation $\frac{x}{2x-24y} = \frac{y}{-x+4y}$ show that the number $\left(\frac{b_2 a_1 - b_1 a_2}{a_1 a_2}\right)^2$ is equal either to 0 or to 25/144.

Solution

$$\frac{x}{2x - 24y} = \frac{y}{-x + 4y} \Leftrightarrow x(-x + 4y) = y(2x - 24y) \Leftrightarrow$$

$$-x^2 + 4xy - 2xy + 24y^2 = 0 \Leftrightarrow 24y^2 + 2xy - x^2 = 0 \overset{(x \neq 0)}{\Longleftrightarrow}$$

$$24\left(\frac{y}{x}\right)^2 + 2\frac{y}{x} - 1 = 0 \qquad (*)$$

Equation (*) is quadratic in the ratio (y/x), so

$$\frac{y}{x} = \frac{-2 \pm \sqrt{4 - 4 \cdot 24 \cdot (-1)}}{2 \cdot 24} = \frac{-2 \pm \sqrt{100}}{48} = \frac{-2 \pm 10}{48}$$

so the possible values of the ratio (y/x) are either $\frac{8}{48} = \frac{1}{6}$, or, $-\frac{12}{48} = -\frac{1}{4}$.

By assumption, the possible values of the quotients $\left(\frac{b_1}{a_1}\right)$ and $\left(\frac{b_2}{a_2}\right)$ are either $\left(\frac{1}{6}\right)$ or $\left(-\frac{1}{4}\right)$.

The number $\left(\frac{b_2 a_1 - b_1 a_2}{a_1 a_2}\right)^2 = \left(\frac{b_2}{a_2} - \frac{b_1}{a_1}\right)^2$. There are two cases:

1) Either $\left(\frac{b_2}{a_2}\right)$ and $\left(\frac{b_1}{a_1}\right)$ represent the same number (either $\frac{1}{6}$ or $\left(-\frac{1}{4}\right)$), and then $\left(\frac{b_2}{a_2} - \frac{b_1}{a_1}\right)^2 = 0$, or

2) One of them is $\left(\frac{1}{6}\right)$ and then, the other will be $\left(-\frac{1}{4}\right)$, and in this case

$$\left(\frac{b_2}{a_2} - \frac{b_1}{a_1}\right)^2 = \left(\frac{1}{6} - \left(-\frac{1}{4}\right)\right)^2 = \left(\frac{1}{6} + \frac{1}{4}\right)^2 = \left(\frac{10}{24}\right)^2 = \left(\frac{5}{12}\right)^2 = \frac{25}{144}$$

Example 9-1-8: a) Under what conditions the roots of $ax^2 + bx + c = 0$ are rational numbers, provided that the coefficients a, b, c are rational? **b)** Show that the roots of $(a - b + c)x^2 + 2cx + (b + c - a) = 0$ are rational, provided that a, b, c are rational.

Solution

a) The roots of $ax^2 + bx + c = 0$ are given by the formula

$$r_{1,2} = \frac{-b \pm \sqrt{b^2 - 4ac}}{2a}$$

The roots will be rational if the discriminant $\Delta = b^2 - 4ac$ is **a perfect square of a rational number**, i.e. when $(b^2 - 4ac) = R^2$, where R is a rational number. Then the square root of Δ will be rational and equal to R, and the roots $r_{1,2}$ shall be rational.

b) The discriminant of the given equation is

$$\Delta = 4c^2 - 4(a - b + c)(b + c - a) = 4\{c^2 - (c + (a - b))(c - (a - b))\}$$

$$\Delta = 4\{c^2 - (c^2 - (a - b)^2)\} = 4(a - b)^2 = \{2(a - b)\}^2$$

and since the discriminant is a perfect square, the roots will be rational.

$$r_{1,2} = \frac{-2c \pm \sqrt{\{2(a-b)\}^2}}{2(a-b+c)} = \frac{-2c \pm 2(a-b)}{2(a-b+c)} \Rightarrow$$

$$r_1 = \frac{a-b-c}{a-b+c}, \quad r_2 = -1$$

(both rational, since by assumption a, b, c are rational).

Example 9-1-9: Provided that the roots of $ax^2 + 2bx + c = 0$, $(b \neq 0)$, are real and distinct, show that the roots of $(a+c)(ax^2 + 2bx + c) = 2(ac - b^2)(x^2 + 1)$ are complex numbers, and conversely.

Solution

The second equation is written equivalently:

$$(a+c)(ax^2 + 2bx + c) - 2(ac - b^2)(x^2 + 1) = 0 \Leftrightarrow$$

$$\big(a(a+c) - 2(ac - b^2)\big)x^2 + 2b(a+c)x + c(a+c) - 2(ac - b^2) = 0 \Leftrightarrow$$

$$(a^2 + 2b^2 - ac)x^2 + 2b(a+c)x + c^2 + 2b^2 - ac = 0 \quad (*)$$

Since $ax^2 + 2bx + c = 0$ has two real and distinct roots, its discriminant must be positive, i.e.

$$4b^2 - 4ac > 0 \Leftrightarrow b^2 - ac > 0 \quad (**)$$

The discriminant of the quadratic equation in (*) is:

$$\Delta = 4b^2(a+c)^2 - 4(a^2 + 2b^2 - ac)(c^2 + 2b^2 - ac) \Leftrightarrow$$

$$\Delta = 4\{b^2(a+c)^2 - (a^2 + 2b^2 - ac)(c^2 + 2b^2 - ac)\}$$

which after some simplifications leads to the following expression, (check it)

$$\Delta = -4(b^2 - ac)\{(a-c)^2 + 4b^2\} \quad (***)$$

Since $b^2 - ac > 0$, from equation (**) and $(a-c)^2 + 4b^2$ is always a positive number (even in the case $a = c$, since $b \neq 0$), the discriminant Δ in

(***) is negative, and this shows that the roots of the quadratic equation in (*) are complex.

Conversely, if the equation in (*) has complex roots, its dicriminant must be negative, i.e. $(b^2 - ac)$ in (***) must be positive, and this shows that the quadratic equation $ax^2 + 2bx + c = 0$ has two real and distinct roots.

PROBLEMS

9-1-1) Solve the equations:

$$x^2 + x - 6 = 0, x^2 - 4x + 5 = 0, 2y^2 - 2y - 1 = 0, t^2 - 4t + 7 = 0$$

(Ans: $(2, -3), (2 \pm i), ((1 \pm \sqrt{3})/2), (2 \pm i\sqrt{3})$).

9-1-2) Solve the equations:

$$3x^2 + 2x + 4 = 0, \qquad 2y^2 + 5y + 3 = 0, \qquad 2x + 3 = x^2 - 7$$

9-1-3) Solve the equations:

$$\frac{1}{x} + \frac{5}{x-3} = 5, \qquad \frac{3}{x-2} - \frac{1}{x+1} = 7, \qquad 5x + \frac{3}{2x} = -2$$

(Ans: $(21 \pm \sqrt{381})/10, (9 \pm \sqrt{613})/14, (-2 \pm i\sqrt{26})/10$).

9-1-4) Find the real roots of the equation: $x^2 - 2x + |5 - 3x| = 0$.

Hint: Consider the two cases, $5 - 3x \geq 0$ and $5 - 3x < 0$.

9-1-5) Show that if $ac < 0$, the equation $ax^2 + bx + c = 0$ has two real and distinct roots. Is the converse statement true?

9-1-6) Solve the following irrational equations:

(a) $\sqrt{7x + 14} = x + 2,$ (b) $\sqrt{\sqrt{5x - 1} - x} = 1$

(c) $x^2 - 3x + 5 - 3\sqrt{x^2 - 3x + 5} = 4,$ (d) $\sqrt{2x + 3} = 6 - x$

Hint: For (c) if we make the substitution $y = x^2 - 3x + 5 \geq 0$, the equation becomes $y - 3\sqrt{y} = 4$, i.e. $y - 4 = 3\sqrt{y}$, square both sides, find y (accepted root $y = 16$, $y = 1$ is rejected (why?)), and then find x, (the correct answer is $x_{1,2} = (3 \pm \sqrt{53})/2$).

9-1-7) Find p and q so that the only root of $3x^2 + 8qx - 24x + 5p = 10$ is the number zero, (i.e. zero is a double root of the equation).

(**Ans:** $p = 2, q = 3$).

9-1-8) Given that a, b, c represent the lengths of the sides of a triangle ABC, show that the roots of the equation $\frac{a^2}{x+1} - \frac{c^2}{x} = b^2$ are complex.

Hint: It suffices to show that the discriminant Δ is negative. Since a, b, c are the lengths of the sides of a triangle, they must satisfy the triangle inequalities, $a < b + c, b < c + a, c < a + b$.

9-1-9) Solve the equation:

$$2\sqrt{8x^2 - 26x + 16} - 1 = \frac{5 - 4\sqrt{8x^2 - 26x + 16}}{\sqrt{8x^2 - 26x + 16}}$$

(**Ans:** $5/2, 3/4$).

Hint: Set $y = \sqrt{8x^2 - 26x + 16} \geq 0$).

9-1-10) For which values of the parameter k the roots r_1, r_2 of the equation $x^2 - 5x + k = 0$, satisfy the relation

$$\left(r_1 - \frac{2}{r_2}\right)\left(r_2 - \frac{2}{r_1}\right) = \frac{8}{3}$$

Hint: See equation (9-1-2).

9-1-11) If the equations $x^2 - ax + ab = 0, x^2 - bx + bc = 0, x^2 - cx + ac = 0$, have roots $(r_1, r_2), (r_2, r_3), (r_3, r_1)$ respectively, show that the quantity $a + b + c = 0$.

Hint: Show first that $r_1 + r_2 + r_3 = (a + b + c)/2$, and then $r_1 = (a - b + c)/2$, etc, and finally consider the products $r_1 r_2$, etc.

9-1-12) If the real numbers a and b satisfy the inequality $|a| - |b| > 1$, ($b \neq 0$), show that the roots of $x^2 + ax + b = 0$ cannot be, both integers.

9-1-13) Form a quadratic equation having roots the reciprocals of the roots of the equation $36x^2 - 13x + 1 = 0$, **(Ans:** $x^2 - 13x + 36 = 0$).

Hint: Apply equation (9-1-3).

9-1-14) If r_1, r_2 are the roots of $x^2 + x - 1 = 0$, form an equation having roots the numbers $(r_1 + 1)/(r_2 + 1), (r_2 + 1)/(r_1 + 1)$.

(Ans: $x^2 + 3x + 1 = 0$).

9-1-15) Assuming that r_1, r_2 are the roots of $ax^2 + bx + c = 0$ and x_1, x_2 are the roots of $Ax^2 + Bx + C = 0$, form an equation having roots the numbers: $\frac{r_1}{x_1} + \frac{r_2}{x_2}, \frac{r_1}{x_2} + \frac{r_2}{x_1}$.

(Ans: $a^2 C x^2 - abBCx + (b^2 AC + B^2 ac - 4acAC) = 0$).

9-1-16) Show that the roots of $abc^2 x^2 + 3a^2 cx + b^2 cx - 6a^2 - ab + 2b^2 = 0$ are rational numbers, provided that a, b, c are rational.

Hint: See Example 9-1-8.

9-1-17) Solve the equation $\{(3-x)^3 + (4+x)^3\}/\{(3-x)^2 + (4+x)^2\} = 7$,   **(Ans:** $r_1 = 3, r_2 = -4$).

9-1-18) If r_1, r_2 are the roots of $kx^2 + 2(k+1)x - 3(k+2) = 0$, find a relation between r_1 and r_2 which is independent of k.

(Ans: $r_1 r_2 = 3 + 3(r_1 + r_2)$).

Hint: In the formulas expressing the sum and the product of two roots, eliminate k.

9-2) Symmetric expressions of the roots r_1 and r_2

a) A function of the two variables r_1 and r_2 is called **symmetric** when it does not change if r_1 is replaced by r_2 and r_2 by r_1. For example, $r_1 + r_2$ and $r_1 r_2$ are the **simplest symmetric expressions** of r_1 and r_2. Also, the expressions

$$r_1^3 + r_2^3, \qquad r_1^5 + 7r_1^3 r_2^3 + r_2^5, \qquad r_1^7 + r_2^7 - r_1^4 r_2^4$$

are symmetric expressions of r_1 and r_2, since they do not change if r_1 becomes r_2 and r_2 becomes r_1. The expressions $r_1^3 - r_2^3$ and $3r_1^2 - r_2 + r_1$, are **not** symmetric.

In general terms, a function $f(r_1, r_2)$ of the two roots **is symmetric** if $f(r_1, r_2) = f(r_2, r_1)$.

An expression which is a symmetric function of the two roots r_1, r_2 of the equation $ax^2 + bx + c = 0$, can be expressed in terms of the coefficients $a, b,$ and c. For example:

$$r_1^2 + r_2^2 = (r_1 + r_2)^2 - 2r_1 r_2 = \left(-\frac{b}{a}\right)^2 - 2\frac{c}{a} = \frac{b^2 - 2ac}{a^2}$$

$$r_1^3 + r_2^3 = (r_1 + r_2)^3 - 3r_1 r_2 (r_1 + r_2) = \left(-\frac{b}{a}\right)^3 - 3\frac{c}{a}\left(-\frac{b}{a}\right)$$
$$= \frac{-b^3 + 3abc}{a^3}$$

b) In general, if we define $S_n \stackrel{\text{def}}{=} r_1^n + r_2^n$, where $n = 3, 4, 5, 6, \cdots$, then the following formula holds true:

$$aS_n + bS_{n-1} + cS_{n-2} = 0 \tag{9-2-1}$$

Indeed, since r_1 and r_2 are the roots of $ax^2 + bx + c = 0$, then

$$\begin{cases} ar_1^2 + br_1 + c = 0 \\ ar_2^2 + br_2 + c = 0 \end{cases} \tag{*}$$

Multiplying the first equation in (*) by r_1^{n-2}, the second by r_2^{n-2} and adding term wise, we obtain,

$$a\underbrace{\left(r_1^n + r_2^n\right)}_{S_n} + b\underbrace{\left(r_1^{n-1} + r_2^{n-1}\right)}_{S_{n-1}} + c\underbrace{\left(r_1^{n-2} + r_2^{n-2}\right)}_{S_{n-2}} = 0$$

and this completes the proof.

Formula (9-2-1) is important, since if, for example, we know S_1 and S_2 we may compute S_3, and then S_4, and then S_5, etc.

c) Symmetric fractional expressions of r_1 and r_2.

Symmetric fractional expressions of the two roots, can similarly be expressed in terms of the coefficients a, b and c. For example,

$$\frac{1}{r_1^n} + \frac{1}{r_2^n} = \frac{r_1^n + r_2^n}{(r_1 r_2)^n} = \frac{S_n}{(r_1 r_2)^n}$$

d) In general, **any symmetric algebraic expression of the two roots can be expressed in terms of the coefficients a, b and c.**

Example 9-2-1: If r_1, r_2 are the roots of $ax^2 + bx + c = 0$, find $r_1^4 + r_2^4$ in terms of the coefficients a, b and c.

Solution

Application of (9-2-1) with $n = 4$ yields, $(S_4 = r_1^4 + r_2^4)$,

$$aS_4 + bS_3 + cS_2 = 0 \Rightarrow aS_4 + b\left(\frac{-b^3 + 3abc}{a^3}\right) + c\left(\frac{b^2 - 2ac}{a^2}\right) = 0 \Rightarrow$$

$$S_4 = \left\{\left(\frac{b}{a}\right)^2 - 2\frac{c}{a}\right\}^2 - 2\left(\frac{c}{a}\right)^2$$

Example 9-2-2: If r_1, r_2 are the roots of $x^2 - 3x + 2 = 0$, find (without solving the equation) the number $A = (1/r_1^3) + (1/r_2^3)$.

Solution

$$A = \frac{1}{r_1^3} + \frac{1}{r_2^3} = \frac{r_1^3 + r_2^3}{(r_1 r_2)^3} = \frac{(r_1 + r_2)^3 - 3r_1 r_2 (r_1 + r_2)}{(r_1 r_2)^3} \Rightarrow$$

$$A = \frac{3^3 - 3 \cdot 2 \cdot 3}{2^3} = \frac{27 - 18}{8} = \frac{9}{8}$$

PROBLEMS

9-2-1) If r_1, r_2 are the roots of $x^2 - 5x + 4 = 0$, find the number $A = r_1^4 r_2^2 - r_1^3 r_2^3 + r_1^2 r_2^4$, without solving the equation, **(Ans: 208)**.

9-2-2) If a and b are the roots of $2x^2 - 3x - 5 = 0$, find the numerical value of the following expressions, without solving the equation:

$$\frac{a^3}{b^3} + \frac{b^3}{a^3}, \quad a^2b^5 + a^5b^2, \quad \frac{2a^2 + 4ab + 2b^2}{a^2 + b^2}$$

9-2-3) Determine k so that the sum of the cubes of the roots of the equation $3x^2 - 3(k-1)x + k^2 = 0$ to be equal to -3, (**Ans:** $-\frac{1}{2}, 2$).

9-2-4) Determine k and λ so that roots of $(k-\lambda)x^2 - 2(k^2-\lambda^2)x + \lambda = 0$, to have sum 12 and product -2.

9-2-5) If r_1, r_2 are the roots of $ax^2 + bx + c = 0$, show that $(r_1 - r_2)^2 = \frac{b^2 - 4ac}{a^2}$.

9-2-6) If k, λ are the roots of $ax^2 + bx + c = 0$, express in terms of a and b the expressions:

$$3k^2\lambda + 3k\lambda^2 + 2k\lambda - k^3 - \lambda^3, \quad \frac{(3k+1)(3\lambda+1)}{(4k-5)(4\lambda-5)}$$

9-2-7) If k, λ are the roots of $ax^2 + bx + c = 0$, show that

$$\frac{2ax + b}{ax^2 + bx + c} = \frac{1}{x-k} + \frac{1}{x-\lambda}$$

9-2-8) If k, λ are the roots of the equation $(x-1)^2 = a(2x-3)$, show that the expression $(2k-3)(2\lambda-3)$ is independent of a.

9-2-9) Show that the difference of the roots of $5x^2 - 2(5k+3)x + 5k^2 + 6k + 1 = 0$, is independent of k, (see Pr. 9-2-5).

9-2-10) If k, λ are the roots of $x^2 - ax + b = 0$ and k^3, λ^3 are the roots of $x^2 - Ax + B = 0$, show that: $A = a(a^2 - 3b)$, $B = b^3$.

9-2-11) For which values of λ and μ the roots of $x^2 + 2(\lambda - \mu)x + 1 - 2\lambda\mu = 0$ are the numbers λ and μ?

(**Ans:** $\lambda = 1/3, \mu = 1$, or, $\lambda = -1/3, \mu = -1$).

9-2-12) Determine a and b, so that when the roots of $x^2 + ax + b = 0$ are increased by one, they become roots of the equation $x^2 - a^2x + ab = 0$.

(**Ans:** $a = 1, b = arbitrary, or, a = -2, b = -1$).

9-2-13) If x_1, x_2 are the roots of $x^2 + ax + 1 = 0$ and r_1, r_2 are the roots of $x^2 + bx + 1 = 0$, determine a and b provided that: **a)** $(x_1 - r_1)(x_2 - r_1)(x_1 + r_2)(x_2 + r_2) = 385$, and **b)** There exists a value of x which satisfies simultaneously the equations $x^2 + ax + 1 = x(52 + x)$ and $x^2 + bx + 1 = (x + 1)(x + 2)$, (**Ans:** $a = 24, b = 31$).

9-2-14) If r_1, r_2 are the roots of $x^2 + ax + b = 0$ and of $x^{2n} + a^n x^n + b^n = 0$ as well, where n is an even positive integer, show that the numbers r_1/r_2 and r_2/r_1 are roots of $x^n + 1 + (x + 1)^n = 0$.

9-3) Conditions between the coefficients of a quadratic equation when the roots satisfy a given relation

Given an equation $ax^2 + bx + c = 0$, we sometimes face the problem of determining the conditions that should exist between the coefficients a, b and c, so that the roots of the equation satisfy **a given relation** of the form $f(r_1, r_2) = 0$, or $f(r_1, r_2) > 0$.

The method of approach is outlined by means of the following illustrative examples.

Example 9-3-1: Determine k so that one root of the equation $3x^2 - 20x + 3k + 1 = 0$ is three times the other.

Solution

If one root is r, the other root shall be $3r$. Then, we have,

$$r + 3r = \frac{20}{3} \quad and \quad r \cdot 3r = \frac{3k + 1}{3} \qquad (*)$$

The first equation in (*) yields, $r = 5/3$, and substituting this value in the second equation in (*) results in

$$3 \cdot \left(\frac{5}{3}\right)^2 = \frac{3k + 1}{3} \Leftrightarrow \frac{25}{3} = \frac{3k + 1}{3} \Leftrightarrow k = 8$$

Indeed, for $k = 8$, the equation becomes $3x^2 - 20x + 25 = 0$. One root is $r_1 = 5/3$ and the other root is $r_2 = 5 = 3r_1$.

Example 9-3-2: Determine k so that the difference of the two roots of the equation $4x^2 + kx - 15 = 0$ to be equal to 4.

Solution

If r_1, r_2 are the roots of the equation we have:

$$\left\{r_1 + r_2 = -\frac{k}{4}, \quad r_1 r_2 = -\frac{15}{4}, \quad r_1 - r_2 = 4\right\} \qquad (*)$$

From the first and the third equations in (*) we get,

$$r_1 = \frac{16 - k}{8}, \quad r_2 = -\frac{16 + k}{8}$$

and substituting in the second equation in (*) results in

$$\frac{16 - k}{8} \cdot \left(-\frac{16 + k}{8}\right) = -\frac{15}{4} \Leftrightarrow (16 - k)(16 + k) = 240 \Leftrightarrow$$

$$16^2 - k^2 = 240 \Leftrightarrow k^2 = 16^2 - 240 = 256 - 240 = 16 \Leftrightarrow k = \pm 4$$

PROBLEMS

9-3-1) Determine k so that one of the roots of $9x^2 - 18(k - 1)x - 8k + 24 = 0$ to be twice the other, **(Ans:** $k = 2$ or $k = -1$).

9-3-2) Determine a and b, given that the difference of the roots of the equation $x^2 + ax + b = 0$ is equal to 3, while the difference of the cubes of the roots is equal to 279.

9-3-3) Determine k so that the two roots of $(k - 1)x^2 - (3k + 4)x + 12k + 3 = 0$ satisfy the relation $4r_1 - 5r_2 = 13$.

(Ans: $k = -1/4$ or $k = 223/125$).

9-3-4) In equations $x^2 - 5x + k + 2 = 0$ and $x^2 - 7x + 2k = 0$ determine k, provided that one root of the second equation is three times of one root of the first.

9-3-5) Let r_1, r_2 be the roots of $3x^2 - 2(a+b+c)x + ab + bc + ca = 0$, $(a, b, c \in \mathbb{R})$. **a)** Show that the roots are always real (for any values of a, b, c) and **b)** Find the relation between a, b, c provided that one roots is three times the other, **(Ans:** $a^2 + b^2 + c^2 = 2(ab + bc + ca)$).

9-3-6) Given that λ and μ are the roots of $ax^2 + 2bx + c = 0$, and $\lambda + \xi$ and $\mu + \xi$ are the roots of $Ax^2 + 2Bx + C = 0$, show that $A^2(b^2 - ac) = a^2(B^2 - AC)$.

9-3-7) For which values of the parameter λ the two roots r_1 and r_2 of the equation $x^2 + \lambda x + 1 = 0$ satisfy the inequalities $r_1 < 3r_2$ **and** $r_2 < 3r_1$.

(Ans: $-\dfrac{4}{\sqrt{3}} < \lambda < -2$).

Hint: The roots must be real, i.e. $\Delta = \lambda^2 - 4 > 0$. Also, the numbers $r_1 - 3r_2$ and $r_2 - 3r_1$ will both be negatives, when their sum is negative and their product is positive, etc.

CHAPTER 10: THE QUADRATIC TRINOMIAL

10-1) General concepts and definitions

Any second degree polynomial in x of the form $f(x) = ax^2 + bx + c$ is called a second degree trinomial in x, or just **a quadratic trinomial in x. The coefficients a, b, c are assumed to be real numbers**. The variable x can take on any real value, i.e. $-\infty < x < \infty$. The trinomial $f(x)$ is therefore a function of x, defined $\forall x \in \mathbb{R}$, provided that the coefficients a, b, c are known.

If x_0 is a given number, then $f(x_0) = ax_0^2 + bx_0 + c$ is the arithmetic value of the trinomial, corresponding to the value x_0 of the variable x.

For example, if $f(x) = x^2 - 2x + 5$, then $f(1) = 1^2 - 2 \cdot 1 + 5 = 4$, $f(-2) = (-2)^2 - 2 \cdot (-2) + 5 = 4 + 4 + 5 = 13$, etc.

A number r is called **a root of the trinomial** $f(x) = ax^2 + bx + c$ if $f(r) = 0$, i.e. if $ar^2 + br + c = 0$. We thus see that the roots of the quadratic trinomial $ax^2 + bx + c$ are the roots of the quadratic equation $ax^2 + bx + c = 0$. **The roots of the trinomial can be either real or complex numbers**, since the roots of the corresponding quadratic equation can be either real or complex, depending on the sign of the discriminant $\Delta = b^2 - 4ac$.

If r_1, r_2 are the roots of the trinomial $f(x) = ax^2 + bx + c$, then as we know, (see equation (9-1-2)), the following equations hold true:

$$\left\{ r_1 + r_2 = -\frac{b}{a}, \quad r_1 r_2 = \frac{c}{a} \right\} \qquad (10-1-1)$$

Any trinomial $f(x) = ax^2 + bx + c$ can be expressed in the following two, equivalent forms:

1) The first form results immediately from equation (*) in section 9-1 and is the following,

$$f(x) = ax^2 + bx + c \equiv a\left\{ \left(x + \frac{b}{2a}\right)^2 + \frac{4ac - b^2}{4a^2} \right\} \qquad (10-1-2)$$

2) The second form expresses the trinomial in terms of its two roots r_1 and r_2 as follows,

$$f(x) = ax^2 + bx + c \equiv a(x - r_1)(x - r_2) \qquad (10-1-3)$$

The proof of (10-1-3) is easy:

$$ax^2 + bx + c = a\left\{x^2 + \frac{b}{a}x + \frac{c}{a}\right\} \xRightarrow{(10-1-1)}$$

$$ax^2 + bx + c = a\{x^2 - (r_1 + r_2)x + r_1 r_2\} = a\{x^2 - r_1 x - r_2 x + r_1 r_2\} \Rightarrow$$

$$ax^2 + bx + c = a\{x(x - r_1) - r_2(x - r_1)\} = a(x - r_1)(x - r_2)$$

Example 10-1-1: Which ones of the following trinomials have real roots and which have complex roots?

$$x^2 - 2x + 3, \qquad 2x^2 + x - 3, \qquad 3x^2 + x + 2, \qquad 4x^2 - 12x + 9$$

Solution

The kind of the roots (real or complex) depends on the discriminant Δ.

1) $x^2 - 2x + 3$: Δ $= (-2)^2 - 4 \cdot 1 \cdot 3 = 4 - 12 = -8 < 0$, complex roots.

2) $2x^2 + x - 3$: Δ $= 1^2 - 4 \cdot 2 \cdot (-3) = 1 + 24 = 25 > 0$, real roots.

3) $3x^2 + x + 2$: Δ $= 1^2 - 4 \cdot 3 \cdot 2 = 1 - 24 = -23 < 0$, complex roots.

4) $4x^2 - 12x + 9$: Δ $= (-12)^2 - 4 \cdot 4 \cdot 9 = 144 - 144 = 0$, real roots.

Example 10-1-2: Express the trinomials $f(x) = 2x^2 - 3x + 1, g(x) = 4x^2 - 12x + 9$ and $w(x) = x^2 + x + 1$ in the forms given by the formulas (10-1-2) and (10-1-3).

Solution

1) $f(x) = 2x^2 - 3x + 1$, $r_{1,2} = \frac{3 \pm \sqrt{9-8}}{4} = \frac{3 \pm 1}{4}$, $r_1 = 1, r_2 = \frac{1}{2}$.

$$f(x) = a\left\{\left(x + \frac{b}{2a}\right)^2 + \frac{4ac - b^2}{4a^2}\right\} = 2\left\{\left(x - \frac{3}{4}\right)^2 - \frac{1}{16}\right\}$$

$$f(x) = a(x - r_1)(x - r_2) = 2(x - 1)\left(x - \frac{1}{2}\right)$$

2) $g(x) = 4x^2 - 12x + 9$, $r_{1,2} = \frac{12 \pm \sqrt{(-12)^2 - 4 \cdot 4 \cdot 9}}{8} = \frac{12 \pm 0}{8}$, $r_1 = r_2 = \frac{3}{2}$.

$$g(x) = a\left\{\left(x + \frac{b}{2a}\right)^2 + \frac{4ac - b^2}{4a^2}\right\} = 4\left(x - \frac{12}{8}\right)^2 = 4\left(x - \frac{3}{2}\right)^2$$

$$g(x) = a(x - r_1)(x - r_2) = 4\left(x - \frac{3}{2}\right)\left(x - \frac{3}{2}\right) = 4\left(x - \frac{3}{2}\right)^2$$

3) $w(x) = x^2 + x + 1$, $r_{1,2} = \frac{-1 \pm \sqrt{1-4}}{2} = \frac{-1 \pm i\sqrt{3}}{2}$, $r_1 = \frac{-1 + i\sqrt{3}}{2}$, $r_2 = \frac{-1 - i\sqrt{3}}{2}$.

$$w(x) = a\left\{\left(x + \frac{b}{2a}\right)^2 + \frac{4ac - b^2}{4a^2}\right\} = \left(x + \frac{1}{2}\right)^2 + \frac{3}{4}$$

$$w(x) = a(x - r_1)(x - r_2) = \left(x + \frac{1 - i\sqrt{3}}{2}\right)\left(x + \frac{1 + i\sqrt{3}}{2}\right)$$

PROBLEMS

10-1-1) Which ones of the following trinomials have real roots and which have complex roots?

$x^2 - x + 10$, $\quad 3x^2 + 7x - 2$, $\quad 2x^2 + 37x - 24$, $\quad x^2 - 3x + 10$

(Ans: Real roots, the second and the third, Complex roots, the first and the fourth).

10-1-2) Express the trinomial $f(x) = (a^2 - 4b^2)x^2 + 2(a^3 + 2b^3)x + a^4 - b^4$ as a product of two, first degree in x factors, i.e. in the form given by formula (10-1-3).

10-2) The sign of the quadratic trinomial as x runs over the set of real numbers

Let us consider the quadratic trinomial $f(x) = ax^2 + bx + c$. We assume that **the coefficients a, b, c are real numbers**, and that the variable x runs over the set of real numbers, i.e. $-\infty < x < \infty$. According to formula (10-1-2), the trinomial may be expressed in the form

$$f(x) = ax^2 + bx + c \equiv a\left\{\left(x + \frac{b}{2a}\right)^2 + \frac{4ac - b^2}{4a^2}\right\} \qquad (*)$$

The following theorem is of great importance in the theory of the quadratic trinomial.

Theorem 10-2-1: a) If the trinomial $f(x)$ has complex roots, then the sign of the trinomial as x varies from $-\infty$ to ∞, coincides with the sign of the coefficient a, i.e. if a is positive, then $f(x) > 0$, $\forall x \in \mathbb{R}$, if a is negative, then $f(x) < 0$, $\forall x \in \mathbb{R}$.

b) If the trinomial $f(x)$ has a double (real) root r, then the sign of the trinomial as x varies from $-\infty$ to ∞, coincides with the sign of the coefficient a, i.e. if a is positive, then $f(x) > 0$, $\forall x \in \mathbb{R} - \{r\}$, if a is negative, then $f(x) < 0$, $\forall x \in \mathbb{R} - \{r\}$, while $f(r) = 0$.

c) The trinomial $f(x)$ has two real and distinct roots r_1 and r_2, (assume $r_1 < r_2$): Then if x takes on values outside the interval $(r_1 \ldots r_2)$, (i.e. $-\infty < x < r_1$ or $r_2 < x < \infty$), the sign of the trinomial coincides with the sign of a, while if x takes on values between the roots (i.e. $r_1 < x < r_2$) the sign of the trinomial is the opposite of the sign of a. At $x = r_1$ or $x = r_2$, then obviously, $f(r_1) = f(r_2) = 0$.

Proof: a) The trinomial $f(x)$ has complex roots when its discriminant Δ is negative, i.e. $\Delta = b^2 - 4ac < 0$, or equivalently $4ac - b^2 > 0$. According to formula (*),

$$f(x) = ax^2 + bx + c \equiv a\left\{\underbrace{\left(x + \frac{b}{2a}\right)^2}_{\geq 0} + \underbrace{\frac{4ac - b^2}{4a^2}}_{>0}\right\} \qquad (**)$$

The quantity inside the braces is always positive, since $\frac{4ac - b^2}{4a^2} > 0$ and $\left(x + \frac{b}{2a}\right)^2 \geq 0$, (actually $\left(x + \frac{b}{2a}\right)^2 > 0$ if $x \neq -\frac{b}{2a}$ and is zero when $x = -\frac{b}{2a}$). If $a > 0$ then $f(x) > 0$, $\forall x \in \mathbb{R}$ while if $a < 0$ then $f(x) < 0$, $\forall x \in \mathbb{R}$.

b) Similar reasoning as in (a).

c) If the trinomial has two real and distinct roots r_1 and r_2, $(r_1 < r_2)$, then we may write (see formula (10-1-3))

$$f(x) = a(x - r_1)(x - r_2) \qquad (***)$$

These types of inequalities were considered in section 3-4.

Fig. 10-1: Sign of the product $(x - r_1)(x - r_2)$.

When $-\infty < x < r_1$ or $r_2 < x < \infty$, the product $(x - r_1)(x - r_2)$ is positive, and hence the trinomial $a(x - r_1)(x - r_2)$ is positive if $a > 0$, or is negative if $a < 0$. When $r_1 < x < r_2$ (x between the two roots), then the product $(x - r_1)(x - r_2)$ is negative, and therefore the trinomial $a(x - r_1)(x - r_2)$ is negative if $a > 0$ or is positive if $a < 0$.

Corollary: There is only **one case** where the trinomial may assume numerical values with sign opposite to the sign of the coefficient a, and this is the case where the trinomial has two real and distinct roots. **So, if we (somehow) find that there exists a number ξ such that $af(\xi) < 0$, then we conclude that the trinomial has two real and distinct roots and that ξ lies between the two roots**.

Example 10-2-1: For what real values of x the trinomial $f(x) = x^2 - 3x + 2$ becomes positive, negative or zero?

Solution

The roots of the trinomial are $r_1 = 1$ and $r_2 = 2$. The coefficient $a = 1$ (positive). According to Theorem 10-2-1, we have:

$$f(x) > 0, \quad \forall x \in (-\infty < x < 1) \cup (2 < x < \infty)$$

$$f(x) < 0, \quad \forall x \in (1 < x < 2)$$

$$f(1) = f(2) = 0$$

Example 10-2-2: Show that $f(x) = -x^2 + 2x - 3$ is always negative, for all real values of x.

Solution

The discriminant of $f(x)$ is $\Delta = 2^2 - 4 \cdot (-1) \cdot (-3) = -8 < 0$, so the trinomial has complex roots, while $a = -1 < 0$. According to Theorem 10-2-1, $f(x) < 0, \forall x \in \mathbb{R}$.

Example 10-2-3: Find the real numbers x and y that satisfy the equation $(x+3)^2(3y^2 - 5y + 9) + (y-2)^{10}(x^2 - x + 1) = 0$.

Solution

The term $(x+3)^2(3y^2 - 5y + 9)$ is **a non-negative number**, (i.e. it can be either positive or zero, cannot be negative). This is so since $(x+3)^2 \geq 0$, while $3y^2 - 5y + 9 > 0, \forall y \in \mathbb{R}$, $(\Delta = 25 - 108 = -83 < 0, a = 3 > 0)$.

Similarly, the term $(y-2)^{10}(x^2 - x + 1)$ is **non-negative**, (check it), and application of Theorem 4-1-2, implies that

$$\begin{cases} (x+3)^2(3y^2 - 5y + 9) = 0 \\ \text{and} \\ (y-2)^{10}(x^2 - x + 1) \end{cases} \Rightarrow \{x = -3, \quad y = 2\}$$

Example 10-2-4: Without solving the equation, show that the trinomial $f(x) = (x-1)(x-5) - 7$ has two real and distinct roots.

Solution

The leading coefficient (the coefficient of x^2) is $a = 1$, and since $af(1) = -7 < 0$, the trinomial has two real and distinct roots, and 1 lies between these roots, (by virtue of the corollary of Theorem 10-2-1).

PROBLEMS

10-2-1) For what values of x the expressions $\sqrt{2x^2 - 7x + 3}$ and $\sqrt{x^2 - x + 2}$ are real numbers? **(Ans:** $x \leq 1/2$ or $x \geq 3, -\infty < x < \infty$).

10-2-2) Determine the domain of definition of the function

$$y = 7\sqrt{x^2 - 4x + 3} - 5\sqrt{-x^2 + 6x + 8}$$

(Ans: $3 - \sqrt{17} \leq x \leq 1$, or, $3 \leq x \leq 3 + \sqrt{17}$).

10-2-3) Consider the trinomial $f(x) = x^2 - 5x + 10$; **a)** Are there real values of x, such that $f(x) < 0$? **b)** Are there complex values of x such that $f(x) < 0$?

(**Ans: a)** No real values of x, since $f(x) > 0, \forall x \in \mathbb{R}$, **b)** Yes, all complex numbers of the form $x = \frac{5}{2} + ki$, where $-\frac{\sqrt{15}}{2} < k < \frac{\sqrt{15}}{2}$).

Hint: Assume that $x = a + ib$ ($b \neq 0$) is such that $f(x) < 0$, i.e. $f(a + ib) < 0$, which after carrying out the calculations assumes the form:

$$a^2 - b^2 - 5a + 10 + i(2ab - 5b) < 0 \quad (*)$$

However, **since inequalities between complex numbers do not exist**, equation (*) will be valid provided that

$$\begin{cases} a^2 - b^2 - 5a + 10 < 0 \\ \text{and} \\ 2ab - 5b = 0 \end{cases} \overset{(b \neq 0)}{\iff} \begin{cases} a^2 - b^2 - 5a + 10 < 0 \\ \text{and} \\ 2a - 5 = 0 \end{cases}$$

10-2-4) Show that the trinomial

$$(a + c)(ax^2 + 2bx + c) - 2(ac - b^2)(x^2 + 1)$$

retains the same sign for all real values of x, provided that the equation $ax^2 + 2bx + c = 0$ has real roots.

10-3) Quadratic inequalities

Using Th. 10-2-1 we may solve inequalities of the form $ax^2 + bx + c > 0$ or $ax^2 + bx + c < 0$, as shown in the following illustrative examples.

Example 10-3-1: Solve the inequality $x^2 - 6x + 5 < 0$.

Solution

Since $a = 1 > 0$, the values of x that make the trinomial negative, according to Theorem 10-2-1, must lie between the roots of the trinomial, which are $r_1 = 1$ and $r_2 = 5$, i.e. the solution of the inequality is the totality of numbers x that satisfy $1 < x < 5$.

Example 10-3-2: Solve the inequality $\frac{2x^2-4x+5}{x^2+2} > 1$.

Solution

$$\frac{2x^2 - 4x + 5}{x^2 + 2} > 1 \Leftrightarrow \frac{2x^2 - 4x + 5}{x^2 + 2} - 1 > 0 \Leftrightarrow$$

$$\frac{2x^2 - 4x + 5 - x^2 - 2}{x^2 + 2} > 0 \Leftrightarrow (x^2 - 4x + 3)(x^2 + 2) > 0 \xLeftrightarrow{(x^2+2>0)}$$

$$x^2 - 4x + 3 > 0 \Leftrightarrow (x-1)(x-3) > 0 \Leftrightarrow x < 1 \text{ or } x > 3$$

PROBLEMS

10-3-1) For which values of k the roots of the trinomial $f(x) = kx^2 + 2(k+1)x - 3(k+2)$ are real?,   **(Ans:** $k \leq \frac{-2-\sqrt{3}}{2}$ or $k \geq \frac{-2+\sqrt{3}}{2}$).

Hint: The roots will be real if the discriminant is positive.

10-3-2) Solve the inequalities:

$$2x^2 + 3x - 1 < 0, \quad 5x^2 + 7x + 10 > 0, \quad \frac{x^2 - 4x + 3}{x + 5} < 0$$

10-3-3) Find the real and integer numbers x and y which satisfy the equation $(y-1)x^2 + (y+1)x + y = 0$.

(Ans: $(0,0), (1,0), (-2,2), (-1,2)$).

Hint: If we consider the given equation as a quadratic equation in x, then

$$x_{1,2} = \frac{-(y+1) \pm \sqrt{(y+1)^2 - 4(y-1)y}}{2(y-1)} = \frac{-(y+1) \pm \sqrt{-3y^2 + 6y + 1}}{2(y-1)}$$

Since x must be real, **the quantity under the radical must be positive or zero**, i.e. $-3y^2 + 6y + 1 \geq 0$, and according to Theorem 10-2-1, y must lie between the roots of $-3y^2 + 6y + 1 = 0$, i.e.

$$\frac{6-\sqrt{48}}{6} \leq y \leq \frac{6+\sqrt{48}}{6} \quad or \quad -0.15... \leq y \leq 2.15...$$

and since we seek **the integer solutions**, the allowed values of y are $\{0, 2\}$, (the value $y = 1$ is rejected since for $y = 1$, the original equation becomes $2x + 1 = 0$, $x = -1/2$, not integer). We set successively, $y = 0$ and $y = 2$ in the original equation and determine the corresponding integer values of x, etc.

10-3-4) Find the domain of definition of the function $y = \sqrt{|x^2 + 8x + 9| - 24}$.

Hint: $|x^2 + 8x + 9| - 24 \geq 0 \Leftrightarrow |x^2 + 8x + 9| \geq 24 \Leftrightarrow (x^2 + 8x + 9)^2 \geq 24^2 \Leftrightarrow (x^2 + 8x + 9)^2 - 24^2 \geq 0 \Leftrightarrow (x^2 + 8x + 9 + 24)(x^2 + 8x + 9 - 24) \geq 0$, etc.

10-4) Position of a real number relative to the roots of a trinomial

Let $f(x) = ax^2 + bx + c$ be a trinomial with **real roots** r_1 and r_2 (assume $r_1 < r_2$), and ξ be **a real number**. There are three cases regarding the position of the number ξ relative to the two roots:

$$\begin{cases} Case\ 1: & -\infty < \xi < r_1 < r_2 \\ Case\ 2: & r_1 < r_2 < \xi < \infty \\ Case\ 3: & r_1 < \xi < r_2 \end{cases} \qquad (*)$$

Each case is characterized by certain conditions between the number and the coefficients of the trinomial.

Let us first consider case 1, where ξ is smaller than the smallest root r_1. Since the roots are real, the discriminant $\Delta = b^2 - 4ac$ must be positive, i.e. $\Delta = b^2 - 4ac > 0$. Also, since ξ lies out of the interval (r_1, r_2), we must have $af(\xi) > 0$, (Theorem 10-2-1). Finally, since $\xi < r_1$ and $\xi < r_2$, we have by term wise addition, $2\xi < (r_1 + r_2)$, or $\xi < (r_1 + r_2)/2$, or even $\xi < -b/2a$.

In summary, if $-\infty < \xi < r_1 < r_2$, we must necessarily have (**necessary conditions**),

$$\left\{ \Delta > 0 \quad af(\xi) > 0 \quad \xi < -\frac{b}{2a} \right\} \qquad (10-4-1)$$

But conditions (10-4-1) are also **sufficient** for ξ to be smaller than $r_1 < r_2$. For, from $\Delta > 0$, the roots are real, from $af(\xi) > 0$ the number ξ lies outside of the interval (r_1, r_2), and from the third condition $\xi < -b/2a$, the number ξ cannot be greater than r_2, (since then we would have $\xi > -b/2a$ (why?), so the only possibility left is to have $-\infty < \xi < r_1$.

Similarly, we may show that **the necessary and sufficient conditions** for case 2 are,

$$\left\{\Delta > 0 \quad af(\xi) > 0 \quad \xi > -\frac{b}{2a}\right\} \qquad (10-4-2)$$

and for case 3,

$$af(\xi) < 0 \qquad (10-4-3)$$

Example 10-4-1: Find the position of the number 7 relative to the roots of the trinomial $x^2 - 6x + 5$, without finding the roots.

Solution

$$\Delta = b^2 - 4ac = 36 - 20 = 16 > 0$$

$$af(7) = 1 \cdot (7^2 - 6 \cdot 7 + 5) = 12 > 0$$

$$-\frac{b}{2a} = -\frac{-6}{2} = 3, \quad \text{and } \xi = 7 > 3$$

so, according to (10-4-2), $\xi > r_2$.

Example 10-4-2: For which values of k both roots of $f(x) = x^2 - (5 + k)x + 3k + \frac{125}{4}$ exceed the number 1?

Solution

If r_1, r_2 the real roots of the trinomial $f(x)$, we must have: $1 < r_1 < r_2$, and the necessary and sufficient conditions for this are given by (10-4-1), i.e.

$$\Delta > 0 \quad af(1) > 0 \quad 1 < -\frac{b}{2a} \qquad (*)$$

The first condition in equation (*) yields,

$$\Delta = (5+k)^2 - 4\left(3k + \frac{125}{4}\right) > 0 \Rightarrow k^2 - 2k - 100 > 0 \Rightarrow$$

$$k < 1 - \sqrt{101} \quad or \quad k > 1 + \sqrt{101} \qquad (**)$$

The second condition in (*) yields,

$$1 \cdot f(1) > 0 \Rightarrow 1 - (5+k) + 3k + \frac{125}{4} > 0 \Rightarrow k > -\frac{109}{8} \qquad (***)$$

The third condition in (*) yields,

$$1 < -\frac{b}{2a} \Rightarrow 1 < -\frac{-(5+k)}{2} \Rightarrow k > -3 \qquad (****)$$

The three inequalities in (**), (***), (****) are satisfied **simultaneously** in the interval $(1 + \sqrt{101} < k < \infty)$.

PROBLEMS

10-4-1) Find the position of the number 7 relative to the roots of the trinomial $x^2 - 9x + 20$, **(Ans:** $r_1 < r_2 < 7$).

10-4-2) Find the values of the parameter k, so that the number 2 lies between the two real roots of the equation $(k+1)x^2 - 3x + k - 8 = 0$.

10-4-3) Find the values of the parameter k, so that the number 3 lies between the two real roots of the equation $(k-2)x^2 - (k-3)x + 15 = 0$.

(Ans: $-1 < k < 2$).

10-4-4) For which values of the parameter λ both real roots of the trinomial $(\lambda - 1)x^2 - (2\lambda + 3)x + 5\lambda$ are smaller than 2?

10-4-5) For which values of the parameter λ both real roots of the trinomial $(\lambda + 1)x^2 + (\lambda - 3)x + \lambda - 3$ are smaller than 2?

(Ans: $-7/3 < \lambda < -1$, or $5/5 < \lambda < 3$).

10-4-6) If r_1, r_2 are the real roots of the trinomial $ax^2 + (b + 2ak)x + ak^2 + bk + c$, show that both roots lie between the roots of the trinomial $ax^2 + bx + c + k(2ax + b)$.

10-4-7) Show that the trinomial $f(x) = \lambda^2(b^2 + x) + \mu^2(a^2 + x) - (a^2 + x)(b^2 + x)$, where $\lambda\mu(a^2 - b^2) \neq 0$, has two real and distinct roots and that one of the two roots lies between $-a^2$ and $-b^2$.

10-5) Position of two real numbers relative to the roots of a trinomial

There are **six different combinations** related to the position of two real numbers ξ and η, relative to the roots r_1 and r_2 of a trinomial $f(x) = ax^2 + bx + c$, (assumed to be **real and distinct**). Assuming $\xi < \eta$ and $r_1 < r_2$ the six different combinations are:

$$\begin{cases} 1)\ \xi < \eta < r_1 < r_2 & 4)\ r_1 < \xi < \eta < r_2 \\ 2)\ \xi < r_1 < \eta < r_2 & 5)\ r_1 < \xi < r_2 < \eta \\ 3)\ \xi < r_1 < r_2 < \eta & 6)\ r_1 < r_2 < \xi < \eta \end{cases} \qquad (10-5-1)$$

For each one of these six combinations there exists an appropriate condition between the coefficients a, b, c of the trinomial.

In general, the position of the two numbers ξ and η relative to the roots r_1 and r_2 depends on the **sign** of the following five quantities:

$$\left\{ \Delta,\ af(\xi),\ af(\eta),\ \xi + \frac{b}{2a},\ \eta + \frac{b}{2a} \right\} \qquad (10-5-2)$$

For example, let us find the conditions under which the position of ξ and η relative to the roots is: $r_1 < r_2 < \xi < \eta$, (case 6 in equation (10-5-1)).

First of all, the roots must be real, meaning that $\Delta > 0$. Also, since ξ and η lie outside the interval $(r_1 \ldots r_2)$, we must have $af(\xi) > 0$ and $af(\eta) > 0$. Finally, since $\xi > r_1$ and $\xi > r_2$, $2\xi > (r_1 + r_2)$, i.e. $\xi > (r_1 + r_2)/2$, i.e. $\xi > -b/2a$ and $\left(\xi + \frac{b}{2a}\right) > 0$, and for similar reasons $\left(\eta + \frac{b}{2a}\right) > 0$.

In summary, provided that

$$\Delta > 0,\quad af(\xi) > 0,\quad af(\eta) > 0,\quad \xi + \frac{b}{2a} > 0,\quad \eta + \frac{b}{2a} > 0$$

the position of ξ and η relative to the roots is: $r_1 < r_2 < \xi < \eta$.

With similar arguments we can find the conditions for the other five combinations in (10-5-1).

Theorem 10-5-1: Provided that $f(\xi)f(\eta) < 0$, the roots of the trinomial $f(x) = ax^2 + bx + c$ are real and distinct and one of the roots lies between ξ and η, and the other root lies outside of the interval $(\xi \ldots \eta)$.

Proof: Without loss of generality, we may assume that the coefficient a is positive, $(a > 0)$.

$$f(\xi)f(\eta) < 0 \Leftrightarrow a^2 f(\xi)f(\eta) < 0 \Leftrightarrow \{af(\xi)\}\{af(\eta)\} < 0 \quad (*)$$

Since the product in (*) is negative, one factor is positive and the other factor is negative. Let us assume that

$$af(\xi) < 0 \quad \text{and} \quad af(\eta) > 0 \quad (**)$$

The first inequality in (*) implies that the roots r_1 and r_2 of $f(x)$ are real and distinct (by virtue of the corollary of Theorem 10-2-1), and ξ lies between the two roots, i.e. $r_1 < \xi < r_2$.

The second inequality in (**) implies that η lies outside the interval of the two roots, i.e. either $\eta < r_1 < \xi < r_2$ or $r_1 < \xi < r_2 < \eta$, and this completes the proof.

Remark: Sometimes, Theorem 10-5-1 may be used to show that the roots of a trinomial are real and distinct, **without computing the discriminant Δ.**

Example 10-5-1: For which values of the parameter k the two real roots of the trinomial $f(x) = (k-20)x^2 - (k-3)x + 8$, lie in the interval $(0,4)$?

Solution

We are in case 3, in equation (10-5-1), $(\xi = 0, \eta = 4)$. The conditions that should be satisfied **simultaneously**, for $0 < r_1 < r_2 < 4$ to hold true, are the following:

$$\left\{\Delta > 0, \quad af(0) > 0, \quad af(4) > 0, \quad 0 + \frac{b}{2a} < 0, \quad 4 + \frac{b}{2a} > 0\right\} \quad (*)$$

where $a = k - 20, b = -(k-3), c = 8$.

Solving the inequalities in (*) and finding the interval where all five inequalities **are satisfied simultaneously**, we find $k > 25$, (let the reader verify the calculations).

Example 10-5-2: For which values of the parameter k, only one of the roots of the trinomial $f(x) = (k-2)x^2 - (k-3)x + 15$, lies within the interval $(-1, 3)$?

Solution

According to Theorem 10-5-1, we must have,

$$f(-1)f(3) < 0 \Leftrightarrow 2(k+5) \cdot 6(k+1) < 0 \Leftrightarrow -5 < k < -1$$

Example 10-5-3: Without the use of the discriminant, show that the equation

$$\frac{b+c}{x-a} + \frac{c+a}{x-b} + \frac{a+b}{x-c} = 0$$

has real and distinct roots, provided that $a^2 < b^2 < c^2$.

Solution

Eliminating the denominators we obtain the equivalent equation

$$\begin{aligned} f(x) = (b+c)(x-b)(x-c) + (c+a)(x-a)(x-c) \\ + (a+b)(x-a)(x-b) = 0 \end{aligned} \quad (*)$$

We notice that $f(a) = (b+c)(a-b)(a-c)$, while $f(b) = (c+a)(b-a)(b-c)$. The product $f(a)f(b)$ is thus equal to:

$$f(a)f(b) = (b+c)(a-b)(a-c)(c+a)(b-a)(b-c) \Rightarrow$$

$$f(a)f(b) = \underbrace{(b^2 - c^2)}_{-} \underbrace{\{-(a-b)^2\}}_{-} \underbrace{(a^2 - c^2)}_{-} < 0$$

and Theorem 10-5-1 implies that $f(x)$ has two real and distinct roots (and one root lies between a and b).

PROBLEMS

10-5-1) For which values of the parameter k the trinomial $f(x) = (k-2)x^2 - 2(k+3)x + 4k$ has one root greater than 3 and the other root smaller than 2 ?, **(Ans: $2 < k < 5$).**

Hint: $af(2) < 0$ and $af(3) < 0$.

10-5-2) For which values of the parameter k one root of the trinomial $f(x) = (k-1)x^2 - (k-5)x + k - 1$ lies between 0 and 2, while the other is greater than 10?

10-5-3) Show that the following equation has two real and distinct roots,

$$\frac{x+1}{x-1} + \frac{x+2}{x-2} + \frac{x+3}{x-3} = 3$$

Hint: Show that the given equation is equivalent to

$$f(x) = 2(x-2)(x-3) + 4(x-1)(x-3) + 4(x-2)(x-3) = 0$$

and then show that $f(1)f(2) < 0$.

10-5-4) Show that the trinomials $g(x) = f(x) + a(2x + p)$ and $w(x) = f(x) + (x+a)(2x+p)$ have roots real and distinct for all real values of a, provided that $f(x) = x^2 + px + q$ has roots real and distinct.

Hint: If r_1, r_2 are the real and distinct roots of $f(x)$, show that the product $g(r_1)g(r_2) < 0$ and similarly $w(r_1)w(r_2) < 0$, and the apply Theorem 10-5-1.

10-5-5) Find the range of variation of the real numbers x and y that satisfy the equation: $y^2 + 9x^2 + 2xy - 92x - 20y + 244 = 0$.

(Ans: $3 \leq x \leq 6$, $1 \leq y \leq 10$).

10-5-6) Show that the roots of the trinomial

$$f(x) = (x-1)(x-2) + (x-2)(x+1) + (x+1)(x-1)$$

are real and distinct, and that $-1 < r_1 < 0 < 1 < r_2 < 2$.

10-5-7) Find the real solutions x, y, z of the equation

$$|x-4|(7y^2 - 3y + 30) + (x - y + 1)^2(x^2 + x + 27)$$
$$+ (x + y - |z|)^2(11z^2 + z + 4) = 0$$

(**Ans:** $(x, y, z) = (4,5,9)$ *or* $(x, y, z) = (4,5,-9)$).

10-5-8) For which values of k the numbers 0 and 2 lie between the roots of the trinomial $x^2 + (2k - 1)x + k + 4$?

10-5-9) Let r_1, r_2 be the roots of the trinomial $(k - 3)x^2 - 4x + k$, $(k \neq 3)$. Find $k \in \mathbb{R}$ so that: **a)** $r_1 < 1 < r_2$, **b)** $1 < r_1 \leq r_2$, **c)** $-2 \leq r_1 < r_2 < 1$, and **d)** $-2 < r_1 < 1 < r_2$.

(**Ans: a)** $3 < k < 7/2$, **b)** $7/2 < k \leq 4$, **c)** $-1 < k \leq 4/5$, **d)** $3 < k < 7/2$).

10-5-10) If a, b, c are real numbers such that $|c| \leq |b| \leq |a| \leq ac$, and r_1, r_2 are the roots of the trinomial $ax^2 + bx + c$, show that $|r_1^3 - r_2^3| < 2$.

10-5-11) Solve the equation $\sqrt{3x + 1} + \sqrt{x - 4} = \sqrt{4x + 5}$, (**Ans:** $x = 5$).

10-5-12) Let $g(x) = (x + k_1)(x + k_2) \cdots (x + k_n)$ where $k_1, k_2, \cdots, k_n$ are numbers outside of the interval $(-2 \ldots 0)$. Show that the equation $ag(2)x^2 + bg(1)x + cg(0) = 0$ has real roots provided that the trinomial $ax^2 + bx + c$ has real roots.

Hint: It suffices to show that $\big(bg(1)\big)^2 - 4acg(0)g(2) \geq 0$, provided that $b^2 - 4ac \geq 0$. Notice that $(1 + k_\lambda)^2 > k_\lambda(k_\lambda + 2) > 0$, $\lambda = 1,2, \ldots, n$, etc.

10-6) Maximum or minimum value of a trinomial

a) We may consider the trinomial $y = ax^2 + bx + c$ (a, b, c are real numbers), **as a function of the real variable x**, where x runs from $x = -\infty$ up to $x = \infty$. Let p be the value of the trinomial, corresponding to **a real value x_0** of the independent variable x, i.e.

$$ax_0^2 + bx_0 + c = p \Leftrightarrow ax_0^2 + bx_0 + c - p = 0 \qquad (*)$$

Formula (*) shows that **the real number x_0 is a root (solution)** of the quadratic equation $ax^2 + bx + c - p = 0$, i.e. this equation admits real roots, and **its discriminant must therefore be greater than or equal to zero**, i.e.

$$\Delta = b^2 - 4a(c-p) \geq 0 \Leftrightarrow 4ap \geq 4ac - b^2 \qquad (**)$$

1) If $a > 0$, formula (**) implies that

$$p \geq \frac{4ac - b^2}{4a} \qquad (10-6-1)$$

and this shows that in this case ($a > 0$), the values of the trinomial **cannot be smaller** than $(4ac - b^2)/4a$. **This number is the minimum value the trinomial can attain**. To find the value of x at which the trinomial attains its minimum value, we assume that the minimum value of the trinomial is obtained at $x = x_1$, i.e.

$$p = \frac{4ac - b^2}{4a} = ax_1^2 + bx_1 + c \Leftrightarrow 4ac - b^2 = 4a^2x_1^2 + 4abx_1 + 4ac \Leftrightarrow$$

$$4a^2x_1^2 + 4abx_1 + b^2 = 0 \Leftrightarrow (2ax_1 + b)^2 = 0 \Leftrightarrow 2ax_1 + b = 0 \Leftrightarrow$$

$$x_1 = -\frac{b}{2a} \qquad (10-6-2)$$

2) If $a < 0$, formula (**) implies,

$$p \leq \frac{4ac - b^2}{4a} \qquad (10-6-3)$$

and this shows that in this case ($a < 0$), the values of the trinomial cannot be greater that $(4ac - b^2)/4a^2$. **This number is the maximum value the trinomial can attain**. Working as in part 1, we find that the maximum value is attained at $x_1 = -b/2a$.

We may summarize our findings in the following theorem.

Theorem 10-6-1: **If $a > 0$ the trinomial $ax^2 + bx + c$ attains a minimum value equal to $(4ac - b^2)/4a$, while if $a < 0$ the trinomial attains its maximum value equal to $(4ac - b^2)/4a$. In both cases, the minimum or the maximum value are attained at $x = -b/2a$, (half the sum of the roots of the trinomial).**

b) With similar arguments, we may find **the possible values (range), of the ratio of two trinomials**. Let us, for example, consider the function

$$y = \frac{ax^2 + bx + c}{Ax^2 + Bx + C} \qquad (***)$$

If, for a given **real value** x_0 of x the corresponding value of y is q, we must have,

$$\frac{ax_0^2 + bx_0 + c}{Ax_0^2 + Bx_0 + C} = q \Leftrightarrow (a - qA)x_0^2 + (b - qB)x_0 + c - qC = 0$$

and this shows that the discriminant Δ must be positive or zero, i.e.

$$(b - qB)^2 - 4(a - qA)(c - qC) \geq 0 \qquad (10-6-4)$$

Formula (10-6-4) is, in general, **an inequality for q (the possible values of the ratio of two trinomials)**. Solving this inequality for q, we find the possible values of y, as expressed in (***), (see Example 10-6-2).

Example 10-6-1: Find the minimum or maximum values of the trinomials:

$$3x^2 + 2x - 1, \quad -2x^2 + 5x + 2$$

Solution

Trinomial $3x^2 + 2x - 1$: $a = 3 > 0$, the trinomial attains **a minimum** value

$$(4ac - b^2)/4a = (4 \cdot 3 \cdot (-1) - 2^2)/(4 \cdot 3) = -\frac{4}{3}$$

The minimum value is attained at $x = -b/2a = -1/3$.

Trinomial $-2x^2 + 5x + 2$: $a = -2 < 0$, the trinomial attains **a maximum** value

$$(4ac - b^2)/4a = (4 \cdot (-2) \cdot 2 - 5^2)/(4 \cdot (-2)) = \frac{41}{8}$$

This maximum value is attained at $x = -b/2a = 5/4$.

Example 10-6-2: Find the possible values (range) of the function $y = (x^2 - x + 1)/(x^2 + x + 1)$.

.

Solution

Let $q = (x^2 - x + 1)/(x^2 + x + 1)$, for real values of x, i.e.

$$\frac{x^2 - x + 1}{x^2 + x + 1} = q \Leftrightarrow (1-q)x^2 - (1+q)x + 1 - q = 0 \quad (*)$$

and since the roots of this equation are real, we must have:

$$\Delta = (1+q)^2 - 4(1-q)(1-q) \geq 0 \Leftrightarrow -3q^2 + 10q - 3 \geq 0 \Leftrightarrow$$

$$3q^2 - 10q + 3 \leq 0 \Leftrightarrow 3\left(q - \frac{1}{3}\right)(q - 3) \leq 0 \Leftrightarrow \frac{1}{3} \leq q \leq 3$$

Thus, the function $y = (x^2 - x + 1)/(x^2 + x + 1)$ is contained between $1/3$ and 3, for all real values of x.

Example 10-6-3: Show that the equation $(x^2 - x + 1)/(x^2 + x + 1) = 5$ is impossible in the set of real numbers.

Solution

Since for any real x, $(x^2 - x + 1)/(x^2 + x + 1) \leq 3$, (as shown in Example 10-6-2), there is **no real value** of x to satisfy the given equation. Of course, there are complex values of x that satisfy the given equation (let the reader determine the complex roots of the equation).

PROBLEMS

10-6-1) Determine the constant k, so that the minimum value of $k(1-x)^2 + (1-k)x^2$ to be the maximum possible, **(Ans: $k = 1/2$).**

10-6-2) Determine k so that the sum of the squares of the roots of $x^2 - (k-2)x - (k+3) = 0$ to be the minimum possible.

10-6-3) If $c_1, c_2, \cdots, c_n$ are given constant numbers, determine the value of x at which the function $y = (x - c_1)^2 + (x - c_2)^2 + \cdots + (x - c_n)^2$ attains its minimum value possible, **(Ans: $x = (c_1 + c_2 + \cdots + c_n)/n$).**

10-6-4) Determine the coefficients of the trinomial $ax^2 + bx + c$, given that $r = 8$ is one root and that the minimum value of the trinomial is -12, attained at $x = 6$.

10-6-5) Show that the function $y = ((x + k)^2 - 4k\lambda)/(2(x - \lambda))$ takes values outside of the interval $(2k \ldots 2\lambda)$, when x varies from $-\infty$ to ∞.

10-6-6) Divide the number 27 into two parts, such that the sum of the square of the first part multiplied by 4 and the square of the second part multiplied by 5 to be the minimum possible, What is this minimum value?

10-6-7) Determine k and λ such that the function $y = (2kx + \lambda)/(x^2 + 1)$ assumes maximum value 4 and minimum value (-1).

(**Ans:** $k = \pm 2$, $\lambda = 3$).

10-6-8) Among all the right triangles having constant perimeter 2τ, which one bounds the maximum area possible? What is this maximum area?

(**Ans:** The isosceles right triangle, max. Area $= (3 - 2\sqrt{2})\tau^2$).

10-6-9) For which value of the positive constant a the maximum value of the function $y = (x - 1)^2/(x^2 + a)$ is equal to 3 ?   (**Ans:** $a = 1/2$).

10-7) When two trinomials have common roots

In this section we shall derive the conditions that should be satisfied, for two given trinomial to have **two common roots** or just **one common root**.

Case 1: Two common roots.

Theorem 10-7-1: **The necessary and sufficient condition for two trinomials to have two common roots is, the coefficients of like powers of x to be proportional.**

Proof: Let us assume that r_1 and r_2 are roots of $f(x) = ax^2 + bx + c$ and $g(x) = Ax^2 + Bx + C$. We shall show that

$$\frac{a}{A} = \frac{b}{B} = \frac{c}{C} \qquad (10-7-1)$$

Indeed, from the first trinomial we have,

$$r_1 + r_2 = -\frac{b}{a}, \quad r_1 r_2 = \frac{c}{a} \qquad (*)$$

while from the second trinomial we have,

$$r_1 + r_2 = -\frac{B}{A}, \quad r_1 r_2 = \frac{C}{A} \qquad (**)$$

From (*) and (**) we find,

$$\frac{b}{a} = \frac{B}{A} \text{ and } \frac{c}{a} = \frac{C}{A} \Rightarrow \begin{cases} \frac{a}{A} = \frac{b}{B} \\ \text{and} \\ \frac{a}{A} = \frac{c}{C} \end{cases} \Rightarrow \frac{a}{A} = \frac{b}{B} = \frac{c}{C}$$

Conversely, if the coefficients satisfy (10-7-1), then

$$\frac{a}{A} = \frac{b}{B} = \frac{c}{C} \stackrel{\text{def}}{=} k \Rightarrow \begin{cases} a = kA \\ b = kB \\ c = kC \end{cases} \Rightarrow f(x) = kg(x)$$

and this shows that both trinomials have the same roots.

Case 2: One common root.

Theorem 10-7-2: **The necessary and sufficient condition for two trinomials $f(x) = ax^2 + bx + c$ and $g(x) = Ax^2 + Bx + C$ to have only one common root r is,**

$$R \stackrel{\text{def}}{=} \begin{vmatrix} a & c \\ A & C \end{vmatrix}^2 - \begin{vmatrix} a & b \\ A & B \end{vmatrix} \begin{vmatrix} b & c \\ B & C \end{vmatrix} = 0 \qquad (10-7-2)$$

and $\begin{vmatrix} a & b \\ A & B \end{vmatrix} \neq 0$. The common root r is given by the formula:

$$r = -\frac{\begin{vmatrix} a & c \\ A & C \end{vmatrix}}{\begin{vmatrix} a & b \\ A & B \end{vmatrix}} \qquad (10-7-3)$$

(For the definition and properties of determinants, see section 4-6).

Proof: If we call r **the only common root** of the two trinomials, we have:

$$f(r) = ar^2 + br + c = 0 \quad \textbf{and} \quad g(r) = Ar^2 + Br + C = 0 \quad (***)$$

Multiplying the first equation in (***) by $(-A)$, the second by a and adding term wise, we obtain easily the expression for r (equation (7-10-3)).

Setting this expression of r in $f(x) = ax^2 + bx + c$, (or in $g(x) = Ax^2 + Bx + C$), we obtain equation (7-10-2), (let the reader check it). The common root is **only one**, since if it were a second common root, then, by virtue of Theorem 10-7-1, we would have, $a/A = b/B$, which however cannot be, since $(aB - Ab) \neq 0$, (by assumption).

Example 10-7-1: Determine k and λ given that the two equations

$$(5k - 52)x^2 - (k - 4)x + 4 = 0 \quad \text{and} \quad (2\lambda + 1)x^2 - 5\lambda x + 20 = 0$$

have the same roots.

Solution

The two trinomials will have the same roots when

$$\frac{5k - 52}{2\lambda + 1} = \frac{-(k-4)}{-5\lambda} = \frac{4}{20} = \frac{1}{5} \Rightarrow \begin{cases} 5(5k - 52) = 2\lambda + 1 \\ \text{and} \\ 5(k - 4) = 5\lambda \end{cases} \quad (*)$$

Solving the system (*) for k and λ, we find: $k = 11$ and $\lambda = 7$. In this case, $3x^2 - 7x + 4 = 0$ and $15x^2 - 35x + 20 = 0$. Verify that both equations have the same roots, $r_1 = 1$ and $r_2 = 4/3$.

Example 10-7-2: Assuming that $abc \neq 0$, find the relation between a, b, c if the two trinomials $f(x) = ax^2 + bx + c$ and $g(x) = cx^2 + bx + a$ have only one root in common.

Solution

The two trinomials will have one common root only if, (equation (7-10-2)):

$$\begin{vmatrix} a & b \\ c & b \end{vmatrix} \neq 0 \quad \textbf{and} \quad \begin{vmatrix} a & c \\ c & a \end{vmatrix}^2 = \begin{vmatrix} a & b \\ c & b \end{vmatrix} \begin{vmatrix} b & c \\ b & a \end{vmatrix} \Leftrightarrow$$

$$b(a-c) \neq 0 \quad \text{and} \quad (a^2-c^2)^2 = b^2(a-c)^2 \overset{(b \neq 0)}{\Longleftrightarrow}$$

$$a - c \neq 0 \quad \text{and} \quad (a-c)^2((a+c)^2 - b^2) = 0 \Leftrightarrow b^2 = (a+c)^2$$

If the coefficients satisfy this equation, the two trinomials will have just one common root.

PROBLEMS

10-7-1) For which values of a and b the two trinomials $x^2 + ax + b$ and $ax^2 + x + b + 1$ have the same roots? Find these roots.

(Ans: $a = -1, b = -1/2, r_{1,2} = (1 \pm \sqrt{3})/2$).

10-7-2) Determine a so that the two trinomials $x^2 + ax + 1$ and $x^2 + x + a$ have: **a)** Two common roots and **b)** only one root in common.

10-7-3) Provided that the trinomials $x^2 + ax + b$ and $x^2 + \lambda x + \mu$ have only one root in common, show that $(b - \mu)^2 = (a\mu - b\lambda)(\lambda - a)$.

10-7-4) Provided that the equations $ax^2 + bx + c = 0$ and $Ax^2 + Bx + C = 0$ have only one common root, show that the discriminant of the equation $ax^2 + bx + c + \lambda(Ax^2 + Bx + C) = 0$ is a perfect square.

10-8) Inequalities with radicals (irrational inequalities)

If in an inequality, the unknown appears under a radical, the inequality is called **an irrational inequality**. For example, inequalities of the form

$$\sqrt{x-3} > 3 + \sqrt{x+2}, \qquad \sqrt{x^2 - 2x + 3} \leq \sqrt{x+7} + 5\sqrt{x-1}$$

are irrational inequalities.

If the inequality contains radicals **of even order, we shall always assume that the quantities under the radical are positive, since inequalities between complex numbers are not defined.** For radicals of odd order, such restrictions are not necessary.

In order to solve an irrational inequality, we isolate the radical (either in the left or in the right side), and then **raise both sides in the suitable power, in**

order to eliminate the radicals. We thus reduce an irrational equation to a rational equation, **coupled with some auxiliary conditions**. This is so, since in general, raising both sides of an inequality to a power, we do not obtain an equivalent inequality.

For example, it is true that $(-2) < (-1)$ but if we square both sides, the inequality $(-2)^2 < (-1)^2$, i.e. $4 < 1$ is **false**. However, if we know that A and B are positive, then $A < B$ implies $A^2 < B^2$, and vice versa. On the other hand, from $A < B$ we conclude that $A^3 < B^3$, and vice versa, with no restrictions on A and B.

The following examples illustrate the method of approach.

Example 10-8-1: Solve the inequality: $\sqrt{x^2 - 1} > \sqrt{x + 1}$.

Solution

First of all, the radicands must be positive, i.e.

$$\begin{Bmatrix} x^2 - 1 \geq 0 \\ \text{and} \\ x + 1 \geq 0 \end{Bmatrix} \Leftrightarrow \begin{Bmatrix} x \leq -1 \ \ or \ \ x \geq 1 \\ \text{and} \\ x \geq -1 \end{Bmatrix} \Rightarrow 1 \leq x < \infty \qquad (*)$$

Second step is to eliminate the radicals, by squaring both sides, (we can safely do so, since both left and right sides are now positive):

$$x^2 - 1 > x + 1 \Leftrightarrow x^2 - x - 2 > 0 \Leftrightarrow (x - 2)(x + 1) > 0 \Rightarrow$$

$$x < -1 \ \ or \ \ x > 2 \qquad (**)$$

The values of x which satisfy both (*) and (**) **simultaneously**, are $2 < x < \infty$, and these values are the solutions of the given inequality.

Example 10-8-2: Solve the inequality: $\sqrt{2x + 1} > 3 + \sqrt{x - 8}$.

Solution

Since the radicands must be positive we must have: $2x + 1 \geq 0$ and $x - 8 \geq 0$, and these two inequalities are satisfied **simultaneously** if $x \geq 8$. Since both sides of the inequality are positive, we obtain an equivalent inequality by squaring both sides, i.e.

$$\sqrt{2x+1} > 3 + \sqrt{x-8} \Leftrightarrow \left(\sqrt{2x+1}\right)^2 > \left(3+\sqrt{x-8}\right)^2 \Leftrightarrow x > 6\sqrt{x-8}$$

and since $x \geq 8$, both sides of the last inequality are positive, and squaring once more, we get,

$$x^2 > 36(x-8) \Leftrightarrow x^2 - 36x + 288 > 0 \Leftrightarrow (x-12)(x-24) > 0$$

and this inequality is satisfied if

$$x < 12 \quad or \quad x > 24 \tag{*}$$

The solution of the original inequality is the totality of the numbers x that satisfy, **simultaneously** the inequalities

$$\begin{cases} x \geq 8 \\ and \\ x < 12 \quad or \quad x > 24 \end{cases} \Rightarrow 8 \leq x < 12 \ \ or \ \ 24 < x < \infty$$

Using set notation we may write:

$$Solution\ set = [8,12) \cup (24, \infty)$$

PROBLEMS

10-8-1) Solve the inequality: $\sqrt{2x+1} < 1 - \sqrt{x+1}$.

(Ans: $-1/2 \leq x < 3 - \sqrt{2}$).

10-8-2) Solve the inequality: $\sqrt{x^2 - 3x + 2} < \sqrt{2x+1}$.

10-8-3) Solve the inequality: $\sqrt{2x+1} < 2(x+1)/(2-x)$.

(Ans: $0 < x < 2$, or $-\frac{1}{2} \leq x < \frac{11-\sqrt{153}}{4}$).

10-8-4) Show that $\sqrt{1+x^2} \geq 1 + \frac{x^2}{2} - \frac{x^4}{8}$, $\forall x \in \mathbb{R}$.

10-8-5) Solve the inequality: $\sqrt[3]{x^2 - 4x + 3} > 1 - x$.

(Ans: $-1 < x < 1$, or $2 < x < \infty$).

CHAPTER 11: EQUATIONS AND INEQUALITIES TRANSFORMABLE TO QUADRATICS

There are equations **of degree higher than two**, which by means of **a suitable transformation (change of the variable)**, can be reduced to quadratics. In this chapter we shall study some types of such equations.

11-1) Biquadratic equations

Biquadratic equation is an equation of the form

$$ax^4 + bx^2 + c = 0, \quad a, b, c \in \mathbb{R} \qquad (11-1-1)$$

A root or a solution of a biqudratic equation is a number (**real or complex**) which satisfies the equation. If we make the change of variable $y = x^2$, equation (11-1-1) becomes

$$ay^2 + by + c = 0 \qquad (11-1-2)$$

which is **quadratic in y**. Solving this quadratic equation we find its two roots y_1 and y_2 and then, from $y = x^2$ we find the corresponding values of x,

$$\{x^2 = y_1 \Rightarrow x = \pm\sqrt{y_1}, \quad x^2 = y_2 \Rightarrow x = \pm\sqrt{y_2}\} \qquad (11-1-3)$$

Any biquadratic equation has four roots (real or complex).

As an example, to solve $x^4 - 10x^2 + 0 = 0$, if we set $y = x^2$, the equation becomes $y^2 - 10y + 9 = 0$. This is a quadratic equation in y and its two roots are $y = 1$ or $y = 9$, i.e. $x^2 = 1$ or $x^2 = 9$, i.e. $x = \pm 1$ or $x = \pm 3$. The **four roots** of the quadratic equation are $\{-1, 1, -3, 3\}$.

The kind or the roots of the biquadratic equation in (11-1-1), (real or complex), obviously depends on the kind of the roots of the quadratic equation in (11-1-2).

a) Conditions for the biquadratic equation to have four real roots:

The biquadratic equation will have four real roots, provided that the roots of (11-1-2) are both real and positive, i.e. provided that $y_1 > 0$ and $y_2 > 0$. If two real numbers are both positive, then their sum and their product are both positive, and this implies the following condition:

The biquadratic equation $ax^4 + bx^2 + c = 0$ has **four real roots** if and only if:

$$\left\{\Delta > 0, \quad y_1 + y_2 = -\frac{b}{a} > 0, \quad y_1 y_2 = \frac{c}{a} > 0\right\} \qquad (11-1-4)$$

b) Conditions for the biquadratic equation to have two real and two complex roots:

The biquadratic equation will have **two real and two complex roots** when the quadratic equation (11-1-2) has one positive and one negative root, and the necessary and sufficient condition for this is simply

$$y_1 y_2 = \frac{c}{a} < 0 \qquad (11-1-5)$$

When (11-1-5) is satisfied, then $\Delta = b^2 - 4ac > 0$, (the equation $ay^2 + by + c = 0$ has real roots) and one of them is positive and the other one is negative, since $y_1 y_2 < 0$. The square root of the positive root yields two real roots for the quadratic, while the square root of the negative root yields to complex roots for the quadratic.

c) Conditions for the biquadratic equation to have four complex roots:

The quadratic equation $ay^2 + by + c = 0$ must have either two negative real roots or two complex roots, and the necessary and sufficient conditions for this are the following:

$$\left\{\begin{array}{l} \Delta > 0, \quad y_1 + y_2 = -\frac{b}{a} < 0, \quad y_1 y_2 = \frac{c}{a} > 0 \\ or \\ \Delta < 0 \end{array}\right\} \qquad (11-1-6)$$

The biqudratic trinomial: Any function of the form $f(x) = ax^4 + bx^2 + c$ is called a biquadratic trinomial. The roots of the biqudratic trinomial are the roots of the corresponding biquadratic equation $ax^4 + bx^2 + c = 0$.

If r_1, r_2, r_3, r_4 are the roots of the biquadratic trinomial $f(x) = ax^4 + bx^2 + c$, the following expression holds true:

$$f(x) = ax^4 + bx^2 + c = a(x - r_1)(x - r_2)(x - r_3)(x - r_4) \quad (11-1-7)$$

This expression results easily from equation (10-1-3) and the fact that $y = x^2$.

Example 11-1-1: Solve the equation $x^4 + 2x^2 - 15 = 0$.

Solution

The change of variable $y = x^2$, reduces the given equation to the following:

$$y^2 + 2y - 15 = 0 \Longrightarrow y_{1,2} = \frac{-2 \pm \sqrt{64}}{2} = \frac{-2 \pm 8}{2} \Longrightarrow \begin{Bmatrix} y_1 = 3 \\ y_2 = -5 \end{Bmatrix} \quad (*)$$

From $y_1 = 3$ we obtain, $x = \pm\sqrt{3}$, while from $y_2 = -5$ we obtain $x = \pm\sqrt{5}\, i$.

Example 11-1-2: Form a biquadratic trinomial having roots: $\pm 2, \pm 3i$.

Solution

Application of formula (11-1-7) yields:

$$f(x) = (x - 2)(x + 2)(x - 3i)(x + 3i) = (x^2 - 4)(x^2 + 9)$$
$$= x^4 + 5x^2 - 36$$

Notice that this biquadratic trinomial **is not unique**, since any trinomial of the form $af(x)$, where $a \neq 0$, does have the same roots.

PROBLEMS

11-1-1) Solve the biquadratic equation: $x^4 - 1 = 0$,   (**Ans:** $\pm 1, \pm i$).

11-1-2) Solve the biquadratic equation: $x^4 + x^2 - 6 = 0$.

11-1-3) Determine the kind of roots (real or complex) of the biquadratic equation $x^4 - 2(k-1)x^2 + k + 1 = 0$, when the real parameter k varies from $-\infty$ to ∞.

(**Ans:** $-\infty < k \leq -1$, two real and two complex roots, $-1 < k < 3$, four complex roots, $3 \leq k < \infty$, four real roots).

11-1-4) Express the roots of the biquadratic trinomial

$$f(x) = x^4 - 2(k^2 + \lambda^2)x^2 + (k^2 - \lambda^2)^2$$

in terms of k and λ.

11-1-5) Solve the equation: $\dfrac{x^2}{(x-1)^2} + \dfrac{x^2}{(x+1)^2} = 3$.

(**Ans:** $\pm\sqrt{\dfrac{8+\sqrt{52}}{2}}, \pm\sqrt{\dfrac{8-\sqrt{52}}{2}}$).

11-2) Biquadratic inequalities

Any inequality of the form $ax^4 + bx^2 + c > 0$, (or < 0), is called a **biquadratic inequality**. If we set $y = x^2$, we obtain **a quadratic inequality in y**. Solving this inequality, we find the range of $y = x^2$, and from this we find the range of x. Let us consider a few examples.

Example 11-2-1: Solve the inequality: $x^4 - x^2 - 2 < 0$.

Solution

If we set $y = x^2$, the given inequality becomes

$$y^2 - y - 2 < 0 \Leftrightarrow (y+1)(y-2) < 0 \Leftrightarrow -1 < y < 2 \qquad (*)$$

(the variable y must lie between the roots (-1) and 2).

Equation (*) implies,

$$-1 < x^2 < 2 \qquad (**)$$

The sought for values of x must satisfy the double inequality in (**). The inequality $(-1) < x^2$ is satisfied for **all real values of x**. The inequality $x^2 < 2$ implies,

$$x^2 < 2 \Leftrightarrow |x|^2 < 2 \Leftrightarrow |x| < \sqrt{2} \Leftrightarrow -\sqrt{2} < x < \sqrt{2} \qquad (***)$$

Thus the values of x that satisfy the double inequality in (**) are $\sqrt{2} < x < \sqrt{2}$.

Example 11-2-2: Solve the inequality: $\frac{x^4 - 2x^2 + 1}{x^2 - 1} > 2$.

Solution

$$\frac{x^4 - 2x^2 + 1}{x^2 - 1} > 2 \iff \frac{x^4 - 2x^2 + 1}{x^2 - 1} - 2 > 0 \iff$$

$$\frac{x^4 - 2x^2 + 1 - 2(x^2 - 1)}{x^2 - 1} > 0 \iff \frac{x^4 - 4x^2 + 3}{x^2 - 1} > 0 \iff$$

$$(x^2 - 1)(x^4 - 4x^2 + 3) > 0 \qquad (*)$$

or, if we set $y = x^2$,

$$(y - 1)(y^2 - 4y + 3) > 0 \iff (y - 1)(y - 1)(y - 3) > 0 \iff$$

$$(y - 1)^2(y - 3) > 0 \xLongrightarrow{((y-1)^2 > 0)} y - 3 > 0$$

and going back to the original variable x,

$$x^2 - 3 > 0 \iff |x|^2 > 3 \iff \begin{cases} x > \sqrt{3} \\ \text{or} \\ x < -\sqrt{3} \end{cases}$$

PROBLEMS

11-2-1) Solve the inequality: $x^4 - 6x^2 + 8 > 0$, $x \in \mathbb{R}$.

(Ans: $x < 2$, or $-\sqrt{2} < x < \sqrt{2}$, or $x > 2$).

11-2-2) Solve the inequality: $x^4 - 1 < 0$, $x \in \mathbb{R}$.

11-2-3) Solve the inequality: $(x^4 + 4x^2 + 7)^2 - 16(x^2 + 2)^2 > 0$, $x \in \mathbb{R}$.

(Ans: $|x| > 1$).

Hint: $(x^4 + 4x^2 + 7)^2 - 16(x^2 + 2)^2 = ((x^2 + 2)^2 + 3)^2 - 16(x^2 + 2)^2$, then set $y = (x^2 + 2)^2$, etc.

11-3) Reciprocal equations

An equation in which the coefficients located at equal distances from the beginning and the end are equal or opposite, is called **a reciprocal equation**.

According to the definition, the following equations are reciprocal:

$$ax^3 + bx^2 + bx + a = 0$$
$$ax^4 + bx^3 + cx^2 + bx + a = 0$$
$$ax^5 + bx^4 + cx^3 + cx^2 + bx + a = 0$$

$$ax^3 + bx^2 - bx - a = 0$$
$$ax^4 + bx^3 - bx - a = 0$$
$$ax^5 + bx^4 + cx^3 - cx^2 - bx - a = 0$$

If a number $r \neq 0$ is a root of a reciprocal equation, then the number $\left(\frac{1}{r}\right)$ (the reciprocal of r) is also a solution. This is a characteristic property of reciprocal equations. For example, let us assume that r is a solution of $f(x) = ax^3 + bx^2 + bx + a = 0$. This means that $f(r) = ar^3 + br^2 + br + a = 0$. We notice that

$$f\left(\frac{1}{r}\right) = a\frac{1}{r^3} + b\frac{1}{r^2} + b\frac{1}{r} + a = \frac{a + br + br^2 + ar^3}{r^3} = \frac{f(r)}{r^3} = 0$$

and this verifies **that $\left(\frac{1}{r}\right)$ is also a root of $f(x) = 0$**. Of course, this characteristic property of the reciprocal equations, **is due to the symmetric structure of its coefficients**.

As we shall see, **the solution of a third, fourth and fifth degree reciprocal equations is reduced to the solution of a quadratic equation.**

a) Reciprocal equation of degree 3: $ax^3 + bx^2 + bx + a = 0$.

$$ax^3 + bx^2 + bx + a = 0 \Longleftrightarrow a(x^3 + 1) + bx(x + 1) = 0 \Longleftrightarrow$$

$$a(x + 1)(x^2 - x + 1) + bx(x + 1) = 0 \Longleftrightarrow$$

$$(x + 1)\{ax^2 + (b - a)x + a\} = 0 \Longleftrightarrow \begin{cases} x + 1 = 0 \\ \text{or} \\ ax^2 + (b - a)x + a = 0 \end{cases}$$

Similarly, we may solve the equation $ax^3 + bx^2 - bx - a = 0$.

b) Reciprocal equation of degree 4: $ax^4 + bx^3 + cx^2 + bx + a = 0$.

Dividing both sides by x^2 ($x \neq 0$), we obtain:

$$ax^2 + bx + c + b\frac{1}{x} + a\frac{1}{x^2} = 0 \Leftrightarrow a\left(x^2 + \frac{1}{x^2}\right) + b\left(x + \frac{1}{x}\right) + c = 0 \quad (*)$$

If we make the substitution:

$$y = x + \frac{1}{x} \Longrightarrow y^2 = x^2 + \frac{1}{x^2} + 2 \Longrightarrow x^2 + \frac{1}{x^2} = y^2 - 2 \quad (**)$$

and substituting in (*), we obtain:

$$a(y^2 - 2) + by + c = 0 \Leftrightarrow ay^2 + by + c - 2a = 0 \quad (***)$$

Solving this quadratic equation we find y, and then from (**) we find the corresponding x.

c) Reciprocal eq. of degree 5: $ax^5 + bx^4 + cx^3 + cx^2 + bx + a = 0$.

$$ax^5 + bx^4 + cx^3 + cx^2 + bx + a = 0 \Leftrightarrow$$

$$a(x^5 + 1) + bx(x^3 + 1) + cx^2(x + 1) = 0 \Leftrightarrow$$

$$a(x+1)(x^4 - x^3 + x^2 - x + 1) + bx(x+1)(x^2 - x + 1) + cx^2(x+1) = 0$$

or, equivalently,

$$(x+1)\underbrace{\{ax^4 + (b-a)x^3 + (a-b+c)x^2 + (b-a)x + a\}}_{Reciprocal\ Eq.of\ degree\ 4} = 0 \quad (****)$$

and this splits in two equations, $x + 1 = 0$, i.e. $x = -1$, and another reciprocal equation of degree 4, which is solved according to part (b).

In a similar fashion, we may solve reciprocal equations of degree higher than 5. For example, to solve **a sixth degree reciprocal equation**, we divide through by x^3, then set $y = x + \frac{1}{x}$, and this yields **a third degree equation in y**, (see Example 11-3-3).

Example 11-3-1: Solve the equation: $6x^4 + 25x^3 + 12x^2 - 25x + 6 = 0$.

Solution

Dividing through by x^2 we obtain,

$$6x^2 + 25x + 12 - \frac{25}{x} + \frac{6}{x^2} = 0 \Leftrightarrow 6\left(x^2 + \frac{1}{x^2}\right) + 25\left(x - \frac{1}{x}\right) + 12 = 0$$

and if we set, $y = x - \frac{1}{x}$ (auxiliary unknown), $y^2 + 2 = x^2 + \frac{1}{x^2}$, the equation becomes,

$$6(y^2 + 2) + 25y + 12 = 0 \Leftrightarrow 6y^2 + 25y + 24 = 0 \Rightarrow \begin{cases} y_1 = -\frac{3}{2} \\ y_2 = -\frac{8}{3} \end{cases}$$

So we have the following equations for x:

$$\left\{ x - \frac{1}{x} = -\frac{3}{2}, \quad x - \frac{1}{x} = -\frac{8}{3} \right\} \tag{*}$$

The first equation in (*) yields,

$$x - \frac{1}{x} = -\frac{3}{2} \Leftrightarrow 2x^2 + 3x - 2 = 0 \Rightarrow x = \frac{1}{2} \text{ or } x = -2$$

and similarly, from the second equation in (*), we find $x = \frac{1}{3}$ or $x = -3$.

The four roots of the original equation are: $\frac{1}{2}, \frac{1}{3}, -2, -3$.

Example 11-3-2: Solve the equation:

$$3x^5 - 16x^4 + 17x^3 + 17x^2 - 16x + 3 = 0$$

Solution

$$3x^5 - 16x^4 + 17x^3 + 17x^2 - 16x + 3 = 0 \Leftrightarrow$$

$$3(x^5 + 1) - 16x(x^3 + 1) + 17x^2(x + 1) = 0 \Leftrightarrow$$

$$3(x+1)(x^4 - x^3 + x^2 - x + 1) - 16x(x+1)(x^2 - x + 1) + 17x^2(x+1) = 0 \Leftrightarrow$$

$$(x+1)\{3x^4 - 19x^3 + 36x^2 - 19x + 3\} = 0 \Leftrightarrow$$

$$\begin{cases} x+1 = 0, & \text{or} \\ 3x^4 - 19x^3 + 36x^2 - 19x + 3 = 0 \end{cases} \qquad (*)$$

The solution of the first equation in (*) is $x = -1$.

The second equation in (*) is **a fourth degree reciprocal equation**. To solve this equation we divide through by x^2, and find:

$$3\left(x^2 + \frac{1}{x^2}\right) - 19\left(x + \frac{1}{x}\right) + 36 = 0 \qquad (**)$$

If we consider the auxiliary unknown, $y = x + \frac{1}{x}$, then $x^2 + \frac{1}{x^2} = y^2 - 2$, and equation (**) becomes:

$$3(y^2 - 2) - 19y + 36 = 0 \Leftrightarrow 3y^2 - 19y + 30 = 0 \Rightarrow \begin{cases} y_1 = \frac{10}{3} \\ y_2 = 3 \end{cases}$$

The values of x are then found from the equation $x + \frac{1}{x} = y$, which is a quadratic equation in x. When $y = y_1 = 10/3$ we find $x = 3$ or $x = 1/3$, while for $y = y_2 = 3$ we find $x = (3 + \sqrt{5})/2$ or $x = (3 - \sqrt{5})/2$.

The five roots of the original equation are: $-1, 3, \frac{1}{3}, \frac{3+\sqrt{5}}{2}, \frac{3-\sqrt{5}}{2}$.

Example 11-3-3: Solve the equation:

$$3x^6 + 5x^5 + 7x^4 + 10x^3 + 7x^2 + 5x + 3 = 0$$

Solution

This is **a sixth degree reciprocal equation**. To solve this equation we divide through by x^3 and find:

$$3x^3 + 5x^2 + 7x + 10 + \frac{7}{x} + \frac{5}{x^2} + \frac{3}{x^3} = 0 \Leftrightarrow$$

$$3\left(x^3 + \frac{1}{x^3}\right) + 5\left(x^2 + \frac{1}{x^2}\right) + 7\left(x + \frac{1}{x}\right) + 10 = 0 \qquad (*)$$

If we consider the auxiliary unknown $y = x + \frac{1}{x}$, then $x^2 + \frac{1}{x^2} = y^2 - 2$.

To express $x^3 + \frac{1}{x^3}$ in terms of y, **we start with $y = x + \frac{1}{x}$ and raise both sides in the third power**:

$$y^3 = \left(x + \frac{1}{x}\right)^3 = x^3 + 3x + \frac{3}{x} + \frac{1}{x^3} = \left(x^3 + \frac{1}{x^3}\right) + 3\underbrace{\left(x + \frac{1}{x}\right)}_{y} \Leftrightarrow$$

$$x^3 + \frac{1}{x^3} = y^3 - 3y$$

Substituting the expressions of $\left(x + \frac{1}{x}\right), \left(x^2 + \frac{1}{x^2}\right), \left(x^3 + \frac{1}{x^3}\right)$ in terms of y in equation (*) we find:

$$3(y^3 - 3y) + 5(y^2 - 2) + 7y + 10 = 0 \Leftrightarrow 3y^3 + 5y^2 - 2y = 0 \Leftrightarrow$$

$$y(3y^2 + 5y - 2) \Rightarrow \begin{cases} y = 0 \\ 3y^2 + 5y - 2 = 0 \end{cases} \Rightarrow \begin{cases} y = 0 \\ y = 1/3 \\ y = -2 \end{cases} \quad (**)$$

The corresponding values of x are obtained from $y = x + \frac{1}{x}$. Let the reader verify that the six roots of the original equation are:
$\pm i, (1 \pm i\sqrt{35})/6, -1$ *(Double root)*.

PROBLEMS

11-3-1) Solve the equation: $2(1 + x^4) = (1 + x)^4$.

(Ans: $1 - \sqrt{3} \pm i\sqrt{2\sqrt{3} - 3}, \ 1 + \sqrt{3} \pm \sqrt{3 + 2\sqrt{3}}$).

11-3-2) Solve the equation: $x^4 - 3x^3 + 4x^2 - 3x + 1 = 0$.

11-3-3) Solve the equation: $6x^5 + 29x^4 + 27x^3 - 27x^2 - 29x - 6 = 0$.

(Ans: $1, -\frac{1}{2}, -2, -\frac{1}{3}, -3$).

11-3-4) Solve the equation: $4x + 17\sqrt[4]{x^3} - 17\sqrt[4]{x} - 4 = 0$.

Hint: Setting $z = \sqrt[4]{x}$, the equation becomes, $4z^4 + 17z^3 - 17z - 4 = 0$.

11-3-5) Solve the equation: $x^3 + \frac{1}{x^3} = 6\left(x + \frac{1}{x}\right)$.

(**Ans**: $\pm i$, $(\pm 3 \pm \sqrt{5})/2$).

11-3-6) Solve the equation: $12x^5 - 8x^4 - 45x^3 + 45x^2 + 8x - 12 = 0$.

11-3-7) Solve the equation: $|x|^3 - 5x^2 - 5|x| + 1 = 0$.

(**Ans**: $|x| = 3 \pm \sqrt{8}$).

11-4) Binomial and trinomial equations

Binomial equation is any equation of the form $ax^k + bx^\lambda = 0$, where $a \neq 0$ and $b \neq 0$ and **k and λ are positive integers or zero**. For example, the equations $3x^5 + 2x^3 = 0, 7x^9 - 4x^6 = 0$ are binomial equations.

Assuming that $k > \lambda > 0$, the binomial equation can be written as

$$x^\lambda(ax^{k-\lambda} + b) = 0$$

or, if we call $n = k - \lambda > 0$, $x^\lambda(ax^n + b) = 0$, from which one root is $x = 0$, while the other roots are the roots of $(ax^n + b) = 0$.

Trinomial equation is any equation of the form $ax^k + bx^\lambda + cx^\mu = 0$, where $abc \neq 0$. The trinomial equation is reduced to binomial equations when $\lambda - k = \mu - \lambda \stackrel{\text{def}}{=} \omega$, since then $\lambda = k + \omega, \mu = \lambda + \omega = k + 2\omega$, and the trinomial becomes

$$ax^k + bx^{k+\omega} + cx^{k+2\omega} = 0 \Leftrightarrow x^k(cx^{2\omega} + bx^\omega + a) = 0$$

from which one root is $x = 0$, while the other roots are the roots of the equation $cx^{2\omega} + bx^\omega + a = 0$. If we set $y = x^\omega$, the last equation becomes $cy^2 + by + a = 0$, which is quadratic in y. Having found y, the corresponding x is then found from $y = x^\omega$.

Example 11-4-1: Solve the equation $x^4 + 1 = 0$.

Solution

Method 1:

$$x^4 + 1 = 0 \Leftrightarrow x^4 + 2x^2 + 1 - 2x^2 = 0 \Leftrightarrow (x^2 + 1)^2 - (\sqrt{2}\,x)^2 = 0 \Leftrightarrow$$

$$\{x^2 + 1 + \sqrt{2}\, x\}\{x^2 + 1 - \sqrt{2}\, x\} = 0 \Leftrightarrow \begin{cases} x^2 + \sqrt{2}\, x + 1 = 0 \\ or \\ x^2 - \sqrt{2}\, x + 1 = 0 \end{cases} \quad (*)$$

Thus, the original equation is split into two quadratic equations as shown in (*). Solving the quadratic equations we find,

$$x = \frac{\pm\sqrt{2} \pm i\sqrt{2}}{2} \quad (four\ different\ roots) \quad (**)$$

Method 2:

The equation $x^4 + 1 = 0$, implies that $x^4 = -1$, i.e. x is the fourth root of the number $z = -1 = 1 \cdot (\cos \pi + i \sin \pi)$, and according to equation (8-7-1) the fourth roots of (-1) are:

$$x_k = \sqrt[4]{1}\left(\cos\left(\frac{\pi}{4} + k\frac{2\pi}{4}\right) + i \sin\left(\frac{\pi}{4} + k\frac{2\pi}{4}\right)\right), k = 0, 1, 2, 3 \quad (***)$$

Let the reader verify that **the values of x in (***) are identical to those found in (**)**.

Remark: Formula (8-7-1), (the nth roots of a complex number) is **the most efficient method** to solve a binomial equation of the form $x^n \pm a = 0$. For example, think how you would solve $x^{38} + 1 = 0$, without using complex numbers?

Example 11-4-2: Solve the equation $x^6 + 5x^3 - 24 = 0$.

Solution

Setting $y = x^3$, the given equation becomes a quadratic in y:

$$y^2 + 5y - 24 = 0 \Leftrightarrow \begin{cases} y_1 = 3 \\ y_2 = -8 \end{cases} \quad (*)$$

From $x^3 = y_1 = 3 = 3(\cos 0 + i \sin 0)$, we find:

$$x_k = \sqrt[3]{3}\left(\cos\left(k\frac{2\pi}{3}\right) + i \sin\left(k\frac{2\pi}{3}\right)\right), k = 0, 1, 2 \quad (**)$$

From $x^3 = y_2 = -8 = 8(\cos \pi + i \sin \pi)$, we find:

$$x_k = \sqrt[3]{8}\left(\cos\left(\frac{\pi}{3} + k\frac{2\pi}{3}\right) + i\sin\left(\frac{\pi}{3} + k\frac{2\pi}{3}\right)\right), \quad k = 0, 1, 2 \quad (***)$$

The six numbers x_k, (three numbers in (**) and three numbers in (***)), constitute the six roots of $x^6 + 5x^3 - 24 = 0$.

PROBLEMS

11-4-1) Solve the equation $x^5 - 1 = 0$, using two methods: **a)** Factor the term $x^5 - 1$ and **b)** Find the fifth roots of 1.

(Ans: $1, \left(-1 \pm \sqrt{5} \pm i\sqrt{10 - 2\sqrt{5}}\right)/4$).

Hint: $x^5 - 1 = (x - 1)(x^4 + x^3 + x^2 + x + 1)$, from which $x = 1$ or x is a root of the second factor. Notice that $x^4 + x^3 + x^2 + x + 1 = 0$, is a reciprocal equation, solved as described in section 11-3. Which is the most efficient method (a) or (b)?

11-4-2) Solve the following equations:

$$2x^4 + 3 = 0, \qquad x^8 + 17x^4 + 16 = 0$$

Hint: the first one is equivalent to $x^4 + \frac{3}{2} = 0$. For the second equation, set $y = x^4$.

11-4-3) a) Solve the equation $x^3 - 1 = 0$, **b)** Show that, regarding the complex roots of this equation, one is equal to the square of the other, i.e. if one complex root is w the other complex root will be w^2.

11-4-4) Show that the sum of the cubic roots of unity, is equal to zero.

11-4-5) If w, w^2 are the complex cubic roots of unity (see Problem 11-4-3), show that:

$$(1 - w)(1 - w^2)(1 - w^4)(1 - w^5) = 9$$
$$(3w^2 + 2w + 1)(2w^2 + 3w + 1) = 3$$
$$(2w^2 + 5w + 2)^6 = (5w^2 + 2w + 2)^6 = 729$$

11-4-6) Solve the equation: $x^8 - 15x^4 - 16 = 0$.

Hint: Set $y = x^4$.

11-4-7) If $k = a + b$, $\lambda = aw + bw^2$, $\mu = aw^2 + bw$, where w, w^2 are the two complex cubic roots of unity, show that:

$$k^3 + \lambda^3 + \mu^3 = 3(a^3 + b^3)$$
$$k^2 + \lambda^2 + \mu^2 = 6ab$$
$$k^3 + \lambda^3 + \mu^3 = 3k\lambda\mu$$

11-5) The general method of substitution

Several times, the solution of an equation of degree higher than 2, is facilitated by means of **a suitable substitution (auxiliary unknown)**, which reduces the original equation to a simpler equation, which can be solved easily.

Unfortunately, there are no standard guidelines as to what substitution is the suitable one, in each case. Experience (gained by working as many problems as possible) and sometimes luck, may help the reader to make the proper substitution.

Let us consider a few illustrative examples.

Example 11-5-1: Solve the equation: $(x^2 - 1)^2 - 5(x^2 - 1) + 6 = 0$.

Solution

Setting $y = x^2 - 1$, the given equation becomes $y^2 - 5y + 6 = 0$, which is a quadratic equation in y, and therefore, easily solvable. We find $y_1 = 2$ and $y_2 = 3$, i.e. $x^2 - 1 = 2$, or $x^2 - 1 = 3$.

From $x^2 - 1 = 2$, we find $x^2 = 3$, i.e. $x = \pm\sqrt{3}$.

From $x^2 - 1 = 3$, we find $x^2 = 4$, i.e. $x = \pm 2$.

The four roots of the original equation are, $\pm\sqrt{3}, \pm 2$.

Example 11-5-2: Solve the equation:

$$(x^2 - x + 1)^4 - 6x^2(x^2 - x + 1)^2 + 5x^4 = 0$$

Solution

We notice that $x = 0$ **is not a root of the equation**, since it does not satisfy the equation. If we therefore divide through both members of the equation by x^4, we find:

$$\left(\frac{x^2 - x + 1}{x}\right)^4 - 6\left(\frac{x^2 - x + 1}{x}\right)^2 + 5 = 0 \qquad (*)$$

and if we make the substitution

$$y = \frac{x^2 - x + 1}{x} \qquad (**)$$

equation (*) is transformed to a biquadratic equation in y:

$$y^4 - 6y^2 + 5 = 0$$

Solving this biquadratic equation (with the substitution $z = y^2$ which transforms the biquadratic in a quadratic in z) we find, $y = \pm 1, \pm\sqrt{5}$, and then from equation (**) we find the corresponding values of x (the roots of the original equation). So the roots of the original equation are obtained from the equations,

$$\frac{x^2 - x + 1}{x} = \pm 1, \quad \frac{x^2 - x + 1}{x} = \pm\sqrt{5} \qquad (***)$$

Each one of the equations in (***) are quadratic in x, and therefore easily solvable. Let the reader complete the calculations.

Example 11-5-3: Solve the equation: $(x + 6)(x - 5)(x - 7)(x + 4) = 504$.

Solution

$$(x + 6)(x - 5)(x - 7)(x + 4) = 504 \Leftrightarrow$$

$$\{(x + 6)(x - 7)\}\{(x - 5)(x + 4)\} = 504 \Leftrightarrow$$

$$\{\{x^2 - x - 42\}\{x^2 - x - 20\}\} = 504 \qquad (*)$$

Making the substitution $y = x^2 - x$, equation (*) becomes a quadratic equation in y, i.e.

$$(y - 42)(y - 20) = 504 \Leftrightarrow y^2 - 42y - 20y + 840 = 504 \Leftrightarrow$$

$$y^2 - 62y + 336 = 0 \Rightarrow y = 6 \text{ or } y = 56$$

When $y = 6$, $x^2 - x = 6$, i.e. $x^2 - x - 6 = 0$, i.e. $x = -4$ or $x = 3$.

When $y = 56$, $x^2 - x = 56$, i.e. $x^2 - x - 56 = 0$, i.e. $x = -7$ or $x = 8$.

The roots of the original equation are, $-7, -4, 3, 8$.

Example 11-5-4: Assuming that $a > 1$, find the positive numbers x that satisfy the equation, (κ and λ are positive integers):

$$2a \sqrt[2\kappa\lambda]{x^{\kappa+\lambda}} = \sqrt[\kappa]{x} + \sqrt[\lambda]{x}$$

Solution

Since **both sides of the equation are positive**, (since $x > 0$), we obtain an **equivalent equation**, by squaring both sides, i.e.

$$2a \sqrt[2\kappa\lambda]{x^{\kappa+\lambda}} = \sqrt[\kappa]{x} + \sqrt[\lambda]{x} \Leftrightarrow \left(2a \sqrt[2\kappa\lambda]{x^{\kappa+\lambda}}\right)^2 = \left(\sqrt[\kappa]{x} + \sqrt[\lambda]{x}\right)^2 \Leftrightarrow$$

$$4a^2 \sqrt[\kappa\lambda]{x^{\kappa+\lambda}} = \left(\sqrt[\kappa]{x} + \sqrt[\lambda]{x}\right)^2 \Leftrightarrow 4a^2 x^{\frac{\kappa+\lambda}{\kappa\lambda}} = \left(x^{\frac{1}{\kappa}} + x^{\frac{1}{\lambda}}\right)^2 \Leftrightarrow$$

$$4a^2 x^{\left(\frac{1}{\kappa}+\frac{1}{\lambda}\right)} = x^{\frac{2}{\kappa}} + x^{\frac{2}{\lambda}} + 2x^{\left(\frac{1}{\kappa}+\frac{1}{\lambda}\right)} \Leftrightarrow$$

$$4a^2 = \frac{x^{\frac{2}{\kappa}} + x^{\frac{2}{\lambda}} + 2x^{\left(\frac{1}{\kappa}+\frac{1}{\lambda}\right)}}{x^{\left(\frac{1}{\kappa}+\frac{1}{\lambda}\right)}} = x^{\left(\frac{1}{\kappa}-\frac{1}{\lambda}\right)} + x^{-\left(\frac{1}{\kappa}-\frac{1}{\lambda}\right)} + 2 \qquad (*)$$

If we make the substitution:

$$y = x^{\left(\frac{1}{\kappa}-\frac{1}{\lambda}\right)} = x^{\frac{\lambda-\kappa}{\kappa\lambda}} = \sqrt[\kappa\lambda]{x^{\lambda-\kappa}} \qquad (**)$$

equation (*) becomes,

$$4a^2 = y + \frac{1}{y} + 2 \Leftrightarrow y^2 + (2 - 4a^2)y + 1 = 0 \qquad (***)$$

Equation (***) is quadratic in y, and its two roots are:

$$y = \frac{4a^2 - 2 \pm \sqrt{(2 - 4a^2)^2 - 4}}{2} = 2a^2 - 1 \pm 2a\sqrt{a^2 - 1} \Leftrightarrow$$

$$y = \left(a \pm \sqrt{a^2 - 1}\right)^2 \qquad (****)$$

Finally, by virtue of (****) equation (**) yields the sough for values of x,

$$y = \sqrt[\kappa\lambda]{x^{\lambda-\kappa}} = \left(a \pm \sqrt{a^2 - 1}\right)^2 \Leftrightarrow x^{\lambda-\kappa} = \left(a \pm \sqrt{a^2 - 1}\right)^{2\kappa\lambda} \Leftrightarrow$$

$$x = \left(a \pm \sqrt{a^2 - 1}\right)^{\frac{2\kappa\lambda}{\lambda-\kappa}}$$

PROBLEMS

11-5-1) Solve the equation: $\left(x + \frac{1}{x}\right)^2 - 2\left(x - \frac{1}{x}\right) = 4$, (set $y = x - \frac{1}{x}$).

(**Ans:** $x = \pm 1,\ 1 \pm \sqrt{2}$).

11-5-2) Solve the equation: $(x^2 + 2x - 3)^2 - 3(x^2 + 2x - 3) + 2 = 0$.

Hint: Set $y = x^2 + 2x - 3$.

11-5-3) Solve the equation: $25\sqrt{x^2 + x + 37} = 7(2x^2 + 2x + 1)$.

(**Ans:** $x = 3\ or\ x = -4$).

Hint: Set $y = x^2 + x + 37$.

11-5-4) Solve the equation: $x^2 - x - 5 + \frac{6}{x^2-x} = 0$.

Hint: Set $y = x^2 - x$.

11-5-5) Solve the equation: $\sqrt[3]{(1+x)^2} + 4\sqrt[3]{(1-x)^2} - 5\sqrt[3]{1 - x^2} = 0$.

(**Ans:** $x = 0\ or\ x = 63/65$).

Hint: See section 7-6.

11-5-6) Solve the equation: $(x+2)(x+3)(x+4)(x+5) = 80$.

Hint: See Example 11-5-3.

11-5-7) Solve the equation: $9(x^3-3) = x^3(x^3-19)$, (set $y = x^3$).

(Ans: $x^3 = 1$ or $x^3 = 27$).

11-5-8) Provided that the rational numbers κ, λ, x, y satisfy the relation

$$(\kappa y - \lambda x)^2 + 4(x - \kappa)(y - \lambda) = 0$$

show that either $x = \kappa$ and $y = \lambda$, or else the number $(1 - \kappa\lambda)$ is the square of a rational number.

11-5-9) Solve the equation: $\sqrt{x^2 + x} + \frac{\sqrt{x-1}}{\sqrt{x^3-x}} = 2$.

(Ans: $x = (-1 \pm \sqrt{5})/2$).

11-5-10) Solve the equation: $5x^2 - (x+1)(x+3) + 3 = 2x^2(x-1)^2$.

(Ans: $x = -1, 0, 1, 2$).

Hint: Simplify the equation, and set $y = x(x-1)$.

11-5-11) Find the real solutions of the equation:

$$\sqrt[n]{1-x^2} = \sqrt[n]{(1+x)^2} - \sqrt[n]{(1-x)^2}$$

Consider two cases, **a)** n odd, and **b)** n even.

(Ans: $n\ odd: x = \left\{(1 \pm \sqrt{5})^n - 2^n\right\}/\left\{(1 \pm \sqrt{5})^n + 2^n\right\}$, $n\ even: x = \left\{(1 + \sqrt{5})^n - 2^n\right\}/\left\{(1 + \sqrt{5})^n + 2^n\right\}$).

Hint: Since $x \neq 1$ (why?), the given equation is equivalent to the following:

$$1 = \sqrt[n]{\frac{(1+x)^2}{1-x^2}} - \sqrt[n]{\frac{(1-x)^2}{1-x^2}} \iff \sqrt[n]{\frac{1+x}{1-x}} - \sqrt[n]{\frac{1-x}{1+x}} = 1$$

or, if we set $y = \sqrt[n]{\frac{1+x}{1-x}}$, $y - \frac{1}{y} = 1$, or $y^2 - y - 1 = 0$, etc.

11-5-12) Find the real solutions of the equation:

$$\sqrt{4x^2 - 17x + 15} = \sqrt{x^2 - 3x} + \sqrt{x^2 - 9}$$

11-5-13) Solve the equation:

$$\frac{x^2 + 4x + 3}{x^2 - 4x + 3} + \frac{x^2 - 4x + 3}{x^2 + 4x + 3} = \frac{x^2 + 6x + 8}{x^2 - 6x + 8} + \frac{x^2 - 6x + 8}{x^2 + 6x + 8}$$

(**Ans:** $x = 0 \ (double \ root), \pm\sqrt{7}, \pm i\sqrt{5}$).

CHAPTER 12: NON LINEAR ALGEBRAIC SYSTEMS

In Chapter 4, we consider linear systems, i.e. systems where the unknowns are raised to the first power. For example, the system

$$\{3x + 2y = 6, \quad 7x - 4x = 19\}$$

is a **linear system**, two equations in two unknowns.

If at least one of the unknowns, either x or y (or both), is raised to a power higher than one, or if the product of two unknowns appears in at least one equation of the system, then the system is called **non linear**. For example, the following systems are non linear.

$$\begin{cases} 2x^2 + y = 9 \\ 2x - 5xy = 7 \end{cases}, \quad \begin{cases} x^3 + y^3 = 27 \\ x^2 - y = 2 \end{cases}, \quad \begin{cases} 3x + 5xy = 1 \\ x^2 - 4y = -8 \end{cases}$$

Solving a non linear system means to find **all the pairs (x, y) of real or complex numbers, which satisfy the system**.

Of course, it is possible to have a non linear system consisting of three equations with three unknowns, for example,

$$\{2xz + 3xy + z^2 = 8, \quad x^2 + y^2 - 4z = 3, \quad x^2 - y - z = 15\}$$

Solving a non linear system is, by far, more difficult as compared to solving a linear system.

In this chapter, we shall consider a few types of non linear systems, which can be solved analytically.

12-1) One of the equations is linear in one of the unknowns

In this case, we solve the equation for the unknown raised to the first power, say for example the unknown y, **and express y in terms of x**, i.e. $y = y(x)$. Then we substitute $y = y(x)$ in the other equation and we thus obtain an equation for x. We solve this equation, find the values of x and from this the corresponding values of y.

Example 12-1-1: Solve the system: $3x + 2y = 7, \quad x^2 - y^2 + 3x = 0$.

Solution

From the first equation (**linear in both unknowns**) we find:
$y = (7 - 3x)/2$, and substituting this expression of y in the second equation (**non linear**), we obtain,

$$x^2 - \left(\frac{7-3x}{2}\right)^2 + 3x = 0 \Leftrightarrow 5x^2 - 54x + 49 = 0$$

which is a quadratic equation in x, having solutions: $x = 1$, or $x = \frac{98}{10}$.

When $x = 1$, $y = \frac{7-3\cdot 1}{2} = 2$.

When $x = \frac{98}{10}$, $y = \frac{7 - 3 \cdot \frac{98}{10}}{2} = -\frac{224}{20} = -\frac{112}{10}$.

The solutions of the system are: $(x, y) = (1, 2)$ or $(x, y) = \left(\frac{98}{10}, -\frac{112}{10}\right)$.

Example 12-1-2: Solve the system: $x + y = 5$, $x^2 + y^2 = 13$.

Solution

From the first equation we find $y = 5 - x$, and substituting in the second equation we obtain:

$$x^2 + (5-x)^2 = 13 \Leftrightarrow x^2 - 5x + 6 = 0 \Rightarrow x = 2 \text{ or } x = 3$$

When $x = 2$, $y = 5 - 2 = 3$.

When $x = 3$, $y = 5 - 3 = 2$.

The solutions of the system are: $(x, y) = (2, 3)$ or $(x, y) = (3, 2)$. Notice that the two solutions are not the same, since (x, y) is **an ordered pair** of numbers.

PROBLEMS

12-1-1) Solve the system: $x - y = 2$, $2x^2 + 3y^2 = 5$.

(Ans: $(x, y) = (1, -1)$ or $(x, y) = \left(\frac{7}{5}, -\frac{3}{5}\right)$).

12-1-2) Solve the system: $y = x^2$, $3x^2 + y^2 = 28$.

12-2) Homogeneous systems

Let us consider the system:

$$\begin{cases} x^2 + 2xy + 3y^2 = 57 \\ 3x^2 + 4xy - 24y^2 = 19 \end{cases} : (S)$$

Each term in the first equation is a monomial of degree 2, with respect to x and y, (the degree of $x^2 = x^2 y^0$ is $2 + 0 = 2$, the degree of xy is $1 + 1 = 2$, and the degree of $y^2 = x^0 y^2$ is $0 + 2 = 2$) and similarly, **each term** in the second equation is also a monomial **of the same degree 2. In this case, we say that the system (S) is an homogeneous system of degree 2**. To solve a homogeneous system **we call λ the ratio y/x**, i.e. we set:

$$\frac{y}{x} = \lambda \Rightarrow y = \lambda x \qquad (12-2-1)$$

We substitute this expression of y in the two equations in (S) and we find:

$$\begin{cases} x^2 + 2\lambda x^2 + 3\lambda^2 x^2 = 57 \\ 3x^2 + 4\lambda x^2 - 24\lambda^2 x^2 = 19 \end{cases} \Rightarrow \begin{cases} x^2(1 + 2\lambda + 3\lambda^2) = 57 \\ x^2(3 + 4\lambda - 24\lambda^2) = 19 \end{cases} \quad (*)$$

and dividing the two equations in (*) term wise, the x^2 is eliminated, and we thus obtain the following **quadratic equation for λ**:

$$\frac{1 + 2\lambda + 3\lambda^2}{3 + 4\lambda - 24\lambda^2} = \frac{57}{19} = 3 \Leftrightarrow 75\lambda^2 - 10\lambda - 8 = 0 \Rightarrow \begin{cases} \lambda_1 = 6/15 \\ or \\ \lambda_2 = -4/15 \end{cases}$$

a) When $\lambda = 6/15$, equation (12-2-1) yields $y = \frac{6}{15} x$, and substituting in the first equation in the system (S) we obtain:

$$x^2 + 2x \cdot \frac{6}{15} x + 3 \left(\frac{6}{15} x \right)^2 = 57 \Leftrightarrow \frac{513}{15^2} x^2 = 57 \Leftrightarrow x^2 = \frac{57 \cdot 15^2}{513} \Leftrightarrow$$

$$x^2 = 25 \Leftrightarrow x = \pm 5 \xrightarrow{\left(y = \frac{6}{15} x \right)} \{ y = 2 \quad or \quad y = -2 \} \qquad (**)$$

b) When $\lambda = -4/15$, equation (12-2-1) yields $y = -\frac{4}{15}x$, and substituting in the first equation in the system (S) we obtain:

$$x^2 + 2x\left(-\frac{4}{15}x\right) + 3\left(-\frac{4}{15}x\right)^2 = 57 \Leftrightarrow \frac{153}{15^2}x^2 = 57 \Leftrightarrow x^2 = \frac{15^2 \cdot 19}{51} \Leftrightarrow$$

$$x = \pm 15\sqrt{\frac{19}{51}} \xRightarrow{\left(y=-\frac{4}{15}x\right)} \left\{ y = -4\sqrt{\frac{19}{51}} \quad \text{or} \quad y = 4\sqrt{\frac{19}{51}} \right\} \quad (***)$$

We have thus found the solutions of the system (S) to be:

$$\left\{ \begin{array}{ll} (x,y) = (5,2) & (x,y) = (-5,-2) \\ (x,y) = \left(15\sqrt{19/51}, -4\sqrt{19/51}\right) & (x,y) = \left(-15\sqrt{19/51}, 4\sqrt{19/51}\right) \end{array} \right\}$$

Let us now consider a homogeneous system consisting of three equations in three unknowns, and outline the method of solution.

Example 12-2-1: Solve the system:

$$\begin{cases} x^2 + y^2 + xy = 37 \\ y^2 + z^2 + yz = 19 \\ z^2 + x^2 + zx = 28 \end{cases} \quad (S)$$

Solution

The given system **is homogeneous of degree 2**. Subtracting the second equation from the first and the third equation from the second, we obtain:

$$\begin{cases} x^2 - z^2 + xy - yz = 18 \\ y^2 - x^2 + yz - zx = -9 \end{cases} \Leftrightarrow \begin{cases} (x-z)(x+z) + y(x-z) = 18 \\ (y-x)(y+x) + z(y-x) = -9 \end{cases} \Leftrightarrow$$

$$\begin{cases} (x-z)(x+y+z) = 18 \\ (y-x)(x+y+z) = -9 \end{cases} \quad (*)$$

and assuming that $(x + y + z) \neq 0$ and dividing the two equations in (*), we obtain,

$$\frac{x-z}{y-x} = -2 \Leftrightarrow x - z = -2(y-x) \Leftrightarrow z = 2y - x \quad (**)$$

Thus far, we have expressed z in terms of x and y, and substituting this expression of z in the second equation in the system (S), we find, $x^2 + 7y^2 - 5xy = 19$. This equation together with the first equation in the system (S) constitute **an homogeneous system, two equations in the two unknowns x and y**, i.e.

$$\begin{cases} x^2 + y^2 + xy = 37 \\ x^2 + 7y^2 - 5xy = -19 \end{cases}: \quad (S_1)$$

This system can be solved by the method described in the preceding example, i.e. we set $\lambda = y/x \Leftrightarrow y = \lambda x$, etc. Once x and y are determined, then z is found from equation (**).

Detailed calculations are left to the reader (see Problem 12-2-1). The solution of the system is found to be:

$$\begin{cases} x = 4 \\ y = 3 \\ z = 2 \end{cases}, \quad \begin{cases} x = -4 \\ y = -3 \\ z = -2 \end{cases}, \quad \begin{cases} x = 10\sqrt{3}/3 \\ y = \sqrt{3}/3 \\ z = -8\sqrt{3}/3 \end{cases}, \quad \begin{cases} x = -10\sqrt{3}/3 \\ y = -\sqrt{3}/3 \\ z = 8\sqrt{3}/3 \end{cases}$$

PROBLEMS

12-2-1) In Example 12-2-1 perform detailed calculations to verify the solution.

12-2-2) Solve the system: $\{x^2 + y^2 = 218, \quad xy - y^2 = 42\}$.

12-2-3) Solve the system:

$$\begin{cases} x^2 + xy + xz = 60 \\ y^2 + yx + yz = 75 \\ z^2 + zx + zy = 90 \end{cases}$$

(Ans: $(x, y, z) = (4, 5, 6) \text{ or } (x, y, z) = (-4, -5, -6)$).

12-2-4) Solve the system:

$$\{x^2 + 2xy - y^2 = y - x, \quad 3x^2 - 4xy + 2y^2 = y + x\}$$

12-2-5) Solve the system:

$$\{x^2 - 3xy + 2y^2 = 2, \quad 2x^2 + xy - y^2 = 20\}$$

(Ans: $(x, y) = (3,1), (-3, -1), (7\sqrt{2}/3, 8\sqrt{2}/3), (-7\sqrt{2}/3, -8\sqrt{2}/3)$).

12-3) Symmetric systems

Symmetric systems are the systems consisting **of symmetric equations**. An equation $f(x, y)$ is said to be symmetric, if it remains unchanged when x and y are mutually interchanged, i.e. when $f(x, y) = f(y, x)$. For example $x^3 + y^3 - 3xy$ is a symmetric expression of x and y, while $x^3 - y^3 - 3xy$ is not.

In a symmetric system, usually we try **to find the sum $x + y$ and the product xy of the two unknowns**, i.e. to find $x + y = S$ (**known**), and $xy = P$ (**known**). Then the x and y will be the roots of the quadratic equation, (see Remark 1 in section 9-1).

$$t^2 - St + P = 0$$

Let us consider a few examples.

Example 12-3-1: Solve the system: $\{x^2 + y^2 + x + y = 18, \quad xy = 6\}$.

Solution

The system is symmetric. We consider the **"auxiliary unknowns"**

$$u = x + y, \quad \text{and} \quad v = xy \qquad (*)$$

The first equation of the system is written as,

$$(x + y)^2 - 2xy + (x + y) = 18 \stackrel{(*)}{\Rightarrow} u^2 - 2v + u = 18 \qquad (**)$$

while the second equation of the system is written as $v = 6$.

In terms of u and v, the original system assumes the following form:

$$\begin{cases} u^2 - 2v + u = 18 \\ v = 6 \end{cases} \Rightarrow u^2 + u - 12 - 18 = 0 \Rightarrow u^2 + u - 30 = 0 \Rightarrow$$

$$\{u = 5 \quad or \quad u = -6\}$$

a) When $u = 5$ and $v = 6$, equation (*) implies: $x + y = 5, xy = 6$, and x and y are the roots of the quadratic equation

$$t^2 - 5t + 6 = 0 \Rightarrow t = 3 \text{ or } t = 2$$

and this means that either $(x = 3, y = 2)$ or $(x = 2, y = 3)$.

b) When $u = -6$ and $v = 6$, equation (*) implies: $x + y = -6, xy = 6$, and x and y are the roots of the quadratic equation

$$t^2 + 6t + 6 = 0 \Rightarrow t = -3 + \sqrt{3} \text{ or } t = -3 - \sqrt{3}$$

and this means that either $\left(x = -3 + \sqrt{3}, y = -3 - \sqrt{3}\right)$ or $\left(x = -3 - \sqrt{3}, y = -3 + \sqrt{3}\right)$.

In summary: The solutions of the system are,

$$(x, y): (3,2), (2,3), \left(-3 + \sqrt{3}, -3 - \sqrt{3}\right), \left(-3 - \sqrt{3}, -3 + \sqrt{3}\right)$$

Example 12-3-2: Solve the system: $\left\{ xy(x + y) = 120, \ \dfrac{1}{x} + \dfrac{1}{y} = \dfrac{8}{15} \right\}$.

Solution

The system is symmetric. In terms of the **auxiliary unknowns** $u = x + y$ and $v = xy$, the system assumes the form:

$$\begin{cases} uv = 120 \\ \dfrac{u}{v} = \dfrac{8}{15} \end{cases} \quad (*)$$

and multiplication of the two equations in (*) yields,

$$u^2 = 64 \Rightarrow u = \pm 8 \stackrel{(*)}{\Rightarrow} \begin{cases} \text{When } u = 8, v = \dfrac{120}{8} = 15 & (a) \\ \text{When } u = -8, v = \dfrac{120}{-8} = -15 & (b) \end{cases} \quad (**)$$

a) $u = 8, v = 15$, i.e. $x + y = 8, xy = 15$, and x, y are the roots of the quadratic equation

$$t^2 - 8t + 15 = 0 \Rightarrow t = 5 \text{ or } t = 3$$

and this means that either $(x, y) = (5,3)$ or $(x, y) = (3,5)$.

b) $u = -8, v = -15$, i.e. $x + y = -8, xy = -15$, and x, y are the roots of the quadratic equation

$$t^2 + 8t - 15 = 0 \Rightarrow t = -4 + \sqrt{31} \text{ or } t = -4 - \sqrt{31}$$

and this means that either $(x, y) = \left(-4 + \sqrt{31}, -4 - \sqrt{31}\right)$ or $(x, y) = \left(-4 - \sqrt{31}, -4 + \sqrt{31}\right)$.

In summary: The solutions of the system are,

$$(x, y): (5,3), (3,5), \left(-4 + \sqrt{31}, -4 - \sqrt{31}\right), \left(-4 - \sqrt{31}, -4 + \sqrt{31}\right)$$

PROBLEMS

12-3-1) Solve the system: $\{x^2 + 2x + 2y + y^2 = 23, \quad x^2 + xy + y^2 = 19\}$

(Ans: $(x, y): (2,3), (3,2), (-5,2), (2, -5)$).

12-3-2) Solve the system: $\{x + y = a, \quad x^3 + y^3 = 2a\}, (a \in \mathbb{R})$. If $|a| < \sqrt{8}$, show that the solutions are real numbers, if $|a| = \sqrt{8}$, then $x = y = \frac{\sqrt{8}}{2}$, while if $|a| > \sqrt{8}$ the solutions of the system are complex numbers.

12-3-3) Solve the system: $\left\{x^2 + y^2 = \frac{97}{9}, \quad x\sqrt{xy} + y\sqrt{xy} = \frac{26}{3}\right\}$, ($x, y$ positive real numbers).

(Ans: $(x, y): \left(3, \frac{4}{3}\right), \left(\frac{4}{3}, 3\right)$).

12-3-4) Solve the system:

$$\begin{cases} \dfrac{1}{x^2} + \dfrac{1}{y^2} + \dfrac{3}{xy} = -1 \\ \dfrac{1}{x} + \dfrac{1}{y} = -1 \end{cases}$$

Hint: Set $X = \dfrac{1}{x}, Y = \dfrac{1}{y}$.

12-3-5) Solve the system:

$$\begin{cases} x^2 + y^2 + x + y = 4 \\ x^2y^2 + x^2y + xy^2 + xy = 4 \end{cases}$$

(Ans: $(x,y): (1,1), (1,-2), (-2,1), (-2,-2)$).

12-3-6) Find the real solutions of the system: $\{x^4 + y^4 = 97, \quad x + y = 5\}$.

(Ans: $(x,y): (2,3), (3,2)$).

12-4) Some general techniques and miscellaneous examples

There is no systematic methods for solving non linear systems. It all depends on the particular system at hand. In this section we shall solve some non linear systems, and in the process of solving we shall outline **some general guidelines** that, potentially, could be used to solve similar types of systems.

Example 12-4-1: Find the real solutions of the system:

$$\{xyz = a, \quad yzw = b, \quad zwx = c, \quad wxy = d\}, \qquad (abcd \neq 0)$$

Solution

Term wise multiplication of the four equations yields,

$$x^3y^3z^3w^3 = abcd \Leftrightarrow (xyzw)^3 = abcd \Leftrightarrow xyzw = \sqrt[3]{abcd} \qquad (*)$$

Dividing equation (*) by the first equation of the system yields,

$$w = \frac{\sqrt[3]{abcd}}{a} = \sqrt[3]{\frac{bcd}{a^2}}$$

and similarly we find x, y and z.

Example 12-4-2: Outline the method of solution of the system:

$$\begin{cases} x^2 + y + z = 9 \\ (y-x)(x+y-5) = 0 \\ (z-2x)(3x+2z-8) = 0 \end{cases} \quad (S)$$

Solution

From the second equation, we have $(y - x) = 0$ or $(x + y - 5) = 0$ and from the third equation we have $(z - 2x) = 0$ or $(3x + 2z - 8) = 0$. Thus, the original system is split into the following four systems:

$$\begin{cases} x^2 + y + z = 9 \\ y - x = 0 \\ z - 2x = 0 \end{cases} \qquad \begin{cases} x^2 + y + z = 9 \\ x + y - 5 = 0 \\ z - 2x = 0 \end{cases}$$

$$\begin{cases} x^2 + y + z = 9 \\ y - x = 0 \\ 3x + 2z - 8 = 0 \end{cases} \qquad \begin{cases} x^2 + y + z = 9 \\ x + y - 5 = 0 \\ 3x + 2z - 8 = 0 \end{cases}$$

To solve, for example, the first system, we express y and z in terms of x, (from the second and third equation, respectively) and substitute into the first equation i.e. $y = x, z = 2x$ and the first equation becomes:

$$x^2 + x + 2x - 9 = 0 \iff x^2 + 3x - 9 = 0 \implies x_{1,2} = \frac{-3 \pm \sqrt{45}}{2}$$

and then find y and z, etc. We repeat the same with the other three systems.

Example 12-4-3: Outline the method of solution of the system:

$$\begin{cases} axy + by = A \\ cyz + dz = B \\ \kappa zx + \lambda x = C \end{cases}, \quad a, b, c, d, \kappa, \lambda, A, B, C \text{ are known constants}$$

Solution

From the third equation we express x in terms of z, and from the second equation we express y in terms of z as well, i.e.

$$x = \frac{C}{\kappa z + \lambda}, \quad y = \frac{B - dz}{cz} \qquad (*)$$

Substituting these expressions of x and y in the first equation, we obtain the following equation, where **the only unknown is z**.

$$a \frac{C}{\kappa z + \lambda} \frac{B - dz}{cz} + b \frac{B - dz}{cz} = A \iff$$

$$\kappa(cA + bd)z^2 + (c\lambda A + ad - bd\lambda - b\kappa B)z - aBC - b\lambda B = 0 \quad (**)$$

Equation (**) is **quadratic in z**, which yields two values of z. Once z is determined, the corresponding values of x and y are obtained from (*).

Example 12-4-4: Outline the method of solution of the system:

$$\begin{cases} x + y = axyz \\ y + z = bxyz \\ z + x = cxyz \end{cases}, \quad a, b, c \text{ are known consatnts } (abc \neq 0)$$

Solution

a) One solution of the system is the obvious solution (**trivial solution**), $x = 0, y = 0, z = 0$. Note that if at least one of x or y or z is zero, then all the unknowns will be zero.

b) We seek **the nontrivial solutions**, $(xyz \neq 0)$. Dividing each equation of the system by $xyz \neq 0$, we obtain:

$$\begin{cases} \dfrac{1}{yz} + \dfrac{1}{zx} = a \\ \dfrac{1}{zx} + \dfrac{1}{xy} = b \\ \dfrac{1}{xy} + \dfrac{1}{yz} = c \end{cases} \quad (*)$$

and if we call, $X = \dfrac{1}{yz}, Y = \dfrac{1}{zx}, Z = \dfrac{1}{xy}$, system (*) is transformed to the following:

$$\begin{cases} X + Y = a \\ Y + Z = b \\ Z + X = c \end{cases} \quad (**)$$

System (*) is **a linear system in X, Y, Z** and can be solved, for example using Cramer's rule. A easier method of solution, is to add term wise the given equations, the result being,

$$2(X + Y + Z) = a + b + c \iff X + Y + Z = \frac{a + b + c}{2}$$

or, since $X + Y = a$, (from the first equation in (**)), $Z = \frac{b+c-a}{2}$, and similarly we find, $X = \frac{a+c-b}{2}$ and $Y = \frac{b+a-c}{2}$, and going back to the original variables x, y, z we have:

$$\left\{\frac{1}{yz} = \frac{a+c-b}{2}, \quad \frac{1}{zx} = \frac{b+a-c}{2}, \quad \frac{1}{xy} = \frac{b+c-a}{2}\right\} \quad (***)$$

or equivalently,

$$\begin{cases} yz = F_1 \stackrel{\text{def}}{=} \dfrac{2}{a+c-b} \\ zx = F_2 \stackrel{\text{def}}{=} \dfrac{2}{b+a-c} \\ xy = F_3 \stackrel{\text{def}}{=} \dfrac{2}{b+c-a} \end{cases} \quad (****)$$

The system in (****) is of the form studied in Example 12-4-1 and is solved according to the method developed in this Example.

PROBLEMS

12-4-1) Solve the system:

$$\{x^2 + xy + xz = 6, \quad y^2 + xy + yz = 12, \quad z^2 + xz + yz = 18\}$$

(Ans: (x, y, z): $(1, 2, 3)$ or $(-1, -2, -3)$).

Hint: Add term wise the three equations to find the sum $(x + y + z)$.

12-4-2) Solve the system: $\{x + y = c, \quad x^2 + y^2 = c^2\}$.

12-4-3) Solve the system:

$$\begin{cases} x + y + z = c \\ x^2 + y^2 + z^2 = c^2 \\ x^3 + y^3 + z^3 = c^3 \end{cases}, \quad c \text{ is a known constant}$$

(Ans: (x, y, z): $(c, 0, 0)$ or $(0, c, 0)$ or $(0, 0, c)$).

12-4-4) Solve the system: $\{xy = 2, \quad yz = 6, \quad zx = 3\}$.

Hint: See Example 12-4-1.

12-4-5) Solve the system:

$$\{x^2 + yz = y + z, \quad y^2 + zx = z + x, \quad z^2 + xy = x + y\}$$

(Ans: (x, y, z): $(0,0,0)$ or $(-1,1,1)$ or $(1,-1,1)$).

Hint: Subtracting the second and the third equation from the first, show that the system reduces to the type studied in Example 12-4-2.

12-4-6) Solve the system: $\left\{x^4 + y^4 = \frac{82}{9}x^2y^2, \quad x + y = 4\right\}$.

Hint: From the first equation we find:

$$\frac{x^4 + y^4}{x^2y^2} = \frac{82}{9} \Leftrightarrow \frac{x^2}{y^2} + \frac{y^2}{x^2} = \frac{82}{9}$$

and if we set $\lambda = \frac{y}{x}$, we obtain a biquadratic equation in λ, etc.

12-4-7) Solve the system:

$$\left\{ \begin{array}{c} 2x^3 = 3y^3 = 4z^3 \\ \frac{1}{x} + \frac{1}{y} + \frac{1}{z} = 8 \end{array} \right\}$$

(Ans: $(x, y, z) = \frac{\left(\sqrt[3]{2}+\sqrt[3]{3}+\sqrt[3]{4}\right)}{8} \left(\frac{1}{\sqrt[3]{2}}, \frac{1}{\sqrt[3]{3}}, \frac{1}{\sqrt[3]{4}}\right)$).

Hint: In the first equation set: $2x^3 = 3y^3 = 4z^3 = \lambda$, i.e. $x = \sqrt[3]{\lambda/2}, y = \sqrt[3]{\lambda/3}$ and $z = \sqrt[3]{\lambda/4}$, which when substituted in the second equation yield an equation for λ, etc.

12-4-8) Solve the system:

$$\left\{ \begin{array}{c} x^2 + (y - z)^2 = 20 \\ y^2 + (z - x)^2 = 26 \\ z^2 + (x - y)^2 = 10 \end{array} \right\}$$

12-4-9) Solve the system:

$$\left\{\frac{x + y - z}{5} = \frac{z + x - y}{11} = \frac{y + z - x}{7} = \frac{xyz}{3}\right\}$$

(**Ans:** (x, y, z): $(2/3, 1/2, 3/4)$ or $(-2/3, -1/2, -3/4)$).

Hint: We set: $\frac{x+y-z}{5} = \frac{z+x-y}{11} = \frac{y+z-x}{7} = \lambda$ and express x, y, z in terms of λ. Then from $\frac{x+y-z}{5} = \frac{xyz}{3}$, we find that $\lambda = \pm\frac{1}{12}$ and since x, y, z have already been expressed in terms of λ, we find the solution of the system.

12-4-10) Solve the system:

$$\{24(y+z) = 7xyz, \quad 24(z+x) = 6xyz \quad 24(x+y) = 5xyz\}$$

Hint: See Example 12-4-4.

12-4-11) Consider a triangle ABC with sides $a = 8\ m, b = 5\ m, c = 4\ m$. **a)** Show that $\hat{A} > 90°$, **b)** When all the sides of the triangle are increased by the same length $x > 0$, the triangle becomes a right triangle, with $\hat{A} = 90°$. Find x.

(**Ans: a)** It suffices to show $a^2 > b^2 + c^2$, **b)** $x = -1 + 2\sqrt{6}\ m$).

12-4-12) Solve the system: (c is a given real number).

$$\{x + y + z = 0, \quad x^3 + y^3 + z^3 = 3xyz, \quad xy + yz + xz = c^2\}$$

(**Ans:** If $c = 0$, then $(x, y, z) = (0,0,0)$, if $c \neq 0$ there are no real solutions.

Hint: See Cauchy's identity, Pr. 7-1-7.

12-4-13) Solve the system:

$$\begin{cases} x(x+y+z) + yz = 15 \\ y(x+y+z) + zx = 18 \\ z(x+y+z) + xy = 30 \end{cases}$$

(**Ans:** (x, y, z): $(1,2,4)$ or $(-1,-2,-4)$).

Hint: Set $y = \lambda x, z = \omega y = \lambda \omega x$, then divide the second equation with the first and the third equation with the second to obtain a system of two equation in the two unknowns λ and ω, etc. As an **alternative solution**, we factor each one of the equations of the system, and if we multiply them together we find that $(x+y)(y+z)(z+x) = \pm 90$, etc.

12-4-14) Solve the system:

$$\{x^2 - yz = 4, \quad y^2 - zx = 9, \quad z^2 - xy = 25\}$$

Hint: Adding the three equations term wise, we find:

$$x^2 + y^2 + z^2 - yz - zx - xy = 38 \Leftrightarrow$$

$$\frac{1}{2}\{(x-y)^2 + (y-z)^2 + (z-x)^2\} = 38 \qquad (*)$$

Subtracting the first equation of the system from the second, we find:

$$y^2 - x^2 + yz - zx = 5 \Leftrightarrow (y-x)(y+x) + z(y-x) = 5 \Leftrightarrow$$

$$(y-x)(y+x+z) = 5$$

and if we call $S = (x + y + z)$, then $(y - x) = 5/S$, and similarly, $(z - x) = 21/S$ and $(z - y) = 16/S$. By virtue of these equations, equation (*) becomes,

$$\frac{1}{2}\frac{5^2 + 21^2 + 16^2}{S^2} = 38$$

The sum $S = x + y + z$ is thus determined, (two values, $S_1 = +\sqrt{S^2}, S_2 = -\sqrt{S^2}$), and the original system is split into two, linear systems,

$$\begin{cases} x + y + z = S_1 \\ y - x = \dfrac{5}{S_1} \\ z - x = \dfrac{21}{S_1} \end{cases}, \qquad \begin{cases} x + y + z = S_2 \\ y - x = \dfrac{5}{S_2} \\ z - x = \dfrac{21}{S_2} \end{cases}$$

12-4-15) Given the three altitudes of a triangle h_a, h_b, h_c find its area E and its three sides a, b and c.

Hint: $ah_a = bh_b = ch_c = 2E$ and $E = \sqrt{\tau(\tau-a)(\tau-b)(\tau-c)}$, where $2\tau = a + b + c$, (the Heron's formula).

12-4-16) Solve the system:
$$\left\{\frac{x^5 + y^5}{x^3 + y^3} = \frac{211}{19}, \quad x^2 + xy + y^2 = 7\right\}$$

CHAPTER 13: POLYNOMIALS IN ONE VARIABLE

13-1) Introduction

If n is a **positive integer**, then any function of the form

$$P(x) = a_n x^n + a_{n-1} x^{n-1} + a_{n-2} x^{n-2} + \cdots + a_2 x^2 + a_1 x + a_0 \quad (13-1-1)$$

is called **a polynomial in x, of degree n**. The numbers $a_n, a_{n-1}, \cdots, a_1, a_0$ are called **the coefficients of the polynomial**, and can be either real numbers or more generally, complex numbers. The number a_n **is the leading coefficient** of the polynomial while the number a_0 **is the constant term** of the polynomial.

Any constant number b may be considered as **a polynomial of degree zero**, since $b = bx^0$.

The polynomial in (13-1-1) is arranged **in descending powers of x**. The same polynomial can be arranged **in ascending powers of x**, as follows:

$$P(x) = a_0 + a_1 x + a_2 x^2 + \cdots + a_{n-2} x^{n-2} + a_{n-1} x^{n-1} + a_n x^n$$

For example, $P(x) = 2x^3 - 3x^2 + 7x + 5$ is a third degree polynomial, arranged in descending powers of x, having leading coefficient 2 and constant term 5. The polynomial $Q(x) = -x^4 + x^2 + 6x + 9$ is a polynomial of degree four, having leading coefficient (-1) and constant term 9. The term x^3 does not appear in the polynomial, but we may assume that x^3 appears in its appropriate position with coefficient $a_3 = 0$, i.e. we may write, $Q(x) = -x^4 + 0 \cdot x^3 + x^2 + 6x + 9$.

The expressions $A(x) = 3x^2 + 2\sqrt{x} + 5$ and $B(x) = x^{3/5} - 2x + 1$ **are not polynomials**, due to the square root $(\sqrt{x})$ in $A(x)$ and the fractional exponent $(3/5)$ in $B(x)$.

Given a polynomial $P(x)$ and a number a, (**real or complex**), the number obtained when the variable x is substituted with a, is called **the arithmetic value of the polynomial corresponding to $x = a$**, and is designated by $P(a)$. For example, if $P(x) = 2x^3 + x^2 - 5x + 1$, then $P(1) = 2 \cdot 1^3 + 1^2 - 5 \cdot 1 + 1 = -1$, $P(-2) = 2 \cdot (-2)^3 + (-2)^2 - 5 \cdot (-2) + 1 = -1$, etc.

A number r such that $P(r) = 0$, is called a root of the polynomial. A root of the polynomial $P(x)$ is a root (or a zero) of the polynomial equation $P(x) = 0$. For example, the numbers 1 and (-1) are the roots of the polynomial $P(x) = x^2 - 1$, while the numbers $\pm i$ are the roots of the polynomial $Q(x) = x^2 + 1$.

For addition and subtraction of polynomials, multiplication of a polynomial by a constant number and multiplication of polynomials, see section 1-3. Division of polynomials is treated in details in the next section.

Example 13-1-1: Given that $P(x) = x^3 + x^2 + 2$ and $Q(x) = 2x^4 + x - 7$, find $2P(x) + 3Q(x)$.

Solution

$$2P(x) + 3Q(x) = 2(x^3 + x^2 + 2) + 3(2x^4 + x - 7) \Rightarrow$$

$$2P(x) + 3Q(x) = 2x^3 + 2x^2 + 4 + 6x^4 + 3x - 21 \Rightarrow$$

$$2P(x) + 3Q(x) = 6x^4 + 2x^3 + 2x^2 + 3x - 17$$

Example 13-1-2: Given $P(x) = x^2 - 1$ and $Q(x) = x^3 - 1$, find the product $P(x)Q(x)$.

Solution

$$P(x)Q(x) = (x^2 - 1)(x^3 - 1) = x^5 - x^3 - x^2 + 1$$

Example 13-1-3: Which of the numbers $0, -2, -3$ is a root of the polynomial $P(x) = x^3 + x^2 + 4$?

Solution

$$P(0) = 0^3 + 0^2 + 4 = 4 \neq 0$$

$$P(-2) = (-2)^3 + (-2)^2 + 4 = -8 + 4 + 4 = 0$$

$$P(-3) = (-3)^3 + (-3)^2 + 4 = -27 + 9 + 4 = -14 \neq 0$$

The number (-2) is a root of $P(x) = x^3 + x^2 + 4$.

PROBLEMS

13-1-1) If $P(x) = x^2 - 1$ and $Q(x) = x^3 - 2x^2 - 3x - 4$, find the polynomial $xP(x) - Q(x)$.

(Ans: $2x^2 + 2x + 4$).

13-1-2) If $P(x) = x + 3$ and $Q(x) = x(x + 2)$, find $(P(x))^2 - 2Q(x)$.

13-1-3) If $P(x) = (x + 1)^3$ and $Q(x) = (x - 1)^3$, find the roots of the polynomial $P(x) - Q(x)$. What is the degree of $P(x) - Q(x)$? The degree of $P(x) + Q(x)$?

(Ans: $x = \pm i(1/\sqrt{3})$, second degree, third degree).

13-2) The division of polynomials

The division of polynomials, is in many respects, analogous to the division of positive integers in arithmetic. In arithmetic, we say that a positive number p is divided **exactly** by the positive number d, if there exists another **integer q** such that when multiplied by d, the result of the multiplication is p, i.e. when $p = qd$. In this case we say that p is divided exactly by d, or that d **divides p**. The number p is called **the dividend**, the number d is called **the divisor** and the number q is called **the quotient** of the division. For example, 2 divides 6, since $6 = 3 \cdot 2$, ($q = 3$), 3 divides 15, since $15 = 5 \cdot 3$, ($q = 5$), while 7 does not divide 20, since **there is not an integer** q such that $20 = q \cdot 7$.

If d does **not** divide p, then as shown in arithmetic, there exist two integers q and r, such that

$$p = qd + r, \quad \text{where} \quad 0 < r < d \qquad (*)$$

For example, if $p = 20, d = 7$, we can write, $20 = 2 \cdot 7 + 6$. In this case, $q = 2$ and $r = 6 < 7 = d$. The number r is called **the remainder** of the division. Notice that always, $0 < r < d$.

Similar concepts, definitions and terminology apply for the division of polynomials, which is studied in considerable depth and details in the sequel of this section.

a) Perfect division of polynomials:

We say that a polynomial $D(x) \neq 0$ divides exactly another polynomial $P(x)$ when there exists another polynomial $Q(x)$ such that the result of the multiplication of $Q(x)$ and $D(x)$ **is identically equal** to $P(x)$, i.e. when

$$P(x) \equiv D(x)Q(x) \qquad (13-2-1)$$

Equation (13-2-1) shows that the polynomial $D(x)$ divides exactly the polynomial $P(x)$, or equivalently, **that $P(x)$ is divided exactly by $D(x)$**. The polynomial $P(x)$ is called **the dividend**, the polynomial $D(x)$ is called **the divisor** and the polynomial $Q(x)$ is called **the quotient** of the perfect (exact) division. Equation (13-2-1) is written equivalently as

$$Q(x) \equiv \frac{P(x)}{D(x)} \qquad (13-2-2)$$

and this shows that the simplest possible form of the fractional expression $P(x)/D(x)$ is $Q(x)$. From (13-2-1) we also conclude that $Q(x)$ divides exactly $P(x)$ and the quotient of this division is $D(x)$.

If the degree of $P(x)$ is n and the degree of $D(x)$ is m, with $n \geq m$, then **the degree of the quotient $Q(x)$ will be $(n - m) \geq 0$.**

For example, if $P(x) = x^3 + 1$ and $D(x) = x + 1$, then (from a known identity),

$$Q(x) = \frac{P(x)}{D(x)} = \frac{x^3 + 1}{x + 1} = \underbrace{x^2 - x + 1}_{Q(x)}$$

The divisor $D(x) = x + 1$, divides exactly the dividend $P(x) = x^3 + 1$, and the quotient of the division is $Q(x) = x^2 - x + 1$. Notice that the degree of the quotient is $2 = 3 - 1$.

Corollary 1: A polynomial $P(x)$ is divided by any constant $c \neq 0$.

Indeed, if $P(x) = a_n x^n + a_{n-1} x^{n-1} + \cdots + a_1 x + a_0$, then we may write that,

$$a_n x^n + a_{n-1} x^{n-1} + \cdots + a_1 x + a_0$$
$$= c \underbrace{\left(\frac{a_n}{c} x^n + \frac{a_{n-1}}{c} x^{n-1} + \cdots + \frac{a_1}{c} x + \frac{a_0}{c} \right)}_{Q(x)}$$

and this shows that the quotient $Q(x)$ of the division $P(x)/c$ is the polynomial inside the parentheses.

b) In general, the division between two polynomials **is not perfect**. Only, under some special conditions, the division will be perfect. One general method to check if a division is perfect, is the following:

Suppose that we want to check if the polynomial $D(x) = x^2 + 1$ divides exactly the polynomial $P(x) = x^3 - 3x^2 + 5$. For the division $P(x)/D(x)$ to be perfect, **a polynomial $Q(x)$ must exist**, such that $P(x) = D(x)Q(x)$. The degree of $Q(x)$ must be $(3-2) = 1$, i.e. $Q(x)$ **must be a first degree polynomial**, having general form $Q(x) = ax + b$, where a, b are numbers to be determined. We have,

$$\frac{P(x)}{D(x)} = Q(x) \Leftrightarrow \frac{x^3 - 3x^2 + 5}{x^2 + 1} = ax + b \Leftrightarrow$$

$$x^3 - 3x^2 + 5 = (x^2 + 1)(ax + b) = ax^3 + ax + bx^2 + b \Leftrightarrow$$

$$x^3 - 3x^2 + 0 \cdot x + 5 = ax^3 + bx^2 + ax + b \qquad (**)$$

The two polynomials in (**) must be **identically equal**, i.e. they must represent **one and the same polynomial**, and this can occur **if and only if the coefficients of the like powers of x are identical**, i.e.

$$\begin{cases} a = 1 \\ b = -3 \\ a = 0 \\ b = 5 \end{cases}$$

But this system **is incompatible**, since a cannot be zero and one, simultaneously, and this shows that there is no polynomial $Q(x)$ such that $P(x) = D(x)Q(x)$, in other words the division $(x^3 - 3x^2 + 5)/(x^2 + 1)$ is **not perfect**.

PROBLEMS

13-2-1) If a polynomial $D(x)$ divides exactly two polynomials $P_1(x)$ and $P_2(x)$, show that it will divide exactly the sum $P_2(x) + P_1(x)$, the difference $P_2(x) - P_1(x)$ and the product $P_2(x) \cdot P_1(x)$. Will $D(x)$ divide $P_1(x)/P_2(x)$ necessarily?

13-2-2) Show that $x^4 - 1$ is not divided exactly by $x^2 + 2$, but it is divided exactly by $x^2 + 1$. What is the quotient in this case?

13-2-3) Determine the constants a and b, so that the polynomial $P(x) = x^3 + ax^2 + bx - 3$ is divided exactly by the polynomial $D(x) = x^2 + x - 1$. What is the quotient of the division?

(**Ans:** $a = 4$, $b = 2$, $Q(x) = x + 3$).

13-3) The identity of the algorithmic division

The following Theorem, known as **Theorem of the algorithmic division**, is perhaps the most important Theorem in the division of polynomials, since many other Theorems and corollaries follow directly from this.

Theorem 13-3-1: **Given two polynomials $P(x)$ of degree n, (the Dividend) and $D(x) \neq 0$ of degree m, with $n \geq m$, (the Divisor), there exist two other polynomials $Q(x)$, (the Quotient of degree $(n - m)$) and $R(x)$, (the remainder, of degree less than m = degree of the Divisor), such that**

$$\begin{cases} P(x) \equiv D(x)Q(x) + R(x) \Leftrightarrow \dfrac{P(x)}{D(x)} \equiv Q(x) + \dfrac{R(x)}{D(x)} \\ \text{Degree of } R(x) < \text{Degree of } D(x) \end{cases} \quad (13-3-1)$$

Some important remarks:

1) If $D(x)$ divides exactly $P(x)$, then the remainder $R(x) \equiv 0$, and then we have $P(x) \equiv D(x)Q(x)$, (perfect division).

2) The degree of the remainder $R(x)$ is **strictly smaller** than the degree of the divisor $D(x)$.

3) The ratio of two polynomials $P(x)$ and $D(x)$, i.e. $P(x)/D(x)$ can be expressed as the sum of a polynomial ($Q(x)$, the quotient) and a rational fraction $(R(x)/D(x))$ where the degree of the numerator **is strictly smaller than the degree of the denominator.**

4) Even in a case where the degree n of $P(x)$ is less than the degree m of $D(x)$, i.e. when $n < m$, we can still consider (13-3-1) as being true, provided that we take, $Q(x) \equiv 0$ and $R(x) \equiv P(x)$. In this particular case, eq. (13-3-1) becomes, $P(x) = D(x) \cdot 0 + P(x)$.

Proof of Theorem 13-3-1: Let $P(x) = a_n x^n + a_{n-1} x^{n-1} + \cdots + a_1 x + a_0$ be a polynomial of degree n, and $D(x) = b_m x^m + b_{m-1} x^{m-1} + \cdots + b_1 x + b_0$ be another polynomial of degree m, with $n \geq m$.

a) First we shall prove the following auxiliary proposition: Given the polynomials $P(x)$ and $D(x)$ there exist **a monomial $P_1(x)$** and **a polynomial $R_1(x)$ of degree less than the degree of $P(x)$**, such that the following identity holds true:

$$P(x) \equiv D(x) P_1(x) + R_1(x) \qquad (13-3-2)$$

Indeed, if we divide the leading term $a_n x^n$ of $P(x)$ by the leading term $b_m x^m$ of $D(x)$, we obtain the monomial $\left(\frac{a_n}{b_m} x^{n-m}\right)$, which we call $P_1(x)$. If we now multiply $P_1(x)$ by $D(x)$ we find:

$$D(x) P_1(x) \equiv (b_m x^m + b_{m-1} x^{m-1} + \cdots + b_1 x + b_0) \left(\frac{a_n}{b_m} x^{n-m}\right) \Rightarrow$$

$$D(x) P_1(x) \equiv a_n x^n + \frac{a_n}{b_m} b_{m-1} x^{n-1} + \cdots + \frac{a_n}{b_m} b_1 x^{n-m+1} + \frac{a_n}{b_m} b_0 x^{n-m}$$

Since the leading term of $P(x)$ is $a_n x^n$, if we subtract $D(x) P_1(x)$ from $P(x)$, we obtain,

$$P(x) - D(x) P_1(x)$$
$$\equiv \underbrace{\left(a_{n-1} - \frac{a_n}{b_m} b_{m-1}\right) x^{n-1} + \left(a_{n-2} - \frac{a_n}{b_m} b_{m-2}\right) x^{n-2} + \cdots}_{\stackrel{\text{def}}{=} R_1(x) \; (Polynomial\ of\ degree\ (n-1))}$$

or, equivalently,

$$P(x) \equiv D(x) P_1(x) + R_1(x), \quad degree\ of\ R_1(x) = n - 1 < n$$

and this completes the proof of equation (13-3-2).

The polynomial $R_1(x)$ is called **the first remainder of the division** of $P(x)$ by $D(x)$.

b) Proof of the main Theorem 13-2-1: According to the auxiliary proposition proved in part (a), given the polynomials $P(x)$ of degree n, (the Dividend) and $D(x)$ of degree m, (the Divisor), with $n \geq m$, there exist a monomial $P_1(x)$ and a polynomial $R_1(x)$ **of degree less that n**, such that

$$P(x) \equiv D(x)P_1(x) + R_1(x)$$

If the degree of $R_1(x)$ is less than the degree of $D(x)$, then the Theorem has been proved. Otherwise, we consider $R_1(x)$ as the dividend and $D(x)$ as the divisor, and again, by virtue of the auxiliary proposition, we may find a monomial $P_2(x)$ and a polynomial $R_2(x)$ **of degree less that the degree of $R_1(x)$**, such that,

$$R_1(x) \equiv D(x)P_2(x) + R_2(x)$$

Notice that in this process, **the degrees of the successive remainders are getting lower**, each time, and sooner or later, we reach a remainder, say $R_k(x)$, whose degree is smaller than the degree of $D(x)$. In such a case, we have the following equalities:

$$\begin{cases} P(x) \equiv D(x)P_1(x) + R_1(x) \\ R_1(x) \equiv D(x)P_2(x) + R_2(x) \\ \cdots \quad \cdots \quad \cdots \\ R_{k-1}(x) \equiv D(x)P_k(x) + R_k(x) \end{cases}, \quad \text{degree of } R_k(x) < \text{degree of } D(x)$$

Adding these equations term wise, we find:

$$P(x) \equiv D(x)\underbrace{\{P_1(x) + P_2(x) + \cdots + P_k(x)\}}_{\stackrel{\text{def}}{=} Q(x)} + \underbrace{R_k(x)}_{\stackrel{\text{def}}{=} R(x)} \Rightarrow$$

$$P(x) \equiv D(x)Q(x) + R(x), \quad \text{degree of } R(x) < \text{degree of } D(x)$$

and this completes the proof of the theorem.

PROBLEMS

13-3-1) For which value of $k \in \mathbb{R}$, the polynomial $x^2 + \sqrt{2}x + k$ divides exactly the polynomial $x^4 + 1$?

(Ans: $k = 1$).

13-3-2) If the dividend $P(x)$ and the divisor $D(x)$ are multiplied by the same polynomial $\phi(x)$, show that the quotient remains unchanged, while the remainder is multiplied by $\phi(x)$.

13-3-3) Given $P(x)$ (the dividend) and $D(x)$ (the divisor), show that the quotient $Q(x)$ and the remainder $R(x)$ of the division $P(x)/Q(x)$, are unique.

13-3-4) Find the quotient of the division: $(x^{9/5} + 1)/(x^{3/5} + 1)$.

(Ans: $x^{6/5} - x^{3/5} + 1$).

Hint: If we set $y = x^{3/5}$, the expression becomes, $\{(y^3 + 1)/(y + 1)\} = y^2 - y + 1$.

13-4) The remainder and factor theorems

Theorem 13-4-1: **The remainder of the division of a polynomial $P(x)$ by $(x - a)$, (a is a given number), is the number $P(a)$.**

Proof: The degree of the divisor $(x - a)$ is 1, and according to Th. 13-3-1, the remainder of the division $\frac{P(x)}{x-a}$ must be of degree less than 1, i.e. the remainder will be of degree zero, in other words **the remainder R must be a constant number independent of x**, and in such a case, equation (13-3-1) implies:

$$P(x) \equiv (x - a)Q(x) + R \qquad (*)$$

Equation (*) is **an identity in x**, and therefore must hold true for all values of x. In particular, if we apply the identity for $x = a$, we find,

$$P(a) = (a - a)Q(a) + R = 0 \cdot Q(a) + R = R$$

and this completes the proof.

Theorem 13-4-2: **The necessary and sufficient condition for a polynomial $P(x)$ to be divisible by $(x - a)$, is $P(a) = 0$.**

Proof: The polynomial $P(x)$ is divisible by $(x - a)$, provided that the remainder of the division is zero, i.e. provided that $R = P(a) = 0$, and this completes the proof.

Theorem 13-4-3: **If the polynomial $P(x)$ is divided exactly by $(x - a)$, $(x - b)$ and $(x - c)$, where a, b and c are numbers different from each other, (i.e. $(a - b)(b - c)(c - a) \neq 0$), then $P(x)$ is divided exactly by the product $(x - a)(x - b)(x - c)$.**

Proof: Since $P(x)$ is divided exactly by $(x - a)$, we have:

$$P(x) \equiv (x - a)Q(x) \qquad (*)$$

The identity in (*) is true for all values of x, and therefore holds true when $x = b$, i.e.

$$P(b) = (b - a)Q(b)$$

But, $P(b) = 0$, since $(x - b)$ divides exactly $P(x)$, and since $(b - a) \neq 0$, the factor $Q(b) = 0$. According to Theorem 13-4-2, $(x - b)$ divides $Q(x)$, i.e. $Q(x) \equiv (x - b)Q_1(x)$, and equation (*) implies,

$$P(x) \equiv (x - a)(x - b)Q_1(x) \qquad (**)$$

Again, applying the identity in (**) for $x = c$, and noting that $P(c) = 0$, (since $(x - c)$ divides exactly $P(x)$) and that $(c - a)(c - b) \neq 0$, we conclude that $Q_1(c) = 0$, which shows that $(x - c)$ divides exactly $Q_1(x)$, i.e. $Q_1(x) \equiv (x - c)Q_2(x)$, and substituting in (**) results in

$$P(x) = (x - a)(x - b)(x - c)Q_2(x)$$

This equation shows that the product $(x - a)(x - b)(x - c)$ divides exactly the polynomial $P(x)$, the quotient of the division being $Q_2(x)$.

Corollary 1: If the polynomial $P(x)$ is divided exactly by $(x - a)$, $(x - b)$, $(x - c)$, $(x - d)$,..., then $P(x)$ is divided **by the product** $(x - a)(x - b)(x - c)(x - d)$ ..., provided that the numbers $a, b, c, d, ...$ are different from each other.

Corollary 2: If the polynomial $P(x)$ vanishes at $x = a, b, c, d, ...$, then $P(x)$ is divided exactly **by the product** $(x - a)(x - b)(x - c)(x - d) ...$, provided that the numbers $a, b, c, d, ...$ are different from each other.

Indeed, if $P(a) = 0$, then $(x - a)$ divides exactly $P(x)$, etc.

Example 13-4-1: Find the remainder of the division of $P(x) = x^4 - 2x^3 - 7x + 1$ by $D(x) = x - 2$.

Solution

Application of Theorem 13-4-1 yields,

$$R = P(2) = 2^4 - 2 \cdot 2^3 - 7 \cdot 2 + 1 = -13$$

Example 13-4-2: The remainder of the division of a polynomial $P(x)$ by $(x^2 - 4)$ is $3x + 4$. Find the remainder of the division of $P(x)$ by $(x + 2)$ or by $(x - 2)$.

Solution

By assumption, we have:

$$P(x) \equiv (x^2 - 4)Q(x) + 3x + 4 \equiv (x + 2)(x - 2)Q(x) + 3x + 4 \quad (*)$$

The remainder of the division of $P(x)$ by $(x + 2)$ is

$$R = P(-2) \stackrel{(*)}{\Rightarrow} R = 0 + 3 \cdot (-2) + 4 = -2$$

Similarly, the remainder of the division of $P(x)$ by $(x - 2)$ is

$$R = P(2) \stackrel{(*)}{\Rightarrow} R = 0 + 3 \cdot 2 + 4 = 10$$

Example 13-4-3: Show that $x + 5$ does not divide exactly the polynomial $P(x) = x^{20} + 5^{20}$.

Solution

The remainder of the division of $P(x)$ by $(x + 5)$ is $R = P(-5) = (-5)^{20} + 5^{20} = 2 \cdot 5^{20} \neq 0$, i.e. the division is not perfect (since $R \neq 0$).

Example 13-4-4: Show that $P(x) = x^8 + x^4 - 2$ is divided by $x^2 + 1$.

Solution

$$P(i) = i^8 + i^4 - 2 = 1 + 1 - 2 = 0, \quad (i = imaginary\ unit)$$

and this shows that $P(x)$ is divided by $(x - i)$. Similarly, since $P(-i) = (-i)^8 + (-i)^4 - 2 = 1 + 1 - 2 = 0$, $P(x)$ is divided by $(x + i)$, and by virtue of Theorem 13-4-3, $P(x)$ is divided by $(x + i)(x - i) = x^2 + 1$.

Example 13-4-5: For what values of a and b the polynomial $P(x) = 2x^4 - ax^3 + bx^2 + 3$ is divided by $(x^2 - 1)$.

Solution

Since $x^2 - 1 = (x - 1)(x + 1)$, the polynomial $P(x)$ will be divided by $(x^2 - 1)$ if it is divided by $(x - 1)$ **and** by $(x + 1)$, (Theorem 13-4-3), i.e. if

$$P(1) = 0 \ and \ P(-1) = 0 \Rightarrow \begin{cases} 2 - a + b + 3 = 0 \\ and \\ 2 + a + b + 3 = 0 \end{cases} \quad (*)$$

Solving the system for a and b in (*) we find, $a = 0$ and $b = -5$.

Example 13-4-6: Show that the remainder of the division of the polynomial $P(x)$ by $(x^2 - k^2)$, where k is a given constant, is given by the formula

$$R(x) = \frac{P(k) - P(-k)}{2k} x + \frac{P(k) + P(-k)}{2}$$

Solution

Since the divisor $x^2 - k^2$ is a second degree polynomial the remainder will be a first degree polynomial, i.e. it will be of the form, $R(x) = ax + b$, and the following identity holds true for all values of x, (see formula 13-3-1):

$$P(x) \equiv (x^2 - k^2)Q(x) + ax + b \quad (*)$$

For $x = k$, identity (*) yields, $P(k) = ak + b$ (**), while for $x = -k$, identity (*) yields, $P(-k) = -ak + b$ (***). Solving the system (**) and (***) for a and b, we find $a = (P(k) - P(-k))/(2k)$ and $b = (P(k) + P(-k))/2$, and this completes the proof.

Example 13-4-7: If n is a positive integer, show that $(x + 1)$ divides the polynomial $P(x) = (x - 1)^{2n} - 4^n$, for every value of n.

Solution

Since $P(-1) = (-1 - 1)^{2n} - 4^n = (-2)^{2n} - 4^n = ((-2)^2)^n - 4^n = 4^n - 4^n = 0$, $(x + 1)$ divides $P(x)$.

PROBLEMS

13-4-1) Find the remainder of the division $(x^4 - 2x^3 + 5x - 7) \div (x - 2)$ and $(x^4 + 2x^3 + 5x + 2) \div (x + 3)$.

(Ans: $3, 14$).

13-4-2) Show that the remainder of the division of $P(x)$ by $(\kappa x + \lambda)$, is $R = P(-\lambda/k)$.

13-4-3) If n is a positive integer, show that $(x + a)$ divides $x^{2n} - a^{2n}$, but it does not divide $x^{2n+1} - a^{2n+1}$, $(a \neq 0)$.

13-4-4) If n is a positive integer, show that $x(x + 1)\left(x + \frac{1}{2}\right)$ divides the polynomial $P(x) = (x + 1)^{2n} - x^{2n} - 2x - 1$.

Hint: Apply Theorem 13-4-3.

13-4-5) If n is a positive integer, show that $(x + 5)$ divides $P(x) = (x + 5)^{2n} + (x + 4)^{2n} - 1$.

13-4-6) The remainder of the division of a polynomial $P(x)$ by $(x^2 + 4x - 6)$ is $R(x) = 7x + 3$. What is the remainder of the division of $P(x)$ by $(x + 5)$? By $(x - 1)$?

Hint: See Example 13-4-2.

13-4-7) If $P(x) = 3x^2 - 5x + 6$, find the polynomial $A(x) = P(3x + 5)$. Show that $(x + a)$ does not divide $A(x)$, $\forall a \in \mathbb{R}$.

(Ans: $A(x) = 27x^2 + 75x + 56$).

13-4-8) If $(x + 3)$ divides $P(x)$, show that $(x - 2)$ divides $P(2x - 7)$.

13-4-9) For which real values of a and b, the polynomial $P(x) = 2x^3 + ax^2 - 13x + b$ is divisible by $(x-2)(x+3)$? **(Ans: $a = 1, b = 6$).**

13-4-10) For which real values of a, b, c the polynomial $P(x) = x^5 + 2x^4 - 3x^3 + ax^2 + bx + c$ is divisible by $(x+2)(x^2-1)$?

13-4-11) Provided that $P(x) = ax^4 + bx^3 + cx^2 + dx + e$ is divided by $(x^2 - k^2)$, $k \neq 0$, show that $ad^2 - bcd + eb^2 = 0$, (assume $abcde \neq 0$).

13-5) Polynomials identically equal

Theorem 13-5-1: If the $n^{\underline{th}}$ degree polynomial $P(x) = a_n x^n + a_{n-1} x^{n-1} + \cdots + a_1 x + a_0$ ($a_n \neq 0$) vanishes for the n different values of x, $r_1, r_2, \cdots, r_n$, then the polynomial is identically equal to the product

$$P(x) \equiv a_n (x - r_1)(x - r_2)(x - r_3) \cdots (x - r_n) \qquad (13-5-1)$$

Equation (13-5-1) is an expression of the polynomial **in terms of its zeros (roots)**. The leading coefficient a_n is assumed to be different from zero, otherwise, the degree of the polynomial would be less than n.

Proof: Since $P(x)$ vanishes at $r_1, r_2, \cdots, r_n$, the terms $(x - r_1), (x - r_2), \ldots, (x - r_n)$ divide $P(x)$, and by virtue of corollary 1 of Theorem 13-4-3, $P(x)$ is divided by the product $(x - r_1)(x - r_2) \cdots (x - r_n)$, i.e.

$$P(x) = (x - r_1)(x - r_2) \cdots (x - r_n) Q(x) \qquad (*)$$

where $Q(x)$ **is the quotient of the division**. However, since the degree of $P(x)$ is n and the degree of the product $(x - r_1)(x - r_2) \cdots (x - r_n)$ is n as well, **the quotient Q must be a constant number**, i.e. $Q(x) = c$ (constant). The coefficient of x^n in the left member in (*) is a_n while the degree of x^n in the right member in (*) is c, and this implies that $c = a_n$, and the proof is completed.

Theorem 13-5-2: If a polynomial $P(x)$ of degree n vanishes for $(n+1)$ different values of x, then all the coefficients of $P(x)$ are equal to zero, i.e.

$$a_n = a_{n-1} = \cdots = a_1 = a_0 = 0 \qquad (13-5-2)$$

Proof: Let us assume that the polynomial $P(x) = a_n x^n + a_{n-1} x^{n-1} + \cdots + a_1 x + a_0$ vanishes for the $(n+1)$ **different** values of x, $r_1, r_2, \cdots, r_n$ and r_{n+1}. According to Theorem 13-5-1, $P(x)$ can be expressed as

$$P(x) \equiv a_n (x - r_1)(x - r_2)(x - r_3) \cdots (x - r_n) \qquad (*)$$

Since r_{n+1} is (by assumption) a root of $P(x)$, $P(r_{n+1}) = 0$, and setting $x = r_{n+1}$ in equation (*) yields,

$$P(r_{n+1}) = 0 = a_n \underbrace{(r_{n+1} - r_1)(r_{n+1} - r_2) \cdots (r_{n+1} - r_n)}_{Term\ A \neq 0} \qquad (**)$$

Since the numbers $r_1, r_2, \cdots, r_n, r_{n+1}$ are different, all the differences $(r_{n+1} - r_1), (r_{n+1} - r_2), \cdots$ are different from zero, and thus the term A in (**) is different from zero. In this case, equation (**) implies that $a_n = 0$, and equation (13-5-1) shows that

$$P(x) \equiv a_n x^n + a_{n-1} x^{n-1} + \cdots + a_1 x + a_0 \equiv 0$$

i.e. $P(x) = 0$, $\forall x$ **real or complex**. This in turn implies that **all the coefficients** $a_n, a_{n-1}, \cdots, a_1, a_0$ **must necessarily be zero**. Since if we assume that **at least one** of the coefficients is not zero, say, **for example** $a_2 \neq 0$, then the polynomial would be $P(x) = a_2 x^2$, and this obviously is not zero if $x \neq 0$, and this contradicts our hypothesis that $P(x) = 0, \forall x$.

Corollary: If a polynomial $P(x)$ vanishes for an infinite number of (different) values of x, then all the coefficients of the polynomial are equal to zero.

Theorem 13-5-3: **If the equality** $A(x) = B(x)$ **between two polynomials holds true for an infinite number of (different) values of x, then the two polynomials are identically equal, i.e.** $A(x) \equiv B(x)$.

Proof: The polynomial $A(x) - B(x)$ vanishes for an infinite number of values of x, and according to the corollary, $A(x) - B(x) \equiv 0$, and this means that $A(x) \equiv B(x)$.

Remark: Two identically equal polynomials are of the same degree and the coefficients of the like powers of x are identical. **Two polynomials of different degrees cannot possibly be identically equal**.

For example, $(x + 1)^2 = x^2 + 2x + 1$ is an identity, since the equality holds true for all values of x, real or complex. **In an identity, one member is equal to the other, written in a different form**.

On the other hand, $x^3 + 2x^2 = x + 2$ **is not an identity**, since it cannot be true for all values of x; instead, this equality is **an equation** which is verified by three values of x, $(x = -1, 1, -2)$, the roots of the equation.

Theorem 13-5-4: **If $A(x) \cdot K(x) \equiv B(x) \cdot K(x)$, where $K(x) \not\equiv 0$, then $A(x) \equiv B(x)$, ($A(x), B(x), K(x)$ are polynomials in x), i.e. the cancellation law holds true for the multiplication of polynomials.**

Proof: The identity $A(x) \cdot K(x) \equiv B(x) \cdot K(x)$ implies that

$$A(x) \cdot K(x) - B(x) \cdot K(x) \equiv 0 \Leftrightarrow K(x) \cdot \big(A(x) - B(x)\big) \equiv 0 \quad (*)$$

Equation (*) shows that the polynomial $K(x) \cdot \big(A(x) - B(x)\big)$ must vanish for an infinite number of values of x, and since $K(x) \not\equiv 0$, the other factor $\big(A(x) - B(x)\big)$ must be identically equal to zero, i.e. $\big(A(x) - B(x)\big) \equiv 0$, i.e. $A(x) \equiv B(x)$.

Remark: A popular method, when working with identically equal polynomials, is the so called "**the method of undetermined coefficients**" which is explained in Example 13-5-2.

Example 13-5-1: Show that the polynomials $A(x) = 3x^3 + bx + 5$ and $B(x) = 7x^3 - ax + 3$ cannot assume same values at four different values of x.

Solution

The polynomial $P(x) = A(x) - B(x)$ is **a third degree** polynomial with leading term $(-4x^3)$. A number r that assigns equal values to $A(x)$ and $B(x)$ is a root of $P(x)$, since $P(r) = A(r) - B(r) = 0$. Since the degree of $P(x)$ is 3, the polynomial cannot have four (different) roots, i.e. $A(x) - B(x)$ cannot be zero at four different values of x, i.e. $A(x) = B(x)$ cannot be satisfied by four different values of x.

Example 13-5-2: Provided that $P(x) = x^4 + ax^3 + bx^2 + cx + d$ is a perfect square, show that $c^2 = a^2 d$ and $(4b - a^2)^2 = 64d$.

Solution

The polynomial whose square is $P(x)$ must be **a second degree polynomial** having general form $\kappa x^2 + \lambda x + \mu$, ($\kappa, \lambda, \mu$ are constant numbers). Then:

$$(\kappa x^2 + \lambda x + \mu)^2 \equiv x^4 + ax^3 + bx^2 + cx + d \Rightarrow$$

$$\kappa^2 x^4 + \lambda^2 x^2 + \mu^2 + 2\kappa\lambda x^3 + 2\kappa\mu x^2 + 2\lambda\mu x \equiv x^4 + ax^3 + bx^2 + cx + d \Rightarrow$$

$$\kappa^2 x^4 + 2\kappa\lambda x^3 + (\lambda^2 + 2\kappa\mu)x^2 + 2\lambda\mu x + \mu^2 \equiv x^4 + ax^3 + bx^2 + cx + d \quad (*)$$

Since the two polynomials in (*) are identically equal, the coefficients of the like powers of x must be identical, i.e.

$$\begin{Bmatrix} \kappa^2 = 1 \\ 2\kappa\lambda = a \\ \lambda^2 + 2\kappa\mu = b \\ 2\lambda\mu = c \\ \mu^2 = d \end{Bmatrix} \quad (**)$$

From the first and the second equations in (**) we have: $4\kappa^2\lambda^2 = a^2$ and since $\kappa^2 = 1$, $\boldsymbol{a^2 = 4\lambda^2}$. From the fourth equation we obtain:

$$c^2 = 4\lambda^2\mu^2 \xrightarrow{(\mu^2=d)(4\lambda^2=a^2)} c^2 = a^2 d$$

Also, $(4b - a^2)^2 = (4\lambda^2 + 8\kappa\mu - 4\lambda^2)^2 = (8\kappa\mu)^2 = 64\kappa^2\mu^2$, or since $\kappa^2 = 1$ and $\mu^2 = d$, we find $(4b - a^2)^2 = 64d$.

The method of equating the coefficients of two polynomials that are equal identically, is known as **"the method of the undetermined coefficients"**.

Example 13-5-3: If a, b, c are three different constant numbers, show that the expression

$$F = (a - b)(x - a)(x - b) + (b - c)(x - b)(x - c)$$
$$+ (c - a)(x - c)(x - a) + (a - b)(b - c)(c - a)$$

is identically equal to zero.

Solution

One possible method to work the problem is to carry out the calculations and show that $F \equiv 0$, i.e. $F = 0 \ \forall x$.

Another method, which saves much time and calculations is the following:

The expression $F = F(x)$ **is a second degree polynomial**, since each one of the first three summands is a quadratic trinomial, while the fourth term is a constant number. We note that, (easy to check)

$$\{F(x = a) = 0, \quad F(x = b) = 0, \quad F(x = c) = 0\}$$

Since the second degree polynomial vanishes for $3 = 2 + 1$, different values of x, $F(x) \equiv 0$, by virtue of Theorem 13-5-2.

Example 13-5-4: Provided that $A(x) = ax^3 + 3bx^2 + 3cx + d$ is divided exactly by $B(x) = ax^2 + 2bx + c$, show that $A(x)$ is a perfect cube and that $B(x)$ is a perfect square.

Solution

Let $Q(x) = x + k$ be the quotient of $A(x) \div B(x)$, i.e.

$$ax^3 + 3bx^2 + 3cx + d \equiv (ax^2 + 2bx + c)(x + k) \qquad (*)$$

where k is to be determined. Notice that the coefficient of x in $Q(x)$ must be 1, so that the coefficients of x^3 in both sides is a. From (*) we obtain:

$$ax^3 + 3bx^2 + 3cx + d \equiv ax^3 + 2bx^2 + cx + akx^2 + 2bkx + ck \iff$$

$$ax^3 + 3bx^2 + 3cx + d \equiv ax^3 + (2b + ak)x^2 + (c + 2bk)x + ck \iff$$

$$\begin{cases} a = a \\ 3b = 2b + ak \\ 3c = c + 2bk \\ d = ck \end{cases} \implies \begin{cases} a = arbitrary \\ b = ak \\ c = bk = ak^2 \\ d = ck = ak^3 \end{cases} \qquad (**)$$

Polynomial $A(x)$:

$$A(x) = ax^3 + 3bx^2 + 3cx + d = ax^3 + 3akx^2 + 3ak^2x + ak^3 \implies$$

$$A(x) = a(x + k)^3 = \left(\sqrt[3]{a}\,(x + k)\right)^3$$

Polynomial $B(x)$:

$$B(x) = ax^2 + 2bx + c = ax^2 + 2akx + ak^2 = \left(\sqrt{a}\,(x+k)\right)^2$$

Example 13-5-5: a) Determine a fourth degree polynomial $P(x)$ such that, $P(x+1) - P(x) \equiv x^3$. **b)** Use the result of part (a), to find the sum $S = 1^3 + 2^3 + 3^3 + \cdots + n^3$.

Solution

a) Let $P(x) = ax^4 + bx^3 + cx^2 + dx + e$. Then $P(x+1) - P(x) \equiv x^3$ implies,

$$a(x+1)^4 + b(x+1)^3 + c(x+1)^2 + d(x+1) + e$$
$$- (ax^4 + bx^3 + cx^2 + dx + e) \equiv x^3 \Leftrightarrow$$

$$4ax^3 + (6a + 3b)x^2 + (4a + 3b + 2c)x + a + b + c + d \equiv x^3 \Leftrightarrow$$

$$\begin{cases} 4a = 1 \\ 6a + 3b = 0 \\ 4a + 3b + 2c = 0 \\ a + b + c + d = 0 \end{cases} \Leftrightarrow \begin{cases} a = 1/4 \\ b = -1/2 \\ c = 1/4 \\ d = 0 \end{cases}$$

The polynomial that satisfies the (difference equation) $P(x+1) - P(x) \equiv x^3$ is:

$$P(x) = \frac{1}{4}x^4 - \frac{1}{2}x^3 + \frac{1}{4}x^2 + e \qquad (*)$$

where e is arbitrary.

b) Making use of the expression in (*) we have:

$$S = 1^3 + 2^3 + 3^3 + \cdots + n^3 \xRightarrow{(P(x+1)-P(x)\equiv x^3)}$$

$$S = \bigl(P(2) - P(1)\bigr) + \bigl(P(3) - P(2)\bigr) + \bigl(P(4) - P(3)\bigr) + \cdots$$
$$+ \bigl(P(n+1) - P(n)\bigr) = P(n+1) - P(1) \xRightarrow{(*)}$$

$$S = \underbrace{\frac{1}{4}(n+1)^4 - \frac{1}{2}(n+1)^3 + \frac{1}{4}(n+1)^2}_{P(n+1)} - \underbrace{\left(\frac{1}{4} - \frac{1}{2} + \frac{1}{4}\right)}_{P(1)} \Rightarrow$$

$$S = 1^3 + 2^3 + 3^3 + \cdots + n^3 = \left(\frac{n(n+1)}{2}\right)^2 \qquad (**)$$

Remark: It can be easily shown that for any positive integer n,

$$1 + 2 + 3 + \cdots + n = \frac{n(n+1)}{2}$$

Formula (**) shows that for **any positive integer n**, we have:

$$1^3 + 2^3 + 3^3 + \cdots + n^3 = (1 + 2 + 3 + \cdots + n)^2 \qquad (***)$$

Let the reader verify (***) for $n = 3,4,5$.

PROBLEMS

13-5-1) Without carrying out calculations show that:

$$(a-b)(x-c)^2 + (b-c)(x-a)^2 + (c-a)(x-b)^2 \\ + (a-b)(b-c)(c-a) \equiv 0$$

(Assume $(a-b)(b-c)(c-a) \neq 0$).

Hint: See Example 13-5-3.

13-5-2) From the identity

$$2x^2 + 5x + 7 \equiv a(x-1)(x+3) + b(x+3) + c$$

determine a, b and c.

13-5-3) Determine a and b, so that the polynomial $A(x) = x^4 + ax^3 + bx^2 - 6x + 1$ is a perfect square of another polynomial.

(**Ans:** $(a = -6, b = 11)$ or $(a = 6, b = 7)$).

13-5-4) Determine the constant c, such that the polynomial $x^3 + ax + b$ to be divisible by $(x-c)^2$.

Hint: Set $y = x - c$, express $x^3 + ax + b$ in terms of y. In order this polynomial to be divisible by y^2, the coefficient of y and the constant term must be zero.

13-5-5) Find the necessary and sufficient condition that the fraction

$$\frac{ax^2 + bx + c}{Ax^2 + Bx + C}$$

takes the same value for three different values of x, ($abcABC \neq 0$).

(Ans: $\frac{a}{A} = \frac{b}{B} = \frac{c}{C}$).

13-5-6) Provided that the polynomial $4x^4 + 12x^3y + kx^2y^2 + 6xy^3 + y^4$ is a perfect square, find k.

13-5-7) Provided that $x^3 + ax^2 + bx + c$ is divisible by $x^2 + dx + f$, show that $c = f(a - d)$ and $(b - f) = d(a - d)$.

13-5-8) Provided that $x^2 + \sqrt{3}\,x + 1$ divides exactly $x^4 + \kappa x^2 + \lambda$, find κ and λ.

13-5-9) Find the general form of fourth degree polynomials $P(x)$ that remain unchanged when x is replaced by $(1 - x)$.

(Ans: $P(x) = \kappa x^4 - 2\kappa x^3 + \lambda x^2 + (\kappa - \lambda)x + \mu$, κ, λ, μ are arbitrary).

13-5-10) Provided that the polynomial $P(x) = (a - 1)x^2 + bx + 1$ satisfies the identity $P(2x + 1) \equiv P(x)$, show that the expression $x^2 + (r_1^3 + r_2^3)x + r_1 r_2$ is a perfect square, where r_1 and r_2 are the roots of the equation $x^2 + ax + b + 1 = 0$.

13-5-11) Provided that $(x - 1)^2$ divides exactly the polynomial $P(x) = x^3 + \kappa x^2 + \lambda x + \mu$, where κ, λ, μ are integers, show that $|\kappa| + |\lambda| + |\mu| \geq 3$.

13-5-12) Provided that the polynomial $P(x) = x^n + ay^n + bz^n$ is divided by the trinomial $x^2 - (\kappa y + \lambda z)x + \kappa \lambda yz$, show that: $\frac{a}{\kappa^n} + \frac{b}{\lambda^n} + 1 = 0$.

Hint: Show that the roots of the trinomial are κy and λz.

13-5-13) Show that the polynomial $P(x, y) = x^{6n} - 2y^{6n} + (xy)^{3n}$ is divisible by $x^2 + xy + y^2$, for every value of the positive integer n.

Hint: It suffices to show that $P(x)$ is divided by $(x - r_1)$ and $(x - r_2)$, where r_1 and r_2 are the roots of $x^2 + xy + y^2$, i.e. $r_{1,2} = ((-1 \pm i\sqrt{3})/2)y$, i.e. to show that $P(r_1) = P(r_2) = 0$. In helps if you express r_1 and r_2 in polar form, and apply De Moivre's Theorem.

13-5-14) Determine κ and λ so that $P(x) = 2x^6 - x^5 + 4x^4 + \kappa x^3 - x^2 + \lambda x - 3$ to be divisible by $x^2 + x + 1$.

CHAPTER 14: GENERAL PROPERTIES OF POLYNOMIAL EQUATIONS

Recall that **a polynomial equation** of degree n, ($n = 1,2,3,\cdots$) is any equation of the form

$$P(x) = a_n x^n + a_{n-1} x^{n-1} + a_{n-2} x^{n-2} + \cdots + a_2 x^2 + a_1 x + a_0 = 0$$

The coefficients $a_n, a_{n-1}, \cdots, a_1, a_0$ could either real or complex numbers, in the general case.

Root of a polynomial equation is any number r, **real or complex**, such that $P(r) = 0$. We say that r is also **a root or a zero** of the polynomial $P(x)$.

At this point two main questions arise:

Question 1: Does every polynomial equation have a root?

The answer is provided by the so called "**The fundamental Theorem of Algebra**", proved rigorously by the great German Mathematician **C. F. Gauss**, in 1799. According to this Theorem, **every polynomial equation $P(x)$ has at least one root**. The proof of this Theorem employs methods and techniques developed in Complex Analysis.

So, the fundamental Theorem of Algebra, guarantees that if $P(x)$ is an arbitrary polynomial with real or complex coefficients, there exists at least one number r, real or complex, such that $P(r) = 0$.

As a consequence of this Theorem, we conclude that **a polynomial of degree n, has exactly n roots**. For, assuming that r is a root of $P(x)$, then $(x - r)$ divides $P(x)$, (Theorem 13-4-2), so we may write $P(x) = (x - r)P_1(x)$ where $P_1(x)$ is now a polynomial of degree $(n - 1)$. Again, by virtue of the fundamental Theorem of Algebra, $P_1(x)$ has a root r_1, so $P_1(x) = (x - r_1)P_2(x)$, where $P_2(x)$ is a polynomial of degree $(n - 2)$, and then

$$P(x) = (x - r)(x - r_1)P_2(x)$$

and this shows that r and r_1 are roots of $P(x)$. With the same reasoning, we finally show that the $n^{\underline{th}}$ degree polynomial has n roots exactly.

For example, the polynomial

$$P(x) = 3x^{27} + 25x^{19} + 9x^{14} - 6x^7 + 2x^2 + 10$$

has 27 roots exactly, (**real or complex**, we do not know), and similarly the polynomial $A(x) = (3-i)x^{65} - 75ix^{37} + (2+i)x^{10} - 13x^3 + 13 - 7i$, has 65 roots exactly, (**real or complex**, we do not know).

Question 2: In the general case, is it always possible to express the roots of a polynomial in terms of its coefficients?

This question has attracted the attention of mathematicians since the sixteen century. For a first and a second degree equation we can, indeed, express the roots of the equation in terms of the coefficients. For a third degree equation, the roots can be expressed by a formula, derived by the Italian mathematician **Girolamo Cardano** in 1535. A little later, a similar formula expressing the roots of a fourth degree equation was derived by **Luigi Ferrari**, another Italian mathematician.

The search for formulas expressing the roots of polynomials of degree 5 or higher, was carried on into the nineteenth century, until two brilliant mathematicians, the Norwegian **N.H. Abel** and the French **E. Galois** proved that, **in the general case, the roots of polynomial equations of degree higher than 4, (i.e. $n = 5, 6, 7, \cdots$), cannot be expressed in terms of the coefficients;** such formulas are not possible.

Of course, for some special types of equations, we may find the roots, for example, we can find the roots of the equation $x^5 - 1 = 0$, (the fifth roots of unity), but for a general fifth degree equation like, for example, $x^5 + 7x^4 - 3x^3 + 2x^2 + 13x - 29 = 0$, we cannot.

14-1) Some general theorems on polynomial equations

Theorem 14-1-1 (The complex conjugate roots theorem):

If the polynomial equation $P(x) = a_n x^n + a_{n-1} x^{n-1} + a_{n-2} x^{n-2} + \cdots + a_2 x^2 + a_1 x + a_0 = 0$, with real coefficients, has a complex root $a + ib$, ($b \neq 0$), then the complex conjugate $a - ib$ will also be a root of $P(x)$.

Proof: Let $r = a + ib$ be a root of $P(x) = 0$. This means that

$$a_n r^n + a_{n-1} r^{n-1} + a_{n-2} r^{n-2} + \cdots + a_2 r^2 + a_1 r + a_0 = 0$$

and the complex conjugate of this number must be zero as well, i.e.

$$\overline{(a_n r^n + a_{n-1} r^{n-1} + a_{n-2} r^{n-2} + \cdots + a_2 r^2 + a_1 r + a_0)} = 0 \quad (*)$$

Exploiting properties of complex conjugate numbers, equation (*) implies:

$$\overline{a_n r^n} + \overline{a_{n-1} r^{n-1}} + \overline{a_{n-2} r^{n-2}} + \cdots + \overline{a_2 r^2} + \overline{a_1 r} + \overline{a_0} = 0 \quad (**)$$

However, $\overline{a_n r^n} = \overline{a_n}\, \overline{r^n}$, and since **the coefficients are real numbers**, $\overline{a_n} = a_n$ and $\overline{r^n} = (\bar{r})^n$, and therefore, $\overline{a_n r^n} = a_n (\bar{r})^n$, and similarly for all the other terms in (**). Equation (**) is now written as

$$a_n (\bar{r})^n + a_{n-1} (\bar{r})^{n-1} + a_{n-2} (\bar{r})^{n-2} + \cdots + a_2 (\bar{r})^2 + a_2 \bar{r} + a_0 = 0$$

and this shows that $\bar{r} = a - ib$ is also a root of the equation $P(x) = 0$.

A different proof of the same theorem:

Let $r = a + ib$ be a root of $P(x)$. This means that $P(r) = 0$. We consider the function

$$\phi(x) = \{x - (a + ib)\}\{x - (a - ib)\} = (x - a)^2 + b^2 \quad (*)$$

If we divide $P(x)$ by $\phi(x)$, the remainder of the division will be a first degree polynomial of the form $R(x) = \kappa x + \lambda$, ($\kappa, \lambda \in \mathbb{R}$), (since the divisor $\phi(x)$ is a second degree polynomial), and according to Theorem 13-3-1, we have,

$$P(x) \equiv \phi(x) Q(x) + \kappa x + \lambda \equiv \{(x-a)^2 + b^2\} Q(x) + \kappa x + \lambda \quad (**)$$

Since $r = a + ib$ is a root of $P(x)$ and $\phi(x)$, if we set $x = r$ in (**) we obtain,

$$0 = 0 \cdot Q(r) + kr + \lambda \Rightarrow \kappa \underbrace{(a + ib)}_{r} + \lambda = 0 \Rightarrow$$

$$(\kappa a + \lambda) + ib\kappa = 0 \Rightarrow \begin{cases} \kappa a + \lambda = 0 \\ \text{and} \\ bk = 0 \end{cases} \overset{(b \neq 0)}{\Longrightarrow} \{\kappa = 0 \quad \text{and} \quad \lambda = 0\}$$

With $\kappa = 0$ and $\lambda = 0$, equation (**) implies that

$$P(x) \equiv \{(x - a)^2 + b^2\} Q(x) \equiv \{x - (a + ib)\}\{x - (a - ib)\} Q(x)$$

and this shows that $P(a - ib) = 0$, i.e. $x = a - ib$ is also a root of $P(x)$.

Notice: The theorem holds true only when **all the coefficients are real**. Even if one of the coefficients is not real, the theorem may or may not hold true.

Corollary: Any **odd** degree polynomial with **real coefficients**, has at least one real root.

This is so, since Theorem 14-1-1 implies that **complex roots appear in pairs** (**a complex root and its conjugate**). For example, any third degree polynomial has **at least one real root**. Of course it could have either three real roots, or have two complex conjugate roots and one real root. It could not have two real roots and one complex.

We now consider polynomial equations with **rational coefficients** (fractions of integer numbers). In such cases, the equation is reduced to an equivalent equation, with **integer coefficients**. Let us, for instance, consider the equation

$$\frac{1}{3}x^3 - \frac{1}{2}x^2 + \frac{5}{6}x - 7 = 0$$

All the coefficients **are rational numbers**. If we multiply through by 6, (**the least common denominator**), we eliminate the denominators, and we obtain **the equivalent equation** $2x^3 - 3x^2 + 5x - 42 = 0$, all coefficients of which are integers.

An equation with integer coefficients may or may not have rational roots. But in case it does, the rational roots are found with the aid of the following theorem.

Theorem 14-1-2 (the rational roots theorem):

If the coefficients of the polynomial equation

$$P(x) = a_n x^n + a_{n-1} x^{n-1} + a_{n-2} x^{n-2} + \cdots + a_2 x^2 + a_1 x + a_0 = 0$$

are all integers and $\left(\frac{p}{q}\right)$ is a rational root, in lowest terms, (p, q integers), then p is a divisor of the constant term a_0 and q is a divisor of the leading term a_n.

Proof: Assuming that $\left(\frac{p}{q}\right)$ is a root of $P(x)$, we have:

$$a_n\left(\frac{p}{q}\right)^n + a_{n-1}\left(\frac{p}{q}\right)^{n-1} + a_{n-2}\left(\frac{p}{q}\right)^{n-2} + \cdots + a_2\left(\frac{p}{q}\right)^2 + a_1\frac{p}{q} + a_0 = 0 \Leftrightarrow$$

$$a_n p^n = -q\{a_{n-1}p^{n-1} + a_{n-2}qp^{n-2} + \cdots + a_1 q^{n-2}p + a_0 q^{n-1}\} \quad (*)$$

Since **all the coefficients are integers and p and q are integers**, the number inside the braces in (*) is an integer, and also the number $a_n p^n$ is an integer. Equation (*) shows that q divides exactly the integer $a_n p^n$, and since p and q are **relatively prime**, q does not divide p, and hence q does not divide p^n, which means that **p must divide a_n**, i.e. **q is a divisor of a_n**.

With similar reasoning, from the equality

$$a_0 q^n = -p\{a_n p^{n-1} + a_{n-1}qp^{n-2} + \cdots + a_2 q^{n-2}p + a_1 q^{n-1}\}$$

we conclude that **p is a divisor of a_0**.

It is of course possible, a polynomial equation **with rational coefficients** to have **an irrational root of the form $a + \sqrt{b}$**, where a and b rational, and b not perfect square of a rational number, i.e. $\sqrt{b}$ is an irrational number. In such a case, we will show that if $a + \sqrt{b}$ is a root, then $a - \sqrt{b}$ is also a root.

Theorem 14-1-3 (Roots of the form $a + \sqrt{b}$):

If the coefficients of the polynomial equation

$$P(x) = a_n x^n + a_{n-1}x^{n-1} + a_{n-2}x^{n-2} + \cdots + a_2 x^2 + a_1 x + a_0 = 0$$

are all rational numbers and $a + \sqrt{b}$ is a root of the equation, where a and b rational numbers and $\sqrt{b}$ irrational, then $a - \sqrt{b}$ is also a root of the equation.

Proof: Let us consider the polynomial

$$\phi(x) \equiv \{x - (a + \sqrt{b})\}\{x - (a - \sqrt{b})\} \equiv (x - a)^2 - b$$

The coefficients of $\phi(x)$ **are rational**, since by assumption, a and b are rational. The remainder of the division of $P(x)$ by $\phi(x)$ will be a first degree polynomial with rational coefficients, of the form $R(x) = \kappa x + \lambda$, where κ and λ are rational numbers. According to Theorem 13-3-1, we have:

$$P(x) \equiv \phi(x)Q(x) + \kappa x + \lambda$$
$$\equiv \{x - (a + \sqrt{b})\}\{x - (a - \sqrt{b})\}Q(x) + \kappa x + \lambda \quad (*)$$

Setting $x = a + \sqrt{b}$ in (*), we obtain,

$$0 = 0 + \kappa(a + \sqrt{b}) + \lambda \Leftrightarrow \kappa a + \lambda + \kappa\sqrt{b} = 0 \quad (**)$$

Equation (**) implies that **$\kappa = 0$ and $\lambda = 0$**. For if we assume that $\kappa \neq 0$, then we would have that $\sqrt{b} = -(\kappa a + \lambda)/k$, i.e. an irrational number would be equal to a rational, which cannot be true. We are therefore forced to conclude that $\kappa = 0$, and then, from (**), $\lambda = 0$.

Having found $\kappa = 0$ and $\lambda = 0$, equation (*) becomes,

$$P(x) \equiv \{x - (a + \sqrt{b})\}\{x - (a - \sqrt{b})\}Q(x)$$

and this shows that $x = a - \sqrt{b}$ is a root, since $P(a - \sqrt{b}) = 0$.

Remark: As we have already mentioned, **a polynomial of degree n has n roots exactly, real or complex**. However, **it may happen that some of the roots are identical**. For example, the polynomial $A(x) = x^2 - 2x + 1$ as a second degree polynomial must have 2 roots. But since $x^2 - 2x + 1 = (x-1)^2$, it seems that $x = 1$ is the only root of $A(x)$. We say that $x = 1$, is **a double root** of $A(x)$, or that $x = 1$ is **a root of multiplicity 2**.

As another example, let $B(x) = (x-2)^3(x+5)^4(x-3)$ be a polynomial of degree 8, and as such, must have 8 roots. Three of the roots are identical and equal to 2, another four roots are identical and equal to -5, while the root $x = 3$ is a single root. We may say that $x = 2$ is a root of multiplicity 3, $x = -5$ is a root of multiplicity 4, while $x = 3$ is a single root.

We may thus say that **a polynomial of degree n has n roots exactly, counting multiplicities**.

Example 14-1-1: Find the roots of the equation $P(x) = x^4 + 3x^3 - 5x^2 - 29x - 30 = 0$, given that one root of the equation is the number $-2 - i$.

Solution

Since the equation has **real coefficients**, the number $-2 + i$ (the conjugate of $-2 - i$) is also a root, and this means that $P(x)$ is divided by $(x + 2 + i)$ and by $(x + 2 - i)$, and by virtue of Theorem 13-4-3, it must be divided by the product $(x + 2 + i)(x + 2 - i) = (x + 2)^2 - i^2 = (x + 2)^2 + 1 = x^2 + 4x + 5$. Dividing $P(x)$ by $(x^2 + 4x + 5)$ results in a quotient $Q(x) = x^2 - x - 6$, and from the identity of the algorithmic division, we have:

$$P(x) = (x^2 + 4x + 5)Q(x) = (x^2 + 4x + 5)(x^2 - x - 6)$$

The other two roots of $P(x) = 0$, are the roots of the quadratic equation $x^2 - x - 6 = 0$, i.e.

$$x_{1,2} = \frac{1 \pm \sqrt{1 + 24}}{2} = \frac{1 \pm 5}{2} \Rightarrow \{x_1 = 3, \quad x_2 = -2\}$$

In summary: The roots of $P(x) = 0$ are, $-2 - i, -2 + i, 3, -2$.

Example 14-1-2: Find the rational roots of the equation

$$P(x) = 3x^3 - x^2 - 3x + 1 = 0$$

Solution

The coefficients of $P(x)$ are all integers. If the equation has a rational root $\left(\frac{p}{q}\right)$, $(p, q$ integers), then p must divide the constant term 1, and q must divide the leading term 3, i.e. the **possible values** of p are ± 1, and the **possible values** of q are $\pm 1, \pm 3$. We have the following combinations for the ratio $\left(\frac{p}{q}\right)$:

$$\frac{p}{q} : \frac{1}{1}, \frac{-1}{1}, \frac{1}{-1}, \frac{-1}{-1}, \frac{1}{3}, \frac{-1}{3}, \frac{1}{-3}, \frac{-1}{-3} \Rightarrow \frac{p}{q} = \pm 1 \text{ or } \pm \frac{1}{3}$$

By direct checking we find the roots: $r_1 = 1, r_2 = -1, r_3 = 1/3$. The number $-1/3$ is not a root.

Example 14-1-3: Find the roots of the equation

$$P(x) = x^4 - 2x^3 - x^2 - 2x - 2 = 0$$

given that one of the roots is the number $(1 + \sqrt{3})$.

Solution

Since the coefficients of $P(x)$ are integers (rational), by Theorem 14-1-3, the number $1 - \sqrt{3}$ is also a root. This means that $P(x)$ is divided by $(x - 1 - \sqrt{3})$ and by $(x - 1 + \sqrt{3})$, and by Theorem 13-4-3, $P(x)$ is divided by the product $(x - 1 - \sqrt{3})(x - 1 + \sqrt{3}) = (x - 1)^2 - 3 = x^2 - 2x - 2$.

Dividing $P(x)$ by $(x^2 - 2x - 2)$ results in a quotient equal to $Q(x) = (x^2 + 1)$, and from the identity of the algorithmic division we have:

$$P(x) = (x^2 - 2x - 2)Q(x) = (x^2 - 2x - 2)(x^2 + 1)$$

and the four roots of $P(x) = 0$ are: $\pm i, 1 + \sqrt{3}, 1 - \sqrt{3}$.

Example 14-1-4: Find the degree of multiplicity of the root $r = 1$, of the polynomial $P(x) = x^3 + x^2 - 5x + 3$.

Solution

The number $r = 1$ is a root of $P(x)$, since $P(1) = 0$. Dividing $P(x)$ by $(x - 1)$ we find $x^2 + 2x - 3$. The roots of $x^2 + 2x - 3 = 0$, are found to be 1 and -3.

In summary, the roots of $P(x)$ are: $1, 1, -3$, and this means that $r = 1$ is a double root, (root of multiplicity 2), while $r = -3$ is a single root. We may write, $x^3 + x^2 - 5x + 3 = (x - 1)^2(x + 3)$, (see equation (13-5-1)).

Example 14-1-5: Find the integer values of a for which the equation $P(x) = x^3 + (a - 1)x + 3a = 0$ has rational roots.

Solution

If the equation has a rational root of the form p/q, (p and q integers), p must divide the constant term $3a$ and q must divide the leading term 1, i.e. $q = \pm 1$, and hence **every rational root (if it exists), will be an integer number.** Let r **(integer)** be such a root. Then,

$$r^3 + (a - 1)r + 3a = 0 \Leftrightarrow r^3 + ar - r + 3a = 0 \Leftrightarrow$$

$$a(r+3) = -(r^3 - r) \Leftrightarrow a = -\frac{r^3 - r}{r + 3} \qquad (*)$$

(Notice in (*) that $(r + 3) \neq 0$, since $r = -3$ is **not a root** of the equation, (check it)). Dividing the polynomial $r^3 - r$ by the polynomial $r + 3$, results in the following identity, (see also equation (13-3-1)),

$$a = -(r^2 - 3r + 8) + \frac{24}{r + 3} \qquad (**)$$

Since r is an integer, the term $(r^2 - 3r + 8)$ is an integer, and since (by assumption) a is also an integer number, **the term $24/(r + 3)$ in (**) must be an integer as well**, and therefore $(r + 3)$ must divide the numerator 24, i.e. the possible values of $(r + 3)$ are the following:

$$\{\pm 1, \quad \pm 2, \quad \pm 3, \quad \pm 4, \quad \pm 6, \quad \pm 8, \quad \pm 12, \quad \pm 24\} \quad (***)$$

Having determined the values of $(k + 3)$, the value of k is obtained, and then from (**) we find **all** the corresponding values of a. For example, if $(r + 3) = -12$, then $r = -15$, and from (**), $a = -280$. Verify directly, that $r = -15$ is a root of $x^3 - 281x - 840 = 0$.

Example 14-1-6: If k and n are integers, with $n \geq 2$, show that the equation $x^n + 2kx + 2 = 0$, does not have rational roots.

Solution

If the equation had a rational root p/q with p and q integers, then according to Theorem 14-1-2, q should divide the leading term 1, i.e. $q = \pm 1$, and p should divide the constant term 2, i.e. $p = \pm 1$ or ± 2, and therefore any rational root r **(if it exists)**, must be one of the integers ± 1 or ± 2.

Can $r = 1$ be a root? If yes, then we would have, $1 + 2k + 2 = 0$, i.e. $k = -\frac{3}{2}$, which **contradicts our assumption that k is an integer**. Similarly we show that $r = -1$ cannot be a root.

Can $r = 2$ be a root? If yes, then we would have, $2^n + 4k + 2 = 0$, i.e. $k = -\frac{1}{2} - 2^{n-2}$. Since by assumption $n \geq 2$, the number 2^{n-2} is integer, and therefore k is a fraction, which again **contradicts our assumption that k is an integer**. Similarly we show that $r = -2$ cannot be a root.

PROBLEMS

14-1-1) Solve the equation $x^4 + x^3 - 5x^2 + x - 6 = 0$, given that $x = i$ is one of its roots,   **(Ans: $\pm i, 2, -3$)**.

14-1-2) Solve the equation $x^6 - 3x^5 + 6x^3 - 3x^2 - 3x + 2 = 0$.

Hint: Determine first the rational roots.

14-1-3) If a, b, c are positive numbers with $a \neq c$, and the two equations $x^4 + ax^3 + bx^2 + cx + 1 = 0$ and $x^4 + cx^3 + bx^2 + ax + 1 = 0$ have one common root, show that $a + c - b = 2$.

14-1-4) Provided that the equations $x^3 + ax + b = 0$ and $3x^2 + a = 0$ have one common root, show that $4a^3 + 27b^2 = 0$.

14-1-5) Solve the equation $x^4 - 2x^3 - x^2 - 2x - 2 = 0$, given that $1 + \sqrt{3}$ is one of its roots,   **(Ans: $\pm i, 1 + \sqrt{3}, 1 - \sqrt{3}$)**.

14-1-6) Form a polynomial equation with integer coefficients having roots the numbers $2 + \sqrt{5}$ and $1 - i$.

14-1-7) Form a polynomial equation with real coefficients having $r = 2 - i$ as a root of multiplicity 2, (double root).

(Ans: $x^4 - 8x^3 + 26x^2 - 40x + 25 = 0$).

Hint: The root $2 + i$ is also a double root.

14-1-8) Solve the equation $6x^4 - 13x^3 - 35x^2 - x + 3 = 0$, given that one of its roots is $2 + \sqrt{3}$.

14-1-9) Solve the equation $x^5 - x^4 + 8x^2 - 9x + 15 = 0$, given that the numbers $\sqrt{3}$ and $1 - 2i$ are two of its roots.

(Ans: $\pm\sqrt{3}, 1 \pm 2i, -1$).

14-1-10) Solve the equation $3x^4 - 10x^3 + 4x^2 - x - 6 = 0$, given that $(1 - i\sqrt{3})/2$ is one of its roots.

14-1-11) Provided that $r = 2$ is a root of the equation

$$x^4 + (k - 2)x^3 - (k + 3)x^2 - 2kx + 3k = 0$$

determine k and find all the other roots.

(**Ans:** $k = 4$, other roots: $-3, -2, 1$).

14-1-12) For what integer values of λ, the equation $x^4 - 4x^3 - \lambda x^2 + 6\lambda x + 9 = 0$ has rational roots?

14-1-13) Show that the number 2 cannot be a multiple root of the equation $x^3 + 3x^2 - 4x - 12 = 0$.

14-1-14) Determine the rational numbers a and b, given that $2 + \sqrt{3}$ is a root of $x^3 + ax^2 + bx + 3 = 0$, (**Ans:** $a = -1, b = -11$).

14-1-15) If in a polynomial $P(x)$ with integer coefficients, both numbers $P(0)$ and $P(1)$ are odd, show that the equation $P(x) = 0$ cannot have integer roots.

Hint: If $x = r$ is an integer root, then $P(x) = Q(x)(x - r)$, where the coefficients of the quotient $Q(x)$ are integers, (since $P(x)$ has integer coefficients by assumption). For $x = 0$ and $x = 1$ we have: $P(0) = -rQ(0)$ and $P(1) = -(r - 1)Q(1)$, where $Q(0)$ and $Q(1)$ are integer numbers. From the first equality we conclude that r divides $P(0)$, and since (by assumption) $P(0)$ is odd, **r must be odd**. Similar reasoning (applied to the second equality) implies that $(r - 1)$ must be odd, and this leads to a contradiction since two consecutive integers, $r - 1$ and r cannot be ,both, integers, and this proves our assertion, i.e. $P(x)$ cannot have integer roots.

14-1-16) For what values of k the polynomial $P(x) = x^4 + kx^2 + k + 15$ has two real and two complex roots?

14-1-17) For what values of k the polynomial $P(x) = x^4 - 5kx^2 + k - 2$ has all its roots real? (**Ans:** $k > 2$).

14-2) Relation between the coefficients of a polynomial and its roots (Vieta's formulas)

Let us consider a polynomial $P(x) = a_n x^n + a_{n-1} x^{n-1} + \cdots + a_1 x + a_0$ having n roots $r_1, r_2, \cdots, r_{n-1}, r_n$, **real or complex**. As we have shown (see equation (13-5-1)), $P(x)$ can be written equivalently as

$$P(x) \equiv a_n (x - r_1)(x - r_2) \cdots (x - r_{n-1})(x - r_n) \qquad (*)$$

Let us call S_1 the **sum of all the roots**, S_2 the **sum of the products of the roots taken by two in all possible ways**, S_3 the **sum of the products of the roots taken by three in all possible ways**,..., S_n the **product of all the roots**, i.e.

$$\left\{ \begin{aligned} S_1 &= r_1 + r_2 + r_3 + \cdots + r_{n-1} + r_n \\ S_2 &= r_1 r_2 + r_1 r_3 + \cdots + r_1 r_n + r_2 r_3 + r_2 r_4 + \cdots + r_{n-1} r_n \\ S_3 &= r_1 r_2 r_3 + r_1 r_2 r_4 + \cdots + r_1 r_2 r_n + r_2 r_3 r_4 + \cdots + r_{n-2} r_{n-1} r_n \\ &\cdots \quad \cdots \quad \cdots \\ S_n &= r_1 r_2 r_3 \cdots r_{n-1} r_n \end{aligned} \right\}$$

$$(14-2-1)$$

Making use of the identity

$$(x - r_1)(x - r_2)(x - r_3) \cdots (x - r_n)$$
$$\equiv x^n - S_1 x^{n-1} + S_2 x^{n-2} - S_3 x^{n-3} + \cdots + (-1)^n S_n \quad (**)$$

equation (*) yields,

$$a_n x^n + a_{n-1} x^{n-1} + a_{n-2} x^{n-2} + a_{n-3} x^{n-3} + \cdots + a_1 x + a_0$$
$$\equiv a_n \{ x^n - S_1 x^{n-1} + S_2 x^{n-2} - S_3 x^{n-3} + \cdots + (-1)^n S_n \}$$

and since **this is an identity**, the coefficients of like powers of x must be the same, i.e.

$$S_1 = -\frac{a_{n-1}}{a_n}, \quad S_2 = \frac{a_{n-2}}{a_n}, \quad S_3 = -\frac{a_{n-3}}{a_n}, \quad \cdots\cdots, \quad S_n = (-1)^n \frac{a_0}{a_n}$$

Formulas (14-2-2) are known as **the Vieta's formulas**.

For example, for a second degree polynomial $ax^2 + bx + c$ having roots r_1 and r_2, we have, $S_1 = r_1 + r_2 = -\frac{b}{a}$ and $S_2 = r_1 r_2 = \frac{c}{a}$. We have derived these two expressions in Chapter 9, equation (9-1-2).

If we consider **a third degree equation** $ax^3 + bx^2 + cx + d = 0$, having roots r_1, r_2 and r_3, the Vieta's formulas yield,

$$S_1 = r_1 + r_2 + r_3 = -\frac{b}{a}$$
$$S_2 = r_1 r_2 + r_1 r_3 + r_2 r_3 = \frac{c}{a}$$
$$S_3 = r_1 r_2 r_3 = -\frac{d}{a}$$

Example 14-2-1: Find the roots of $x^3 + 4x^2 + x - 6 = 0$, and verify Vieta's formulas.

Solution

If the equation has rational roots of the form (p/q) then $q = \pm 1$ and $p = \pm 1, \pm 2, \pm 3, \pm 6$. By direct substitution we find that the roots of the equation are $r_1 = 1, r_2 = -2, r_3 = -3$.

$$S_1 = r_1 + r_2 + r_3 = 1 - 2 - 3 = -4 = -\frac{4}{1}$$

$$S_2 = r_1 r_2 + r_1 r_3 + r_2 r_3 = -2 - 3 + 6 = 1 = \frac{1}{1}$$

$$S_3 = r_1 r_2 r_3 = 6 = -\frac{-6}{1}$$

Example 14-2-2: Solve the equation $8x^4 - 2x^3 - 27x^2 + 6x + 9 = 0$, given that two of its roots are opposite numbers.

Solution

Let $r_1 = r, r_2 = -r, r_3, r_4$ be the four roots of the equation. Application of Vieta's formulas yield:

$$S_1 = r - r + r_3 + r_4 = -\frac{-2}{8} \Rightarrow r_3 + r_4 = \frac{1}{4} \qquad (*)$$

$$S_2 = -r^2 + rr_3 + rr_4 - rr_3 - rr_4 + r_3 r_4 = -\frac{27}{8} \Rightarrow$$

$$S_2 = -r^2 + r(r_3 + r_4) - r(r_3 + r_4) + r_3 r_4 = -\frac{27}{8} \Rightarrow$$

$$S_2 = -r^2 + r_3 r_4 = -\frac{27}{8} \qquad (**)$$

$$S_3 = -r^2 r_3 - r^2 r_4 + r r_3 r_4 - r r_3 r_4 = -\frac{6}{8} = -\frac{3}{4} \Rightarrow$$

$$S_3 = -r^2(r_3 + r_4) = -\frac{3}{4} \qquad (***)$$

$$S_4 = -r^2 r_3 r_4 = \frac{9}{8} \qquad (****)$$

From (*) and (***) we obtain:

$$-r^2 \cdot \frac{1}{4} = -\frac{3}{4} \Rightarrow r^2 = 3 \Rightarrow r = \pm\sqrt{3} \Rightarrow \begin{Bmatrix} r_1 = \sqrt{3} \\ r_2 = -\sqrt{3} \end{Bmatrix}$$

From equation (****), with $r^2 = 3$, we find: $r_3 r_4 = -\frac{3}{8}$, (we should have obtained the same equation had we used (**) instead of (****)).

Since we know the sum and the product of the two numbers r_3 and r_4, r_3 and r_4 are the roots of the quadratic equation (see Remark 1 in section 9-1),

$$t^2 - \frac{1}{4}t - \frac{3}{8} = 0 \Leftrightarrow 8t^2 - 2t - 3 = 0 \Rightarrow \begin{Bmatrix} r_3 = 3/4 \\ r_4 = -1/2 \end{Bmatrix}$$

In summary: The four roots of the equation are, $\pm\sqrt{3}, 3/4, -1/2$.

Example 14-2-3: What should be the relation between a and b, if one root of the equation $x^3 + ax + b = 0$ is to be equal to the product of the other two?

Solution

If $r_1, r_2, r_3 = r_1 r_2$ are the roots of the given equation, Vieta's formulas yield:

$$S_1 = r_1 + r_2 + r_1 r_2 = 0 \qquad (*)$$

$$S_1 = r_1 r_2 + r_1(r_1 r_2) + r_2(r_1 r_2) = a \Leftrightarrow$$

$$r_1 r_2 (1 + r_1 + r_2) = a \qquad (**)$$

$$r_1 r_2 (r_1 r_2) = -b \Leftrightarrow (r_1 r_2)^2 = -b \qquad (***)$$

Solving (*) for $r_1 + r_2$, substituting in (**) and taking into account (***) results in the following relation between a and b : $(a - b)^2 = -b$.

Example 14-2-4: If a, b, c are the roots of $x^3 + \kappa x^2 + \lambda x + \mu = 0$, express the quantity $A = a^2 + b^2 + c^2 + 5abc$ in terms of κ, λ, μ.

Solution

$$A = a^2 + b^2 + c^2 + 5abc$$
$$= (a + b + c)^2 - 2(ab + bc + ca) + 5abc \xrightarrow{(Vieta\ formulas)}$$
$$A = (-\kappa)^2 - 2\lambda - 5\mu = \kappa^2 - 2\lambda - 5\mu$$

PROBLEMS

14-2-1) Solve the equation $x^3 - 7x + 6 = 0$, given that one of its roots is twice one of the others, **(Ans: 1, 2, −3).**

14-2-2) If a, b, c are the roots of $x^3 + \kappa x^2 + \lambda x + \mu = 0$, express in terms of κ, λ, μ the quantities:

$$A = a^{-2} + b^{-2} + c^{-2}, \qquad B = (a+b)(b+c)(c+a)$$

14-2-3) Verify that the coefficients of the equation $x^3 - 2x - 4 = 0$ satisfy the condition established in Example 14-2-3, and then solve the equation.

(Ans: $-1 + i$, $-1 - i$, 2).

14-2-4) Provided that the roots of $x^5 - 5x^4 - 5x^3 + 25x^2 + 4x - 20 = 0$ are of the form $r_1, -r_1, r_2, -r_2, r_3$, solve the equation.

14-2-5) Without solving the equation $x^3 - 5x^2 + 3x - 7 = 0$, find the sum of the cubes of its roots, **(Ans: 101).**

Hint: Make use of Cauchy's identity:

$$a^3 + b^3 + c^3 - 3abc = (a + b + c)(a^2 + b^2 + b^2 - ab - bc - ca)$$

14-2-6) If r_1, r_2, r_3 are the roots of $x^3 + ax^2 + bx + c = 0$ and $S_1 = r_1 + r_2 + r_3$, show that $(S_1 - r_1)(S_1 - r_2)(S_1 - r_3) = c - ab$.

14-2-7) If a, b, c are the roots of $x^3 + \kappa x^2 + \lambda x + \mu = 0$, form a third degree equation having roots the numbers a^2, b^2, c^2.

(Ans: $x^3 + (2\lambda - \kappa^2)x^2 + (\lambda^2 - 2\kappa\mu)x + \mu^2 = 0)$.

14-3) Some general remarks in solving polynomial equations

1) The main idea to solve a polynomial equation $P(x) = 0$ is to factor $P(x)$ into simpler terms, and then set each factor equal to zero. This results in equations much simpler than $P(x) = 0$, (see section 2-4, Factored Equations).

If we can, somehow, find one root r of the polynomial, then $(x - r)$ divides exactly $P(x)$, i.e. $P(x) = (x - r)Q(x)$, where $Q(x)$ is a polynomial of degree $(n - 1)$, if $P(x)$ is of degree n. We now try to find a root r_1 of $Q(x)$, and if we succeed, following the same procedure, $Q(x) = (x - r_1)Q_1(x)$, where $Q_1(x)$ is of degree $(n - 2)$, (simpler than $Q_1(x)$), and similarly, we repeat the same procedure until we find all the roots of the original polynomial.

The Theorems developed in section 14-1 are particularly useful when solving polynomial equations.

If a complex number $(a + ib)$ is a root **of multiplicity m**, of a polynomial equation with **real coefficients**, then the complex conjugate of the root, i.e. $(a - ib)$ is also a root **of the same multiplicity**.

If the irrational number $(a + \sqrt{b})$ (a rational, b rational, $\sqrt{b}$ irrational) is a root of **multiplicity m**, of a polynomial with **rational coefficients**, then $(a - \sqrt{b})$ is also a root **of the same multiplicity**.

To locate the rational roots of an equation **with integer coefficients**, (if there exist any), we make use of Theorem 14-1-2 (**the rational roots Theorem**).

In the general case, it is very difficult to find the roots of a polynomial equation. The degree of difficulty increases with the degree of the equation. In cases where analytic computation of the roots is not possible, there are methods to compute the roots **approximately**. These methods are studied in Numerical Analysis courses.

2) Equations with complex coefficients: Note that **when the coefficients of an equation are complex numbers, then the conjugate roots Theorem, may not hold true**. This means that $(a + ib)$ may be a root, but $(a - ib)$ may not be a root.

Let us for example consider **a quadratic equation with complex coefficients**,

$$az^2 + bz + c = 0, \quad a, b, c \in \mathbb{C} \; (set\; of\; complex\; numbers)$$

Following the procedure developed in section 9-1, we find that the roots of the quadratic equation are given by the same formula, i.e.

$$z_{1,2} = \frac{-b \pm \sqrt{b^2 - 4ac}}{2a} \qquad (14-3-1)$$

where a, b, c are now complex numbers. **For the computation of the square root of a complex number see section 8-7 and for an alternative computation, see Example 8-7-4.**

The sum and the product of the two roots, is given by the same formulas as in the case with real coefficients, i.e.

$$\left\{ z_1 + z_2 = -\frac{b}{a}, \quad z_1 z_2 = \frac{c}{a} \right\} \qquad (14-3-2)$$

Using properties of the absolute values of complex numbers, eq. (14-3-2) implies,

$$\left\{ \frac{|b|}{|a|} = |z_1 + z_2| \le |z_1| + |z_2|, \quad |z_1 z_2| = \frac{|c|}{|a|} \right\} \qquad (14-3-3)$$

Example 14-3-1: Solve the equation $z^2 - (2 - i)z + (3 - i) = 0$.

Solution

The discriminant is:

$$\Delta = (2 - i)^2 - 4(3 - i) = 4 + i^2 - 4i - 12 + 4i = 4 - 1 - 12 = -9$$

$$z_{1,2} = \frac{(2-i) \pm \sqrt{-9}}{2} = \frac{(2-i) \pm 3i}{2} \Rightarrow \begin{cases} z_1 = 1+i \\ z_2 = 1-2i \end{cases}$$

We see that, **since the coefficients of the trinomial are not real**, the roots z_1 and z_2 **are not** complex conjugate.

Example 14-3-2: If p and q are real numbers, such that $pq = 1$, show that the equation $x^3 + px^2 + qx + 1 = 0$ has one real and two imaginary roots.

Solution

Since $pq = 1, q = \frac{1}{p}$ and the equation becomes:

$$x^3 + px^2 + \frac{1}{p}x + 1 = 0 \Leftrightarrow px^3 + p^2x^2 + x + p = 0 \Leftrightarrow$$

$$px^2(x+p) + (x+p) = 0 \Leftrightarrow (x+p)(px^2 + 1) = 0 \Leftrightarrow \begin{cases} x+p = 0 \\ \text{or} \\ px^2 + 1 = 0 \end{cases} \Leftrightarrow$$

$$\{x = -p \quad \text{or} \quad x = \pm i\sqrt{1/p}\}$$

and this shows that the equation has one real root ($x = -p$) and two imaginary roots ($\pm i\sqrt{1/p}$).

Example 14-3-4: Let $p_1, p_2, \cdots, p_n$ be real numbers ($p_1 p_2 \cdots p_n \neq 0$), $q_1, q_2, \cdots, q_n$ be also n distinct real numbers, (i.e. different from each other), and F be any real number $\neq 0$. Show that all the roots of the equation

$$\frac{p_1^2}{x - q_1} + \frac{p_2^2}{x - q_2} + \cdots + \frac{p_n^2}{x - q_n} = F$$

are real numbers.

Solution

If we eliminate the denominators, we see that the given equation reduces to a polynomial equation with real coefficients, of degree n. If the equation has a complex root $r = a + ib$, (a, b real), then according to Theorem 14-1-4, $\bar{r} = a - ib$ must also be a root, i.e. **both equalities must hold simultaneously:**

$$\frac{p_1^2}{r-q_1} + \frac{p_2^2}{r-q_2} + \cdots + \frac{p_n^2}{r-q_n} = F \qquad (*)$$

$$\frac{p_1^2}{\bar{r}-q_1} + \frac{p_2^2}{\bar{r}-q_2} + \cdots + \frac{p_n^2}{\bar{r}-q_n} = F \qquad (**)$$

and subtracting (*) from (**) yields,

$$p_1^2 \frac{r-\bar{r}}{(r-q_1)(\bar{r}-q_1)} + p_2^2 \frac{r-\bar{r}}{(r-q_2)(\bar{r}-q_2)} + \cdots + p_n^2 \frac{r-\bar{r}}{(r-q_n)(\bar{r}-q_n)} = 0$$

or, since $r - \bar{r} = 2ib$ and $(r-q_1)(\bar{r}-q_1) = (a-q_1)^2 + b^2$, etc,

$$2ib\left\{\frac{p_1^2}{(a-q_1)^2 + b^2} + \frac{p_2^2}{(a-q_2)^2 + b^2} + \cdots + \frac{p_n^2}{(a-q_n)^2 + b^2}\right\} = 0 \quad (***)$$

We notice that the quantity inside the braces **is always positive**, and therefore, equation (***) implies that $b = 0$, i.e. $r = a + ib = a + i \cdot 0 = a$ real, and this completes the proof.

Example 14-3-5: If r is a root of $x^n + a_{n-1}x^{n-1} + \cdots + a_1 x + a_0 = 0$, show that $|r| < 1 + |a_{n-1}| + |a_{n-2}| + \cdots + |a_1| + |a_0|$.

Solution

Assuming that $x = r$ is a root of the equation, we have:

$$r^n + a_{n-1}r^{n-1} + a_{n-2}r^{n-2} + \cdots + a_1 r + a_0 = 0 \Leftrightarrow$$

$$r^n = -a_{n-1}r^{n-1} - a_{n-2}r^{n-2} - \cdots - a_1 r - a_0 \Leftrightarrow$$

$$r = -a_{n-1} - \frac{a_{n-2}}{r} - \cdots - \frac{a_1}{r^{n-2}} - \frac{a_0}{r^{n-1}} \qquad (*)$$

Taking the absolute values of both sides of equation (*) and applying the well known property of absolute values, $|z_1 + z_2 + \cdots| \leq |z_1| + |z_2| + \cdots$, **(valid for real and complex numbers)**, we obtain,

$$|r| = \left|a_{n-1} + \frac{a_{n-2}}{r} + \cdots + \frac{a_1}{r^{n-2}} + \frac{a_0}{r^{n-1}}\right|$$

$$\leq |a_{n-1}| + \frac{|a_{n-2}|}{|r|} + \cdots + \frac{|a_1|}{|r|^{n-2}} + \frac{|a_0|}{|r|^{n-1}} \qquad (**)$$

We now consider the two cases, either $|r| \geq 1$, or $|r| < 1$.

Case 1: If $|r| \geq 1$, then: $\frac{1}{|r|} \leq 1$, $\frac{1}{|r|^2} \leq 1, \cdots, \frac{1}{|r|^{n-2}} \leq 1$, $\frac{1}{|r|^{n-1}} \leq 1$, and equation (**) implies,

$$|r| \leq |a_{n-1}| + |a_{n-2}| + \cdots + |a_1| + |a_0| \Rightarrow$$

$$|r| < 1 + |a_{n-1}| + |a_{n-2}| + \cdots + |a_1| + |a_0|$$

Case 2: If $|r| < 1$, then $|r| < 1 + |a_{n-1}| + |a_{n-2}| + \cdots + |a_1| + |a_0|$.

So, in both cases, $|r| < 1 + |a_{n-1}| + |a_{n-2}| + \cdots + |a_1| + |a_0|$, and this completes the proof.

PROBLEMS

14-3-1) Solve the equation: $x^2 - 2(3+i)x - (8+6i) = 0$.

(**Ans:** $3 + 3\sqrt{2} + i(1+\sqrt{2})$, $3 - 3\sqrt{2} + i(1-\sqrt{2})$).

14-3-2) Solve the equation: $x^4 + x^3 - \frac{1}{4}x^2 + \frac{1}{4}x + \frac{1}{16} = 0$.

Hint: Divide through by x^2, (since $x = 0$ is **not** a root), and make the substitution $y = x + \frac{1}{4x}$, (**Ans:** $\frac{1 \pm i\sqrt{3}}{4}$, $\frac{-3 \pm \sqrt{5}}{4}$).

14-3-3) Making use of the "**change of variable**" $y = x + \frac{p+q}{2}$, show that the equation $(x+p)^4 + (x+q)^4 = k$, $(p, q, k$ given numbers), reduces to the following biquadratic equation in y, which in principle can be solved:

$$y^4 + 6\left(\frac{p-q}{2}\right)^2 y^2 + \left(\frac{p-q}{2}\right)^4 - \frac{k}{2} = 0$$

14-3-4) Determine the integer value of the parameter p, for which the equation $x^4 - 3x^3 + px^2 - 4x - 1 = 0$, admits a rational root.

14-3-5) Solve the equation: $5x^5 - 3x^4 - 10x^3 + 6x^2 - 40x + 24 = 0$.

(**Ans:** $\pm 2, 3/5, \pm i\sqrt{2}$).

14-3-6) Solve the equation: $x^2 + \frac{4x^2}{(x+2)^2} = 5$.

(**Ans**: $2, -1, (-5 \pm i\sqrt{15})/2$).

14-3-7) Solve the equation: $4x^4 - 4x^3 + x^2 - 4x - 3 = 0$.

(**Ans**: $3/2, -1/2, \pm i$).

14-3-8) Solve the equation: $(x^2 + 3x - 1)^2 - 4(x^2 + 3x + 5) = 21$.

(**Ans**: $2, -5, (-3 \pm i\sqrt{7})/2$).

Hint: If we set $y = x^2 + 3x$, the equation is transformed to a quadratic in y.

14-3-9) Solve the equation: $(x - 3)(x - 4)(x + 5)(x + 4) = 84$.

(**Ans**: $2, -3, (-1 \pm \sqrt{105})/2$).

Hint: Combine the first with the fourth factor, and the second with the third, and then make the substitution $y = x^2 + x$, etc.

14-3-10) Solve the equation: $\sqrt[6]{(1+x)^2} - \sqrt[6]{(1-x)^2} = \sqrt[6]{1-x^2}$.

(**Ans**: $\{(1+\sqrt{5})^6 - 2^6\}/\{(1+\sqrt{5})^6 + 2^6\}$).

14-4) Algebraic numbers

A number b (real or complex) is called algebraic, if it is a solution of a polynomial equation $a_n z^n + a_{n-1} z^{n-1} + \cdots + a_1 z + a_0$, where all the coefficients $a_n, a_{n-1}, \ldots, a_1, a_0$ are integers.

For example, $x = 3$ is an algebraic number, since it satisfies the equation $x - 3 = 0$. The number $x = \sqrt{2}$ is algebraic (even though irrational), since it satisfies $x^2 - 2 = 0$, and similarly, $x = \sqrt[5]{3}$ is algebraic, since it satisfies $x^5 - 3 = 0$. The complex number $z = i$ is algebraic, since it satisfies $z^2 + 1 = 0$, etc.

There are numbers that are not algebraic, i.e., do not satisfy any polynomial equation with integer coefficients. These numbers are called "**transcendental numbers**". It can be shown that the number $\pi = 3.1415 \ldots$ is transcendental. Another transcendental number is the number $e = 2.718281 \ldots$, (see section

17-1). Note that it is not, yet known, whether the numbers $(e + \pi)$ and $(e \cdot \pi)$ are transcendental or algebraic.

Example 14-4-1: Show that the number $z = \sqrt{2} + 3i$ is algebraic.

Solution

$$z = \sqrt{2} + 3i \Rightarrow z - 3i = \sqrt{2} \Rightarrow (z - 3i)^2 = \left(\sqrt{2}\right)^2 \Rightarrow$$

$$z^2 - 6iz + (3i)^2 = 2 \Rightarrow z^2 - 6iz - 9 = 2 \Rightarrow z^2 - 11 = 6iz$$

and squaring both sides, once more, we obtain,

$$(z^2 - 11)^2 = (6iz)^2 \Rightarrow z^4 - 22z^2 + 11^2 = -36z^2 \Rightarrow z^4 + 14z^2 + 121 = 0$$

and this shows that the number $z = \sqrt{2} + 3i$ is algebraic, since it satisfies a polynomial equation with integer coefficients.

PROBLEMS

14-4-1) Show that the numbers $a = \sqrt{3} + \sqrt{5}$ and $b = -\frac{1}{2} + \sqrt{7}\, i$, are algebraic numbers.

14-4-2) Show that the numbers $x = \sqrt[3]{2} + \sqrt{5}$ and $y = -3 + \sqrt{7}\, i$, are algebraic numbers.

14-4-3) Show that the numbers $z_1 = \cos\frac{\pi}{3} + i \sin\frac{\pi}{3}$, $z_2 = 2\left(\cos\frac{\pi}{4} + i\sin\frac{\pi}{4}\right)$, $w = z_1 z_2$ and $u = z_1/z_2$ are algebraic numbers.

CHAPTER 15: POLYNOMIALS IN SEVERAL VARIABLES

15-1) Introduction

Any function of the form $cx^n y^m$, where c is a constant and n and m are natural numbers, is called **a monomial in the two variables x and y**. For example, the expressions $3x^2 y^3, (-7)x^5 y^4, (1+\sqrt{2})x^8 y$ are monomials in two variables. **The degree of the monomial $cx^n y^m$ in x and y is the sum of the exponents of x and y, i.e. is equal to $(m+n)$.** For example the degree of $2x^3 y^4$ is $(3+4) = 7$, and the degree of $5x^7 y$ is $(7+1) = 8$.

The sum of monomials in x and y, is called a polynomial in x and y, (see section 1-5 (c) and (d)). **The degree of a polynomial in x and y is the degree of the monomial that has the highest degree.** Le us, for example, consider the following polynomial,

$$P(x,y) = 3x^3 y^4 + 2x^2 y^3 - 6x^2 y^2 + 7xy + 5 \qquad (*)$$

The degree of the polynomial in x and y is $(3+4) = 7$. Of course, it is possible to arrange the polynomial **in descending powers of x**, and write it as

$$P(x,y) = (3y^4)x^3 + (2y^3 - 6y^2)x^2 + (7y)x + 5 \qquad (**)$$

or, we may arrange it **in descending powers of y**, in which case we write,

$$P(x,y) = (3x^3)y^4 + (2x^2)y^3 - (6x^2)y^2 + (7x)y + 5 \qquad (***)$$

The degree of the polynomial in x in (**) is 3, while the degree of the polynomial in y in (***) is 4. The degree of the polynomial in both x and y in (*) is $7 = 4 + 3$.

Two polynomials $P(x,y)$ and $Q(x,y)$ are identically equal (and we write $P(x,y) \equiv Q(x,y)$) when they consist of equal monomials. **Two identically equal polynomials assume the same arithmetic values for any ordered pair of numbers (x_0, y_0).** The difference of two identically equal polynomials, is the zero polynomial, i.e. a polynomial all the coefficients of which are zero.

As an example, if $x^3 y^2 + ax^2 y^2 + 5xy \equiv bx^3 y^2 - 7x^2 y^2 + cxy$, we conclude that $b = 1, a = -7, c = 5$.

Similar definitions apply for polynomials $P(x, y, z)$ in three variables, x, y and z, for polynomials $P(x, y, z, w)$ in four variables x, y, z and w, etc.

Example 15-1-1: For what values of a, b and c the two polynomials $P(x, y, z) = 2x^3y^2z + 3axy^2z + 7xyz$ and $Q(x, y, z) = bx^3y^2z - 2xy^2z + cxyz$ are identically equal?

Solution

$$P(x, y, z) \equiv Q(x, y, z) \Leftrightarrow b = 2, a = -\frac{2}{3}, c = 7$$

PROBLEMS

15-1-1) Show that: $(x + y)^2 \equiv x^2 + 2xy + y^2$.

15-1-2) Show that:

$$x^3 + y^3 + z^3 - 3xyz \equiv \frac{1}{2}(x + y + z)\{(x - y)^2 + (y - z)^2 + (z - x)^2\}$$

15-1-3) Assuming that x, y, z are real numbers, show that if $x^3 + y^3 + z^3 = 3xyz$, then either $(x + y + z) = 0$ or $x = y = z$.

15-2) The division of polynomials in several variables

The perfect division of polynomials in several variables is defined as in the case of polynomials in one variable. We say that the polynomial $D(x, y, z, ...)$ divides exactly the polynomial $P(x, y, z, ...)$ if there exists a polynomial $Q(x, y, z, ...)$ such that

$$P(x, y, z, ...) \equiv Q(x, y, z, ...)D(x, y, z, ...) \qquad (15 - 2 - 1)$$

The polynomial $P(x, y, z, ...)$ is called **the dividend**, $D(x, y, z, ...)$ is called **the divisor** and $Q(x, y, z, ...)$ is called **the quotient** of the division. Notice that from (15-2-1) if $D(x, y, z, ...)$ divides $P(x, y, z, ...)$, then $Q(x, y, z, ...)$ divides $P(x, y, z, ...)$, the quotient of the division being $D(x, y, z, ...)$.

As an example, $x + y$ divides exactly $x^3 + y^3$, since

$$\frac{x^3 + y^3}{x + y} \equiv x^2 - xy + y^2 \Leftrightarrow x^3 + y^3 \equiv (x + y)(x^2 - xy + y^2)$$

Theorem 15-2-1: **If the polynomial $P(x, y, z, ...)$ vanishes when x is replaced by y, then the polynomial is divided by $(x - y)$, and conversely.**

Proof: If we arrange the polynomial in descending powers of x, and divide by $(x - y)$, the remainder of the division must be a constant (a polynomial of degree zero), and according to the identity of the algorithmic division (equation 13-3-1), we have,

$$P(x, y, z, ...) \equiv (x - y)Q(x, y, z, ...) + c \qquad (*)$$

Applying the identity in (*) for $x = y$, we have, $0 = 0 + c$, (since $P(y, y, z, ...) = 0$, by assumption), or $c = 0$, and equation (*) yields,

$$P(x, y, z, ...) \equiv (x - y)Q(x, y, z, ...)$$

which shows that the polynomial $P(x, y, z, ...)$ is divided by $(x - y)$.

The converse statement is obvious.

Theorem 15-2-2: **If the polynomial $P(x, y, z)$ is divided by $(x - y)$ and by $(x - z)$, then it will be divided by the product $(x - y)(x - z)$ as well.**

Proof: Since $P(x, y, z, ...)$ is divided by $(x - y)$, we have the identity,

$$P(x, y, z) \equiv (x - y)Q(x, y, z) \qquad (*)$$

Applying the identity in (*) for $x = z$, we have, (since $P(z, y, z) = 0$, by assumption),

$$0 \equiv (z - y)Q(z, y, z) \Leftrightarrow Q(z, y, z) \equiv 0$$

and by virtue of Theorem 15-2-1, we conclude that $(x - z)$ divides $Q(x, y, z)$, i.e. $Q(x, y, z) = (x - z)Q_1(x, y, z)$, and substituting in (*) we obtain,

$$P(x, y, z) \equiv (x - y)(x - z)Q_1(x, y, z) \qquad (**)$$

Equation (**) shows that $P(x, y, z)$ is divided by the product $(x - y)(x - z)$ and this completes the proof.

Theorem 15-2-3: **(a)** If the polynomial $P(x, y, z)$ is divided by $(x - y)$ and by $(y - z)$ and by $(z - x)$, then the polynomial will be divided and by the product $(x - y)(y - z)(z - x)$.

(b) If the polynomial $P(x, y, z)$ is divided by $(x + y)$ and by $(y + z)$ and by $(z + x)$, then it will be divided and by the product $(x + y)(y + z)(z + x)$.

Proof: The proof is identical to the proof of Theorem 15-2-2.

Example 15-2-1: Show that the polynomial

$$P(x, y, z) = x^n(y - z) + y^n(z - x) + z^n(x - y)$$

is divided by the product $(x - y)(y - z)(z - x)$, for all values of the natural number $n \geq 2$.

Solution

If x is replaced by y, we find: $P(y, y, z) = y^n(y - z) + y^n(z - y) + 0 = 0$, and by Theorem 15-2-1, $P(x, y, z)$ is divided by $(x - y)$. Similarly we show that the polynomial is divided by $(y - z)$ and by $(z - x)$, and by virtue of Theorem 15-2-3, it will divided by the product $(x - y)(y - z)(z - x)$.

PROBLEMS

15-2-1) Provided that n is an odd positive integer, $(n \geq 3)$, show that the polynomial $P(x, y, z) = (x + y + z)^n - x^n - y^n - z^n$ is divided by the product $(x + y)(y + z)(z + x)$.

Hint: Since $P(x = -y, y, z) = 0$, the polynomial is divided by $(x + y)$, and similarly show that is divided by $(y + z)$ and by $(z + x)$, and then apply Theorem 15-2-3 (b).

15-2-2) Show that the polynomial

$$P(a, b, c) = a^n(b^k - c^k)^{2m+1} + b^n(c^k - a^k)^{2m+1} + c^n(a^k - b^k)^{2m+1}$$

is divided by the product $(a - b)(b - c)(c - a)$, for all values of the positive integers k, m, n.

Hint: Show that $P(a = b, b, c) = 0$, which shows that $P(a, b, c)$ is divided by $(a - b)$, etc.

15-3) Symmetric polynomials

Polynomials which remain unchanged by **a cyclic permutation** of their variables, are called **symmetric** polynomials.

For example, the following polynomials in **two** variables x and y are symmetric:

$$x + y - 3, \quad x^2 + xy + y^2, \quad x^3 + y^3 + x^2 + y^2 + 7xy$$

These polynomials remain unchanged when x is replaced by y, and y by x.

Also, the following polynomials in **three** variables are symmetric:

$$x^2 + y^2 + z^2 + xy + yz + zx, \qquad x^3 + y^3 + z^3 - 3xyz$$

In general, if $P(x, y, z) \equiv P(y, z, x)$, then $P(x, y, z)$ is symmetric.

The most general form of **first degree symmetric** polynomials in x, y, z is:

$$a(x + y + z) + b \qquad (15 - 3 - 1)$$

The most general form of **second degree symmetric** polynomials in x, y, z is:

$$a(x^2 + y^2 + z^2) + b(xy + yz + zx) + c(x + y + z) + d \quad (15 - 3 - 2)$$

The most general form of **third degree symmetric** polynomials in x, y, z is:

$$a(x^3 + y^3 + z^3) + b(x^2y + y^2z + z^2x) + c(xy^2 + yz^2 + zx^2) + dxyz \\ + \kappa(x^2 + y^2 + z^2) + \lambda(xy + yz + zx) + \mu(x + y + z) + \nu$$

Properties of symmetric polynomials:

Theorem 15-3-1: The product of two symmetric polynomials is a symmetric polynomial.

Proof: Let $A(x, y, z)$ and $B(x, y, z)$ be two symmetric polynomials, i.e.

$$A(x, y, z) = A(y, z, x) \text{ and } B(x, y, z) = B(y, z, x) \qquad (*)$$

Then their product $P(x, y, z) = A(x, y, z)B(x, y, z)$ is also symmetric, since
$P(y, z, x) = A(y, z, x)B(y, z, x) = A(x, y, z)B(x, y, z) = P(x, y, z)$.

Theorem 15-3-2: In a perfect division, if the dividend and the divisor are symmetric polynomials, then the quotient will be also symmetric.

Proof: Similar to the proof of Theorem 15-3-1.

Theorem 15-3-3: (a) If the symmetric polynomial $P(x, y, z)$ is divided by $(x - y)$, then it will be divided and by the product $(x - y)(y - z)(z - x)$.

(b) If the symmetric polynomial $P(x, y, z)$ is divided by $(x + y)$, then it will be divided and by the product $(x + y)(y + z)(z + x)$.

Proof: (a) Since (by assumption), $(x - y)$ divides $P(x, y, z)$, we have:

$$P(x, y, z) \equiv (x - y)Q(x, y, z) \qquad (*)$$

and by a cycling permutation we obtain,

$$P(y, z, x) \equiv (y - z)Q(y, z, x)$$

or since $P(x, y, z)$ is symmetric,

$$P(x, y, z) \equiv (y - z)Q(y, z, x)$$

and this shows that $P(x, y, z)$ is divided by $(y - z)$ as well. Similarly, we show that $P(x, y, z)$ is divided by $(z - x)$, and then, from Theorem 15-2-3, the polynomial $P(x, y, z)$ is divided and by the product $(x - y)(y - z)(z - x)$.

(b) Similar to the proof in part (a).

The elementary symmetric polynomials:

If we consider the n variables, $x_1, x_2, \cdots, x_n$ and call $S_1 = x_1 + x_2 + \cdots + x_n$ (the sum of all the variables), $S_2 = x_1 x_2 + x_1 x_3 + \cdots + x_{n-1} x_n$, (the sum of the products of the variables, taken by two, in all possible ways),

$$S_3 = x_1 x_2 x_3 + x_1 x_2 x_4 + \cdots + x_{n-2} x_{n-1} x_n$$

(the sum of the products of the variables, taken by three, in all possible ways),....., and $S_n = x_1 x_2 x_3 \cdots x_n$ (the product of all the variables), we notice that $S_1, S_2, S_3, \cdots, S_n$ are **symmetric polynomials in the variables** $x_1, x_2, \cdots, x_n$, and are called **the elementary symmetric polynomials**. For example, the elementary symmetric polynomials of two variables x and y, are $S_1 = x + y$ and $S_2 = xy$. The elementary symmetric polynomials of x, y, z are,

$$S_1 = x + y + z, \qquad S_2 = xy + yz + zx, \qquad S_3 = xyz$$

Theorem 15-3-4: **Every symmetric polynomial can be expressed in terms of the elementary symmetric polynomials** $S_1, S_2, S_3, \ldots$.

Example 15-3-1: Express $P(x, y) = x^3 + y^3 - 3xy^2 - 3x^2y + 2(x^2 + y^2)$ in terms of the elementary symmetric polynomials $S_1 = x + y$ and $S_2 = xy$.

Solution

$$P(x,y) = \underbrace{(x+y)^3 - 3xy(x+y)}_{x^3+y^3} - 3xy(x+y) + 2\underbrace{\{(x+y)^2 - 2xy\}}_{x^2+y^2} \Rightarrow$$

$$P(x,y) = S_1^3 - 3S_2 S_1 - 3S_2 S_1 + 2(S_1^2 - 2S_2) \Rightarrow$$

$$P(x,y) = S_1^3 + 2S_1^2 - 6S_1 S_2 - 4S_2$$

Example 15-3-2: Express $P(x, y, z) = x^3 + y^3 + z^3$ in terms of the elementary symmetric polynomials $S_1 = x + y + z, S_2 = xy + yz + zx$ and $S_3 = xyz$.

Solution

From Problem 15-1-2 we have, (**Cauchy's identity**):

$$x^3 + y^3 + z^3 - 3xyz = (x + y + z)(x^2 + y^2 + z^2 - xy - yz - zx) \Rightarrow$$

$$\begin{aligned} x^3 + y^3 + z^3 = {} & 3xyz \\ & + (x+y+z)\big((x+y+z)^2 - 2(xy+yz+zx) \\ & - (xy+yz+zx)\big) \Rightarrow \end{aligned}$$

$$x^3 + y^3 + z^3 = 3S_3 + S_1(S_1^2 - 3S_2) = S_1^3 - 3S_1 S_2 + 3S_3$$

PROBLEMS

15-3-1) Express $x^4 + y^4 + z^4$ in terms of the elementary symmetric polynomials, (**Ans:** $S_1^4 + 2S_2^2 - 4S_1^2 S_2 + 4S_1 S_3$).

15-3-2) Express $x^3 + y^3$ in terms of $(x + y + 3)$ and $(xy - x - y)$.

15-3-3) Express $x^2y^2z^2 + y^2z^2w^2 + z^2w^2x^2 + w^2x^2y^2$ in terms of the elementary symmetric polynomials, **(Ans: $S_3^2 - 2S_2S_4$).**

15-3-4) Express xyz in terms of $\phi_1 \equiv x + y + z + 3$, $\phi_2 \equiv xy + yz + zx$ and $\phi_3 \equiv x^2(y + z) + y^2(z + x) + z^2(x + y)$.

15-3-5) Assuming that the equation $x^2 + kx + m = 0$ has two positive roots r_1 and r_2, express $\sqrt[4]{r_1} + \sqrt[4]{r_2}$ in terms of k and m.

(Ans: $\sqrt[4]{-k + 6\sqrt{m} + 4\sqrt[4]{m}\sqrt{-k + 2\sqrt{m}}}$ **).**

15-4) Homogeneous polynomials

A polynomial $P(x, y, z, ...)$ is called **homogeneous** when **all its terms (monomials) are of the same degree in the variables $x, y, z,$ ...** For example, the polynomial $x^2y + xy^2 - 3x^3 + 7y^3$ is **a third degree homogeneous polynomial in x and y**, $2ab - a^2 + b^2$ is **a second degree homogeneous polynomial in the variables a and b**, while $x^2y + z^3 - y^3$ is **a third degree homogeneous polynomial in x, y and z**. The polynomials $x - 2y^2$ and $xyz + x^2y^2$ are **not** homogeneous.

A polynomial in three variables $P(x, y, z)$ consists of the sum of monomials of the form $cx^\kappa y^\lambda z^\mu$. If the polynomial is **homogeneous**, then we must have, $\kappa + \lambda + \mu = constant \stackrel{\text{def}}{=} n$ (a positive integer), where **n is the degree of the homogeneity of the polynomial**. If p is an arbitrary number, and each one of the variable x, y, z is multiplied by p, then the monomial $cx^\kappa y^\lambda z^\mu$ becomes, $c(px)^\kappa(py)^\lambda(pz)^\mu = cp^{\kappa+\lambda+\mu}x^\kappa y^\lambda z^\mu = p^n(cx^\kappa y^\lambda z^\mu)$, i.e. the monomial is multiplied by p^n. We thus conclude that **if $P(x, y, z)$ is a homogeneous polynomial of degree n, then for any constant p, we must have:**

$$P(px, py, pz) \equiv p^n P(x, y, z) \qquad (15 - 4 - 1)$$

A polynomial may be symmetric and homogeneous at the same time. The general form of such polynomials is obtained from the general form of symmetric polynomials if we delete the terms that destroy the homogeneity.

For example, **the symmetric and homogeneous** polynomials in x, y, z are the following:

(a): First degree: $a(x + y + z)$

(b): Second degree: $a(x^2 + y^2 + z^2) + b(xy + yz + zx)$

(c): Third degree:

$$a(x^3 + y^3 + z^3) + b(x^2y + y^2z + z^2x) + c(xy^2 + yz^2 + zx^2) + dxyz$$

where a, b, c, d are given constants.

Notice that all the elementary symmetric polynomials are symmetric and homogeneous (S_1 first degree, S_2 second degree, S_3 third degree, etc).

The following Theorem is proved easily.

Theorem 15-4-1: **(a) The product of two homogeneous polynomials of degree n and m, is a homogeneous polynomial of degree $(n + m)$.**

(b) In the case of perfect division of two homogeneous polynomials of degree n and m, ($n \geq m$), the quotient is a homogeneous polynomial of degree $(n - m)$.

Proof: (a) Let $P(x, y, z)$ be the product of two homogeneous polynomials $A(x, y, z)$ and $B(x, y, z)$ of degrees n and m respectively. If p is any constant, then

$$P(px, py, pz) \equiv A(px, py, pz)B(px, py, pz) \equiv p^n A(x, y, z) p^m B(x, y, z) \Rightarrow$$

$$P(px, py, pz) \equiv p^{n+m} A(x, y, z) B(x, y, z) \equiv p^{n+m} P(x, y, z)$$

and this shows that $P(x, y, z)$ is a homogeneous polynomial of degree $(n + m)$.

Similarly, we prove part (b).

Properties of polynomials that **are symmetric and homogeneous simultaneously,** may be used to prove easily, complicated algebraic identities, as shown in the following examples.

Example 15-4-1: Factor the expression

$$P \equiv x^3(y-z) + y^3(z-x) + z^3(x-y)$$

Solution

The given expression is symmetric since it remains unchanged by the cyclic permutation $\{x \to y, y \to z, z \to x\}$. We notice that when $x = y$, $P \equiv 0$, since $P(y,y,z) = y^3(y-z) + y^3(z-y) + z^3(y-y) \equiv 0$, and this shows that $P(x,y,z)$ is divided by $(x-y)$, and since P is symmetric, by virtue of Theorem 15-3-3, $P(x,y,z)$ is divided by the product $(x-y)(y-z)(z-x)$. At the same time, $P(x,y,z)$ a homogeneous polynomial of degree 4, while the product $(x-y)(y-z)(z-x)$ is a homogeneous polynomial of degree 3, and by Theorem 15-4-1, the quotient bust be a homogeneous polynomial of degree $1 = (4-3)$, i.e. must be of the form $a(x+y+z)$. This leads to the identity,

$$x^3(y-z) + y^3(z-x) + z^3(x-y) \equiv a(x-y)(y-z)(z-x)(x+y+z)$$

This identity must be true for any values of x, y, z. For example, if we apply it for $x = 1, y = 2, z = 0$, we obtain, $2 - 8 + 0 = 6a$, i.e. $a = -1$, and finally,

$$x^3(y-z) + y^3(z-x) + z^3(x-y) \equiv -(x-y)(y-z)(z-x)(x+y+z)$$

This identity is true for any x, y, z.

Example 15-4-2: Factor the polynomial

$$P(x,y,z) \equiv (x+y+z)^5 - x^5 - y^5 - z^5$$

Solution

The polynomial $P(x,y,z)$ is symmetric and homogeneous of degree 5. Since $P(-y,y,z) = (-y+y+z)^5 - (-y)^5 - y^5 - z^5 \equiv 0$, $P(x,y,z)$ is divided by the product $(x+y)(y+z)(z+x)$, (Theorem 15-3-3 (b)). Also, since $P(x,y,z)$ and $(x+y)(y+z)(z+x)$ are homogeneous polynomials of degree 5 and 3 respectively, the quotient will be symmetric and homogeneous of degree $2 = (5-3)$, i.e. it will have the form $a(x^2+y^2+z^2) + b(xy+yz+zx)$, i.e.

$$(x+y+z)^5 - x^5 - y^5 - z^5$$
$$\equiv (x+y)(y+z)(z+x)\{a(x^2+y^2+z^2)$$
$$+ b(xy+yz+zx)\} \qquad (*)$$

The identity in (*) holds true for any value of x, y, z. Therefore, it is true, for example, for $x = 1, y = 1, z = 1$ and for $x = 1, y = 1, z = 1$. In the first case, (*) implies $2a + b = 15$, while in the second case we obtain $a + b = 10$. Solving the system for a and b we find, $a = 5, b = 5$, and finally, equation (*) implies,

$$(x+y+z)^5 - x^5 - y^5 - z^5$$
$$\equiv 5(x+y)(y+z)(z+x)\{(x^2+y^2+z^2) + (xy+yz+zx)\}$$

Example 15-4-3: Find the expansion of $P(x, y, z) \equiv (x+y+z)^3$.

Solution

The polynomial $P(x, y, z)$ is a symmetric and homogeneous polynomial of degree 3, and therefore must be of the general form

$$(x+y+z)^3 \equiv (x^3+y^3+z^3) + b(x^2y+y^2z+z^2x)$$
$$+ c(xy^2+yz^2+zx^2) + dxyz \qquad (*)$$

Since the identity in (*) is true for all values of x, y, z, let us apply it for some (relatively small values of x, y, z), so as to keep the calculations as simple as possible.

For $x = 1, y = 1, z = 0$, we find: $b + c = 6$,

For $x = 1, y = -1, z = 0$, we find: $c - b = 0$,

For $x = 1, y = 1, z = 1$, we find: $3b + 3c + d = 24$,

and solving the system for a, b, c results in: $b = 3, c = 3, d = 6$, and the identity in (*) becomes,

$$(x+y+z)^3 \equiv (x^3+y^3+z^3) + 3(x^2y+y^2z+z^2x+xy^2+yz^2+zx^2)$$
$$+ 6xyz$$

PROBLEMS

15-4-1) Simplify the expression $P(x, y, z)$:

$$P \equiv (x+y)(y+z)(z+x) - yz(y+z) - zx(z+x) - xy(x+y)$$

(**Ans:** $P \equiv 2xyz$).

Hint: Since $P(x = 0, y, z) = yz(y+z) - yz(y+z) \equiv 0$, P is divided by $x - 0 = x$, and since is symmetric is divided by xyz, i.e. $P = cxyz$, etc.

15-4-2) Factor the expressions:

$$P \equiv x^2(y+z) + y^2(z+x) + z^2(x+y) + 2xyz$$
$$A \equiv a(b^3 - c^3) + b(c^3 - a^3) + c(a^3 - b^3)$$
$$B \equiv (x+y)^9 - x^9 - y^9$$
$$K \equiv (xy + yz + zx)^3 - x^3y^3 - y^3z^3 - z^3x^3$$

15-4-3) If the symmetric polynomial $P(x, y)$ is divided by $(x - y)$, show that it will be divided by $(x - y)^2$ as well.

15-4-4) Show the identity:

$$x^4(y-z) + y^4(z-x) + z^4(x-y)$$
$$\equiv -(x-y)(y-z)(z-x)(x^2 + y^2 + z^2 + xy + yz + zx)$$

15-4-5) Simplify the expression:

$$K \equiv \frac{x^3(y+z)}{(x-y)(x-z)} + \frac{y^3(z+x)}{(y-z)(y-x)} + \frac{z^3(x+y)}{(z-x)(z-y)}$$

(**Ans:** $xy + yz + zx$).

15-4-6) Simplify the fractions:

$$\frac{a^3(b-c) + b^3(c-a) + c^3(a-b)}{a^2(b-c) + b^2(c-a) + c^2(a-b)}$$

$$\frac{(x-y)^5 + (y-z)^5 + (z-x)^5}{(x-y)^3 + (y-z)^3 + (z-x)^3}$$

15-4-7) Simplify the expression:

$$P \equiv \frac{2x}{x+y} + \frac{2y}{y+z} + \frac{2z}{z+x} + \frac{(x-y)(y-z)(z-x)}{(x+y)(y+z)(z+x)}$$

(**Ans:** $P = 3$).

15-4-8) Simplify the expression:

$$P \equiv \frac{x^4}{(x-y)(x-z)} + \frac{y^4}{(y-z)(y-x)} + \frac{z^4}{(z-x)(z-y)}$$

15-4-9) Factor the expression:

$$P \equiv (x-a)^3(b-c)^3 + (x-b)^3(c-a)^3 + (x-c)^3(a-b)^3$$

(**Ans:** $3(x-a)(x-b)(x-c)(a-b)(b-c)(c-a)$).

15-4-10) Simplify the expression:

$$A \equiv \frac{x^3(y-z) + y^3(z-x) + z^3(x-y)}{(y-z)^3 + (z-x)^3 + (x-y)^3}$$

CHAPTER 16: PROGRESSIONS

In Algebra we study three types of progressions, **the arithmetic, the harmonic and the geometric progression**.

16-1) Arithmetic progressions

If we arrange n numbers in a series $a_1, a_2, a_3, \ldots, a_{n-1}, a_n$, we obtain a **succession** of numbers. The first number is a_1, the second number is $a_2,\ldots$, the last number is a_n, (or the nth number).

A succession of numbers is called **an arithmetic progression**, if each number is obtained from its preceding number by adding **one and the same number ω, which is called the ratio of the arithmetic progression**. This means that the difference of two successive terms (a number minus the preceding one), is always constant, and equal to ω, i.e. $(a_2 - a_1) = (a_3 - a_2) = \cdots = (a_n - a_{n-1}) \stackrel{def}{=} \omega$.

For example, the numbers 3,7,11,15,19,23 form an arithmetic progression with first term 3 and ratio 4. Notice that **the difference of any two successive terms is 4, the ratio of the progression**.

Also, the numbers, 35,33,31,29,27,25 form an arithmetic progression with first term 35 and ratio (-2).

If we call a is the first number of the progression and ω the ratio of the progression, then the second term of the progression will be $a_2 = (a + \omega)$, the third term of the progression will be $a_3 = (a + \omega) + \omega = a + 2\omega$, and in general, the nth term of the progression will be

$$a_n = a + (n-1)\omega \qquad (16-1-1)$$

The progression is completely determined by the first term a and the ratio ω. For example, if the first term is $a = 5$ and $\omega = 3$, then the seventh term of the progression is $a_7 = 5 + (7-1)3 = 5 + 30 = 35$, and the twentieth term is $a_{20} = 5 + (20-1)3 = 5 + 57 = 62$.

The sum of the first n terms of an arithmetic progression is given by the following Theorem.

Theorem 16-1-1: The sum of the first n terms of an arithmetic progression is expressed in terms of the first tem $a_1 = a$, the ratio ω and the number of terms n, by means of the formula

$$S_n \stackrel{\text{def}}{=} a_1 + a_2 + \cdots + a_n = \frac{2a + (n-1)\omega}{2} \cdot n \qquad (16-1-2)$$

Proof: Let $S_n \stackrel{\text{def}}{=} a_1 + a_2 + \cdots + a_n$. Expressing the terms of the progression in terms of the first term a and the ratio ω, we have:

$$S_n = a + (a + \omega) + (a + 2\omega) + \cdots + (a + (n-1)\omega) \qquad (*)$$

The sum remains of course unchanged if we reverse the order in the summation, i.e.

$$S_n = (a + (n-1)\omega) + (a + (n-2)\omega) \cdots + (a + \omega) + a \qquad (**)$$

Adding (*) and (**) we find:

$$2S_n = \{a + (a + (n-1)\omega)\} + \{a + \omega + (a + (n-2)\omega)\} + \cdots$$
$$+ \{a + (n-1)\omega + a\} \qquad (***)$$

Notice that **inside each one of the braces, the sum of the terms is the same and equal to $2a + (n-1)\omega$**, and since we have n braces, equation (***) yields,

$$2S_n = \{2a + (n-1)\omega\}n \Leftrightarrow S_n = \frac{2a + (n-1)\omega}{2} \cdot n$$

Corollary: If we take into account formula (16-1-1), the sum S may be written equivalently as

$$S_n \stackrel{\text{def}}{=} a_1 + a_2 + \cdots + a_n = \frac{a_1 + a_n}{2} \cdot n \qquad (16-1-3)$$

The problem of interpolation: Given to numbers A and B, $(A < B)$, we seek to determine k numbers $x_1, x_2, x_3, \cdots, x_k$, ($k$ **a given positive integer**), which together with the first number A and the last number B to form an arithmetic progression. This problem is called **the problem of arithmetic interpolation**. Since we know the first term A of the progression, the problem will be solved if we determine the ratio ω of the progression. In the progression

$\{A, x_1, x_2, \cdots, x_k, B\}$ the first term is A, the last term is B, and the number of terms is $(k+2)$, and according to formula (16-1-1) we have:

$$B = A + ((k+2) - 1)\omega \Leftrightarrow B = A + (k+1)\omega \Leftrightarrow \omega = \frac{B-A}{k+1} \quad (16-1-4)$$

The sought for numbers are given by the formulas,

$$x_1 = A + \frac{B-A}{k+1}, x_2 = A + 2\frac{B-A}{k+1}, x_3 = A + 3\frac{B-A}{k+1}, \cdots \quad (16-1-5)$$

Representation of numbers in an arithmetic progression: In various problems we have to work with three or four or five numbers, etc, which are **successive terms** of an arithmetic progression. In such cases, it is sometimes convenient to use the following **symmetric representation:**

Three terms: $x - \omega, x, x + \omega$, (Ratio ω),

Four terms: $x - 3\omega, x - \omega, x + \omega, x + 3\omega$, (Ratio 2ω),

Five terms: $x - 2\omega, x - \omega, x, x + \omega, x + 2\omega$, (Ratio ω).

Using these symmetric expressions, we are working with two variables, x and ω, while at the same time, the numbers form an arithmetic progression.

Theorem 16-1-2: **The necessary and sufficient condition that three numbers a, b, c form an arithmetic progression is $2b = a + c$.**

Proof: If a, b, c form an arithmetic progression, then

$$c - b = b - a (= ratio\ of\ the\ progression) \Leftrightarrow a + c = 2b \Leftrightarrow b = \frac{a+c}{2}$$

The number b is called the "**mean arithmetic**" of a and c.

Example 16-1-1: Find the sum of the first n positive integers.

Solution

$$S_n = 1 + 2 + 3 + \cdots + (n-1) + n \quad (*)$$

The numbers $1, 2, 3, \ldots, (n-1), n$ are in arithmetic progression, with first term 1 and last term n, and application of formula (16-1-3) yields,

$$S_n = 1 + 2 + 3 + \cdots + (n-1) + n = \frac{1+n}{2} \cdot n = \frac{n(n+1)}{2}$$

Example 16-1-2: Find the sum of the first k odd positive integers.

Solution

The ratio of the progression is $\omega = 2$, and application of formula (16-1-2) yields:

$$S = \underbrace{1 + 3 + 5 + 7 + \cdots}_{k \text{ terms}} = \frac{2 \cdot 1 + (k-1) \cdot 2}{2} \cdot k = \frac{2k}{2} \cdot k = k^2$$

The sum of the first k odd positive integers, is equal the square of their number. For example, the sum of the first five odd integers is: $1 + 3 + 5 + 7 + 9 = 25 = 5^2$.

Example 16-1-3: Find the sum of the squares of the first n positive integers.

Solution

We want to compute the sum $S = 1^2 + 2^2 + 3^2 + \cdots + n^2$. In order to evaluate S, we start with the identity,

$$(x+1)^3 = x^3 + 3x^2 + 3x + 1 \qquad (*)$$

Applying this identity for $x = 1, 2, 3, \ldots, n$, and adding term wise, we find:

$$\left. \begin{array}{l} 2^3 = 1^3 + 3 \cdot 1^2 + 3 \cdot 1 + 1 \\ 3^3 = 2^3 + 3 \cdot 2^2 + 3 \cdot 2 + 1 \\ 4^3 = 3^3 + 3 \cdot 3^2 + 3 \cdot 3 + 1 \\ \cdots \quad \cdots \quad \cdots \\ n^3 = (n-1)^3 + 3 \cdot (n-1)^2 + 3 \cdot (n-1) + 1 \\ (n+1)^3 = n^3 + 3 \cdot n^2 + 3 \cdot n + 1 \end{array} \right\} \text{(Term wise addition)} \Longrightarrow$$

$$(n+1)^3 = 1^3 + 3 \underbrace{(1^2 + 2^2 + \cdots + n^2)}_{S} + 3 \underbrace{(1 + 2 + \cdots + n)}_{n(n+1)/2} + n \Rightarrow$$

$$3S = 3(1^2 + 2^2 + \cdots + n^2) = (n+1)^3 - 1 - 3\frac{n(n+1)}{2} - n \Rightarrow$$

$$S = 1^2 + 2^2 + \cdots + n^2 = \frac{n(n+1)(2n+1)}{6}$$

We notice that in the term wise addition of the n equalities, the term 2^3 in the first equation cancels the term 2^3 in the second equation, the term 3^3 in the second equation cancels the term 3^3 in the third equation, etc. The only surviving term in the left hand side is the term $(n+1)^3$.

Remark: Let us define $S_k \stackrel{\text{def}}{=} 1^k + 2^k + 3^k + \cdots + (n-1)^k + n^k$, i.e. the sum of the $k^{\underline{th}}$ powers of the first n positive integers, $(k = 1,2,3,\ldots)$. We have found, for example, that

$$S_1 = 1 + 2 + 3 + \cdots + (n-1) + n = \frac{n(n+1)}{2}$$

$$S_2 = 1^2 + 2^2 + \cdots + (n-1)^2 + n^2 = \frac{n(n+1)(2n+1)}{6}$$

The evaluation of the sums S_k for various values of the positive integer k, is an interesting problem in Algebra. The technique used to evaluate S_2 may be used to evaluate $S_3, S_4, S_5, \ldots$

For example, starting with the identity

$$(x+1)^4 = x^4 + 4x^3 + 6x^2 + 4x + 1$$

and applying for $x = 1,2,3,\cdots,(n-1),n$, adding term wise the resulting equalities and using the expressions for S_1, S_2 already found, we find

$$S_3 = 1^3 + 2^3 + 3^3 + \cdots + (n-1)^3 + n^3 = \left(\frac{n(n+1)}{2}\right)^2$$

(For the evaluation of S_3 see also Example 13-5-5).

Thus far, we have evaluated the sums S_1, S_2 and S_3. Starting with the identity,

$$(x+1)^5 = x^5 + 5x^4 + 10x^3 + 10x^2 + 5x + 1$$

applying for $x = 1,2,3,\ldots,n$, adding term wise and using the found expressions for S_1, S_2 and S_3, we may evaluate $S_4 = 1^4 + 2^4 + 3^4 + \cdots + n^4$. Having found S_4 we may then find S_5, then S_6, etc.

Example 16-1-4: Provided that x^2, y^2, z^2 form an arithmetic progression, show that the numbers $1/(y+z), 1/(z+x), 1/(x+y)$ also form an arithmetic progression.

Solution

Since by assumption x^2, y^2, z^2 form an arithmetic progression, we must have (Theorem 16-1-2), $2y^2 = x^2 + z^2$.

To show that $1/(y+z), 1/(z+x), 1/(x+y)$ form an arithmetic progression, it suffices to show that:

$$\frac{2}{z+x} = \frac{1}{y+z} + \frac{1}{x+y} \Leftrightarrow \frac{2}{z+x} = \frac{x+2y+z}{(y+z)(x+y)} \Leftrightarrow$$

$$2(y+z)(x+y) = (z+x)(x+2y+z) \Leftrightarrow$$

$$2yx + 2zx + 2y^2 + 2zy = zx + x^2 + 2zy + 2xy + z^2 + xz \Leftrightarrow$$

$$2yx + 2zx + 2y^2 + 2zy = 2xy + 2zx + 2zy + x^2 + z^2 \Leftrightarrow$$

$$2y^2 = x^2 + z^2$$

and since this is true, the proof follows immediately.

Example 16-1-5: Given that the expression

$$a(b-c)x^2 + b(c-a)xy + c(a-b)y^2$$

is a perfect square, show that the numbers $1/a, 1/b, 1/c$ form an arithmetic progression, $(abc \neq 0)$.

Solution

$$a(b-c)x^2 + b(c-a)xy + c(a-b)y^2$$
$$= y^2 \left\{ a(b-c)\left(\frac{x}{y}\right)^2 + b(c-a)\frac{x}{y} + c(a-b) \right\} \quad (*)$$

Equation (*) shows that the given expression will be a perfect square, when the quantity inside the braces, which is **a trinomial in** $\left(\frac{x}{y}\right)$, will be a perfect square, and this occurs when its discriminant is zero, i.e. when

$$\{b(c-a)\}^2 - 4\{a(b-c)\}\{c(a-b)\} = 0 \Leftrightarrow$$

$$b^2c^2 + b^2a^2 + 4a^2c^2 + 2acb^2 - 4a^2bc - 4abc^2 = 0 \Leftrightarrow$$

$$(bc + ab - 2ac)^2 = 0 \Leftrightarrow bc + ab - 2ac = 0$$

and dividing both sides by $abc \neq 0$, we obtain, $\frac{2}{b} = \frac{1}{a} + \frac{1}{c}$, and this shows that the numbers $\frac{1}{a}, \frac{1}{b}, \frac{1}{c}$ form an arithmetic progression, (Theorem 16-1-2).

PROBLEMS

16-1-1) Find k so that the numbers $(3k^2 - 2k + 3)$, $(2k^2 - 7k + 5)$, $(k^2 + 4k - 9)$, form an arithmetic progression. What is the ratio of the progression?   (**Ans:** $k = 1, \omega = -4$).

16-1-2) Interpolate 9 numbers between $A = 5$ and $B = 75$, so that the eleven numbers $\{A, x_1, x_2, \ldots, x_9, B\}$ to form an arithmetic progression.

16-1-3) In an arithmetic progression, the sum of the first n terms is N, the sum of the first k terms is K and the sum of the first m terms is M. Show that

$$\frac{K}{k}(n - m) + \frac{N}{n}(m - k) + \frac{M}{m}(k - n) = 0$$

16-1-4) Find the coefficients and the roots r_1 and r_2 of the trinomial $ax^2 + bx + c$, given that the numbers $\{r_1, a, b, c, r_2\}$ form an arithmetic progression.

Hint: It facilitates the calculations, if we set, $r_1 = k - 2\omega, a = k - \omega, b = k, c = k + \omega, r_2 = k + 2\omega$. Then the equation becomes:

$$(k - \omega)x^2 + kx + (k + \omega) = 0$$

and then, $r_1 + r_2 = 2k = -\frac{k}{k-\omega}$ (*), and $r_1 r_2 = k^2 - 4\omega^2 = \frac{k+\omega}{k-\omega}$ (**). Equations (*) and (**) form a system in k and ω, etc.

16-1-5) Show that the sum of any number of consecutive odd positive integers cannot be a prime number.

Hint: Prime numbers are the ones that are divide only by themselves and one, for example 2,3,5,7,11,13 are **the first prime numbers**. The numbers 8 and 9 are not prime, since 8 is divided by 2, while 9 is divided by 3. To solve the problem, see Example 16-1-2.

16-1-6) Express in terms of n the sums:

$$A = 1\cdot 2 + 2\cdot 3 + 3\cdot 4 + \cdots + n(n+1)$$

$$B = 1\cdot 2\cdot 3 + 2\cdot 3\cdot 4 + 3\cdot 4\cdot 5 + \cdots + n(n+1)(n+2)$$

Hint: the general term $n(n+1)$ in the sum A, is equal to $n^2 + n$, so

$$A = 1^2 + 1 + 2^2 + 2 + 3^2 + 3 + \cdots + n^2 + n \Rightarrow$$

$$A = \underbrace{(1^2 + 2^2 + \cdots + n^2)}_{S_2} + \underbrace{(1 + 2 + 3 + \cdots + n)}_{S_1}$$

The sums S_1 and S_2 have been evaluated in Example 16-1-3. By a similar method we find B, (the sum S_3 was obtained also in Example 16-1-3).

16-1-7) Find the positive integer n, given that the quadratic equation

$$x(x+1) + (x+1)(x+2) + (x+2)(x+3) + \cdots + (x+n-1)(x+n) - 10n = 0$$

admits as roots two consecutive integer numbers, **(Ans:** $n = 11$).

16-1-8) If A is the sum of the first n terms of an arithmetic progression, B the sum of the first $2n$ terms and C the sum of the first $3n$ terms, show that $C = 3(B - A)$.

16-1-9) Show that in any arithmetic progression $\{x_1, x_2, \ldots, x_n\}$ of positive numbers, we have:

$$\frac{1}{\sqrt{x_1}+\sqrt{x_2}} + \frac{1}{\sqrt{x_2}+\sqrt{x_3}} + \cdots + \frac{1}{\sqrt{x_{n-1}}+\sqrt{x_n}} = \frac{n-1}{\sqrt{x_1}+\sqrt{x_n}}$$

16-1-10) Determine the constant k provided that the roots of the equation $x^4 - (3k+5)x^2 + (k+1)^2 = 0$ form an arithmetic progression.

Hint: Set $r_1 = a - 3\omega, r_2 = a - \omega, r_3 = a + \omega, r_4 = a + 3\omega$

16-1-11) If a, b, c, d are successive terms in an arithmetic progression, show that $bc > ad$.

16-1-12) Provided that $x_1, x_2, \ldots, x_n$ are successive terms of an arithmetic progression and that for every n, $x_1 + x_2 + \cdots + x_n = n^2$, show that the arithmetic progression is a progression of odd integers.

16-1-13) Can the numbers $\sqrt{3}, \sqrt{5}, \sqrt{7}$ be successive terms of one and the same arithmetic progression? Can these numbers be terms, of any order, of one and the same arithmetic progression?

(**Ans:** In both cases the answer is no).

16-1-14) Determine a if the cubes of the roots of $ax^4 + 2x^2 + 1 = 0$ form an arithmetic progression, (**Ans:** $4\sqrt[3]{3} = a\left(1 + \sqrt[3]{5}\right)^2$).

16-1-15) If the positive numbers x, y, z, w form an arithmetic progression, show that $(x + w)/2 > \sqrt[4]{xyzw}$.

Hint: Consider $x = a - 3\omega, y = a - \omega, z = a + \omega, w = a + 3\omega$.

16-2) Harmonic progressions

We say that the n numbers $a_1, a_2, a_3, \ldots, a_n$ form **an harmonic progression** if their reciprocals $\frac{1}{a_1}, \frac{1}{a_2}, \frac{1}{a_3}, \ldots, \frac{1}{a_n}$ form an arithmetic progression.

If three numbers a, b, c form an harmonic progression, then, by definition, $\frac{1}{a}, \frac{1}{b}, \frac{1}{c}$ form an arithmetic progression, and therefore

$$\frac{2}{b} = \frac{1}{a} + \frac{1}{c} \Longleftrightarrow b = \frac{2ac}{a+c} \qquad (16-2-1)$$

Equation (16-2-1) is **the necessary and sufficient condition** for three numbers $\frac{1}{a}, \frac{1}{b}, \frac{1}{c}$ to form an harmonic progression. The number b, as defined in (16-2-1), is called the "**mean harmonic**" of a and b.

The problem of interpolation: Given two numbers A and B, we seek to determine k numbers, $y_1, y_2, \ldots, y_k$, (k is **a given positive integer**), such that the numbers $\{A, y_1, y_2, \ldots, y_k, B\}$ to form an harmonic progression. This is the problem of the "**harmonic interpolation**". The numbers $\{A, y_1, y_2, \ldots, y_k, B\}$ form an harmonic progression, if and only if the numbers $\left\{\frac{1}{A}, \frac{1}{y_1}, \frac{1}{y_2}, \ldots, \frac{1}{y_k}, \frac{1}{B}\right\}$ form an arithmetic progression, and thus **the harmonic interpolation reduces to the arithmetic interpolation**.

Example 16-2-1: Find the eighth term of the harmonic progression $\frac{1}{1}, \frac{1}{3}, \frac{1}{5}, \ldots$

Solution

The reciprocal of the terms of the harmonic progression, are, 1,3,5,7,9,11,13,15,17, ..., which form an arithmetic progression, and the reciprocal of those, are:

$$\frac{1}{1}, \frac{1}{3}, \frac{1}{5}, \frac{1}{7}, \frac{1}{9}, \frac{1}{11}, \frac{1}{13}, \frac{1}{15}, \frac{1}{17}, \ldots \qquad (*)$$

The numbers in (*) are in harmonic progression, and its eighth term is $\frac{1}{15}$.

Example 16-2-2: Given that $\frac{x-y}{xy} = \frac{y-z}{yz} = \frac{z-w}{zw}$ show that the numbers x, y, z, w form an harmonic progression, and that $\frac{5x-3y}{xy} = \frac{z+w}{zw}$.

Solution

The given equality is expressed equivalently as,

$$\frac{1}{y} - \frac{1}{x} = \frac{1}{z} - \frac{1}{y} = \frac{1}{w} - \frac{1}{z}$$

and this shows that the numbers $\frac{1}{x}, \frac{1}{y}, \frac{1}{z}, \frac{1}{w}$ form an arithmetic progression, i.e. x, y, z, w form an harmonic progression.

Since the numbers $\frac{1}{x}, \frac{1}{y}, \frac{1}{z}, \frac{1}{w}$ form an arithmetic progression, if we call ω the ratio of the progression we have:

$$\frac{1}{y} = \frac{1}{x} + \omega, \qquad \frac{1}{z} = \frac{1}{x} + 2\omega, \qquad \frac{1}{w} = \frac{1}{x} + 3\omega \qquad (*)$$

From the first equation in (*), we obtain: $\frac{5}{y} = \frac{5}{x} + 5\omega$ and subtracting $\frac{3}{x}$ from both sides, results in

$$\frac{5}{y} - \frac{3}{x} = \frac{5}{x} + 5\omega - \frac{3}{x} \Leftrightarrow \frac{5x - 3y}{xy} = \frac{2}{x} + 5\omega \quad (**)$$

Also, adding term wise the second and the third equations in (*) we find:

$$\frac{1}{z} + \frac{1}{w} = \frac{2}{x} + 5\omega \Leftrightarrow \frac{z+w}{zw} = \frac{2}{x} + 5\omega \quad (***)$$

Comparing (**) and (***) we find $\frac{5x-3y}{xy} = \frac{z+w}{zw}$, and this completes the proof.

Example 16-2-3: What equation should be satisfied by the coefficients of the equation $x^3 + ax^2 + bx + c = 0$, given that its roots form an harmonic progression?

Solution

If the three roots r_1, r_2, r_3 of the equation form an harmonic progression, we must have

$$r_2 = \frac{2r_1 r_3}{r_1 + r_3} \quad (*)$$

Also, the roots satisfy **Vieta's formulas**:

$$r_1 + r_2 + r_3 = -a, \quad r_1 r_2 + r_2 r_3 + r_3 r_1 = b, \quad r_1 r_2 r_3 = -c \quad (**)$$

From the second equation in (**) we find:

$$r_2(r_1 + r_3) + r_1 r_3 = b \overset{(*)}{\Rightarrow} 2r_1 r_3 + r_1 r_3 = b \Rightarrow 3r_1 r_3 = b \overset{(**)}{\Rightarrow}$$

$$3\frac{-c}{r_2} = b \Rightarrow r_2 = -\frac{3c}{b} \quad (***)$$

and since r_2 is a root of the equation, **it must satisfy the equation**, i.e.

$$r_2^3 + ar_2^2 + br_2 + c = 0 \overset{(***)}{\Longrightarrow}$$

$$\left(-\frac{3c}{b}\right)^3 + a\left(-\frac{3c}{b}\right)^2 + b\left(-\frac{3c}{b}\right) + c = 0 \Rightarrow 27c^2 + 2b^3 = 9abc$$

PROBLEMS

16-2-1) If the numbers x, y, z form an harmonic progression, show that the numbers $y - x, y, y - z$ also form an harmonic progression.

16-2-2) Provided that the numbers $\{x_1, x_2, \ldots, x_{n-1}, x_n\}$ form an harmonic progression, show that $x_1 x_2 + x_2 x_3 + \cdots + x_{n-1} x_n = (n-1) x_1 x_n$.

16-2-3) If the positive numbers x, y, z form an harmonic progression, show that

$$\frac{x+y}{2x-y} + \frac{y+z}{2z-y} > 4$$

Is the inequality valid if x, y, z are all negative? Can we assert the validity of the inequality if no information is given about the sign of the numbers?

Hint: $y = 2xz/(x+z)$; also, for any positive number p it is true that $p + \frac{1}{p} \geq 2$, (why?).

16-2-4) Find the number x, if the numbers $(x+1), (x+5), (x+11)$ form an harmonic progression, **(Ans: 19)**.

16-2-5) Provided that the numbers x, y, z, w form an arithmetic progression, show that the numbers yzw, zwx, wxy, xyz form an harmonic progression.

16-2-6) If the numbers x, y, z form an harmonic progression, show that the numbers yz, zx, xy form an arithmetic progression.

16-3) Geometric progressions

A succession of numbers is called **a geometric progression, if the quotient of any number by its preceding number is constant. This constant quotient, is called the ratio of the geometric progression.**

For example, if the numbers x, y, z, w, p, q form a geometric progression, then

$$\frac{y}{x} = \frac{z}{y} = \frac{w}{z} = \frac{p}{w} = \frac{q}{p} \stackrel{\text{def}}{=} \omega \ (the\ ratio\ of\ the\ progression)$$

An equivalent definition is the following: **A succession of numbers forms a geometric progression when each number(except the first one), is obtained from its preceding number by multiplication by one and the same number, the ratio ω of the progression.** In the aforesaid example, notice that

$$y = x\omega, z = y\omega, w = z\omega, p = w\omega, q = p\omega$$

If $|\omega| > 1$, the geometric progression is called **increasing**, while if $|\omega| < 1$, the progression is called **decreasing**. In an increasing geometric progression, the terms keep increasing in absolute value, while in a degreasing progression, the terms keep decreasing in absolute value.

The necessary and sufficient condition that three numbers a, b, c (in this order) form a geometric progression is

$$b^2 = ac \qquad (16-3-1)$$

The number b as defined in (16-3-1) is called the **"mean geometric"** of the other two numbers a and c.

Expression of the $n\underline{^{th}}$ term in terms of the first number and the ratio:

Let us consider the geometric progression $\{a_1, a_2, a_3, \ldots, a_{n-1}, a_n\}$. If we call a the first term, $(a_1 = a)$, and ω the ratio of the progression, then, $a_2 = a\omega, a_3 = a_2\omega = (a\omega)\omega = a\omega^2, a_4 = a_3\omega = (a\omega^2)\omega = a\omega^3$, and in general,

$$a_n = a\omega^{n-1} \qquad (16-3-2)$$

The geometric progression, is completely determined by the first term a and the ratio ω. For example, if the first term is $a = 2$ and $\omega = 3$, then the sixth term will be $a_6 = a\omega^5 = 2 \cdot 3^5$.

The sum of the first n terms of a geometric progression is given by the following Theorem.

Theorem 16-3-1: The sum of the first n terms of a geometric progression is expressed in terms of the first term $a_1 = a$, the ratio ω and the number n of terms, by means of the formula

$$S_n \stackrel{\text{def}}{=} a + a\omega + a\omega^2 \ldots + a\omega^{n-1} = a\frac{\omega^n - 1}{\omega - 1} \qquad (16-3-3)$$

Proof: If $a_1 = a$ is the first term and ω is the ratio of the geometric progression, then $a_2 = a\omega, a_3 = a\omega^2,\ldots$, and the sum S_n will be,

$$S_n = a + a\omega + a\omega^2 + \cdots + a\omega^{n-2} + a\omega^{n-1} \qquad (*)$$

Multiplying both sides in (*) by ω, we find,

$$\omega S_n = a\omega + a\omega^2 + \cdots + a\omega^{n-1} + a\omega^n \qquad (**)$$

and subtracting (*) from (**) results in the following equation:

$$(\omega - 1)S_n = a\omega^n - a \stackrel{(\omega \neq 1)}{\Longrightarrow} S_n = a\frac{\omega^n - 1}{\omega - 1}$$

and this completes the proof. Notice that in deriving the expression for S_n, we have assumed that $\omega \neq 1$, since the case $\omega = 1$ corresponds to the trivial case $a_1 = a, a_2 = a, \ldots, a_n = a$.

The sum S_n in (16-3-3) may be written equivalently as

$$S_n = \frac{a_n \omega - a}{\omega - 1} \qquad (16-3-4)$$

Indeed,

$$S_n = a\frac{\omega^n - 1}{\omega - 1} = \frac{(a\omega^{n-1})\omega - a}{\omega - 1} = \frac{a_n \omega - a}{\omega - 1}$$

since $a_n = a\omega^{n-1}$. Formula (16-3-4) expresses **the sum of the terms of a geometric progression in terms of the first term a, the last term a_n and the ratio ω**.

Theorem 13-6-2: If $a_1 = a$ is the first term of a geometric procession and a_n the last term, then the product P of the first n terms of the progression is given by the formula,

$$P^2 = (a_1 a_2 a_3 \cdots a_{n-1} a_n)^2 = (a_1 a_n)^n \qquad (16-3-5)$$

Proof: The terms of the geometric progression are

$$a_1 = a, a_2 = a\omega, a_3 = a\omega^2, \ldots, a_{n-1} = a\omega^{n-2}, a_n = a\omega^{n-1}$$

The quantity P^2 is equal to

$$P^2 = P \cdot P = (a_1 a_2 a_3 \cdots a_{n-1} a_n)(a_n a_{n-1} \cdots a_3 a_2 a_1) \Rightarrow$$

$$P^2 = (a_1 a_n)(a_2 a_{n-1})(a_3 a_{n-2}) \cdots (a_{n-1} a_2)(a_n a_1) \Rightarrow$$

$$P^2 = (a^2 \omega^{n-1})(a^2 \omega^{n-1})(a^2 \omega^{n-1}) \cdots (a^2 \omega^{n-1})(a^2 \omega^{n-1}) \Rightarrow$$

$$P^2 = \{a^2 \omega^{n-1}\}^n = \{a(a\omega^{n-1})\}^n = \{a_1 a_n\}^n$$

The problem of geometric interpolation: Given two numbers A and B, ($A < B$), we seek to determine k numbers $a_1, a_2, \ldots, a_{k-1}, a_k$ (k is **a fixed positive integer**), which together with the first number A and the last number B, form a geometric progression. This means that the succession of $(k+2)$ numbers $\{A, a_1, a_2, \ldots, a_{k-1}, a_k, B\}$ forms a geometric progression, with first term A and last term B. According to formula (16-3-2), if ω is the ratio of the progression, we must have:

$$B = A\omega^{k+1} \Leftrightarrow \omega = \sqrt[k+1]{B/A} \qquad (16-3-6)$$

The sought for numbers $a_1, a_2, \ldots, a_k$, (which together with the first number A and the last term B form a geometric progression), are:

$$a_1 = A \sqrt[k+1]{B/A}, a_2 = A \left(\sqrt[k+1]{B/A}\right)^2, a_3 = A \left(\sqrt[k+1]{B/A}\right)^3, \ldots$$

The sum of the infinite number of terms of a decreasing geometric progression: Let $a, a\omega, a\omega^2, a\omega^3, \ldots, a\omega^n, a\omega^{n+1}, \ldots$ are the terms of an infinite, decreasing geometric progression, ($|\omega| < 1$). It can be shown that, the sum S of **the infinite number of terms** of the **decreasing** geometric progression, **is finite**, and given by the formula,

$$S = a + a\omega + a\omega^3 + \cdots + a\omega^n + a\omega^{n+1} + \cdots = \frac{a}{1-\omega} \qquad (16-3-7)$$

A rigorous proof of formula (16-3-7) requires knowledge of sequences and infinite series. However, in Example 16-3-4, we give an "**intuitive**" proof of this formula.

Representation of numbers in a geometric progression: In various problems we have to work with three or four or five, etc numbers which are successive terms of a geometric progression. In such cases it is convenient to use the following **symmetric representation**:

Three numbers: $\frac{x}{\omega}, x, x\omega$, (Ratio ω).

Four numbers: $\frac{x}{\omega^3}, \frac{x}{\omega}, x\omega, x\omega^3$, (Ratio ω^2).

Five numbers: $\frac{x}{\omega^2}, \frac{x}{\omega}, x, x\omega, x\omega^2$, (Ratio ω).

Example 16-3-1: Given that the numbers $(x+2), (x+4), (x+7)$ form a geometric progression, find x.

Solution

Application of the necessary and sufficient condition (16-3-1) yields:

$$(x+4)^2 = (x+2)(x+7) \Leftrightarrow x^2 + 8x + 16 = x^2 + 9x + 14 \Leftrightarrow x = 2$$

The three numbers are: $\{4,6,9\}$ which form a geometric progression since $\frac{6}{4} = \frac{9}{6} = \omega$, (the ratio of the progression).

Example 16-3-2: The sum of the first six terms of a geometric progression with ratio $\omega = 2$, is $S = 189$. Find the geometric progression.

Solution

Since a geometric progression is determined completely by the first term and the ratio, and since the ratio $\omega = 2$ is given, it suffices to find the first term a. Application of formula (16-3-3) with $n = 6$, yields:

$$S = 189 = a\frac{2^6 - 1}{2 - 1} \Leftrightarrow 189 = a\frac{64 - 1}{1} \Leftrightarrow a = \frac{189}{63} = 3$$

The geometric progression is: $3, 3 \cdot 2, 3 \cdot 2^2, 3 \cdot 2^3, \ldots$, i.e. 3,6,12,24,48,96, ...

Verify by direct calculation that $3 + 6 + 12 + 24 + 48 + 96 = 189$.

Example 16-3-3: The sum of three numbers a, b, c is 72. Given that a, b, c form an arithmetic progression, while $c, b, a + 2$ form a geometric progression, find the three numbers.

Solution

By assumption we have:

$$\{a + b + c = 72, \quad 2b = a + c, \quad b^2 = c(a + 2)\} \tag{*}$$

Formula (*) constitutes **a system of three equations in three unknowns**, the solution of which will determine a, b, c. By virtue of the second equation in (*), the first equation yields, $2b + b = 72$, i.e. $3b = 72$, i.e. $\boldsymbol{b = 24}$. The third equation in (*), by virtue of the second equation becomes,

$$24^2 = (2 \cdot 24 - a)(a + 2) \Leftrightarrow a^2 - 46a + 480 = 0 \Rightarrow \begin{cases} a = 30 \\ or \\ a = 16 \end{cases}$$

If $a = 30, b = 24, c = 18$, while if $a = 16, b = 24, c = 32$. **The problem has two solutions**. Verify that both solutions satisfy the requirements of the problem.

Example 16-3-4: Consider a segment AB of length 1. Let M_1 be the midpoint of AB, M_2 the midpoint of M_1B, M_3 the midpoint of M_2B, etc. This process continues **indefinitely**.

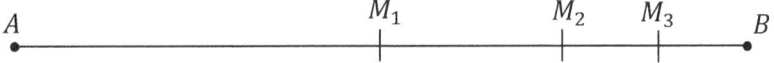

Fig. 16-1: The sum of an infinite number of terms of a degreasing G.P.

It is intuitively clear that, since $\boldsymbol{AB = 1}$,

$$AB = AM_1 + M_1M_2 + M_2M_3 + \cdots = 1 \tag{*}$$

$$AM_1 = M_1B = \frac{1}{2}AB = \frac{1}{2} \cdot 1 = \frac{1}{2}$$
$$M_1M_2 = M_2B = \frac{1}{2}M_1B = \frac{1}{2} \cdot \frac{1}{2} = \frac{1}{2^2}$$
$$M_2M_3 = M_3B = \frac{1}{2} \cdot M_2B = \frac{1}{2} \cdot \frac{1}{2^2} = \frac{1}{2^3}$$
$$\cdots \quad \cdots \quad \cdots$$

Taking into account these equalities, equation (*) implies that,

$$\frac{1}{2} + \frac{1}{2^2} + \frac{1}{2^3} + \frac{1}{2^4} + \cdots = 1 \qquad (**)$$

and this shows that **the sum of the infinite number of terms of the decreasing geometric series** $\left\{\frac{1}{2}, \frac{1}{2^2}, \frac{1}{2^3}, \frac{1}{2^4}, \cdots\right\}$, $(a = \frac{1}{2}, \omega = \frac{1}{2})$, is finite and must be equal to 1.

Notice that application of formula (16-3-7) yields the same result, $(S = \frac{a}{1-\omega} = \frac{1/2}{1-(1/2)} = \frac{1/2}{1/2} = 1)$.

Example 16-3-5: Between the numbers $A = 3$ and $B = 192$, interpolate 5 geometric means, i.e. five numbers which together with the numbers $A = 3$ and $B = 192$, form a geometric progression.

Solution

The first number of the progression is $A = 3$. The ratio of the progression is given by formula (16-3-6), with $A = 3, B = 192, k = 5$, i.e. $\omega = \sqrt[6]{192/3} = \sqrt[6]{64} = 2$, i.e. the sought for progression is: 3,6,12,24,48,96,192.

Example 16-3-6: In the decimal numeral system (see Notice in Ex. 4-13-1), the integer X has $2n$ digits equal to 1, the integer Y has $(n + 1)$ digits equal to 1 and the number Z has n digits equal to 6. Show that the number $(X + Y + Z + 8)$ is a perfect square.

Solution

In the decimal system, the numbers X, Y and Z are the following:

$$X = 1 \cdot 10^{2n-1} + 1 \cdot 10^{2n-2} + \cdots + 1 \cdot 10^1 + 1 \cdot 10^0 \xRightarrow{(16-3-4)}$$

$$X = \frac{10^{2n-1} \cdot 10 - 1}{10 - 1} = \frac{10^{2n} - 1}{9} \qquad (*)$$

$$Y = 1 \cdot 10^n + 1 \cdot 10^{n-1} + \cdots + 1 \cdot 10^1 + 1 \cdot 10^0 \xRightarrow{(16-3-4)}$$

$$Y = \frac{10^n \cdot 10 - 1}{10 - 1} = \frac{10^{n+1} - 1}{9} \qquad (**)$$

$$Z = 6 \cdot 10^{n-1} + 6 \cdot 10^{n-2} + \cdots + 6 \cdot 10^1 + 6 \cdot 10^0 \xRightarrow{(16-3-4)}$$

$$Z = \frac{6 \cdot 10^{n-1} \cdot 10 - 6}{10 - 1} = \frac{6 \cdot 10^n - 6}{9} \qquad (***)$$

By virtue of (*), (**) and (***) the number $X + Y + Z + 8$ assumes the following form:

$$X + Y + Z + 8 = \frac{10^{2n} - 1}{9} + \frac{10^{n+1} - 1}{9} + \frac{6 \cdot 10^n - 6}{9} + 8 \Rightarrow$$

$$X + Y + Z + 8 = \frac{10^{2n} + 10^{n+1} + 6 \cdot 10^n - 8}{9} + 8 \Rightarrow$$

$$X + Y + Z + 8 = \frac{10^{2n} + 10^{n+1} + 6 \cdot 10^n - 8 + 72}{9} \Rightarrow$$

$$X + Y + Z + 8 = \frac{10^{2n} + 10^n(10 + 6) + 64}{9} = \frac{10^{2n} + 16 \cdot 10^n + 8^2}{9} \Rightarrow$$

$$X + Y + Z + 8 = \left(\frac{10^n + 8}{3}\right)^2$$

PROBLEMS

16-3-1) Given three numbers a, b, c, find a number x such that the numbers $x + a, x + b, x + c$ form a geometric progression.

(**Ans:** $x = (b^2 - ac)/(a + c - 2b)$).

16-3-2) If the numbers a, b, c form a geometric progression, show that the numbers $(a + b), (2b), (b + c)$ form an harmonic progression.

16-3-3) If the mean arithmetic of two numbers a and b is twice the mean geometric of the same number, i.e. if $(a+b)/2 = 2\sqrt{ab}$ find the ratio a/b.

(Ans: $7 + 4\sqrt{3}$ or $7 - 4\sqrt{3}$).

16-3-4) Given that the numbers a, b, c form an arithmetic progression and that b, c, a form a geometric progression, show that either $a = b = c$ or $a = 4b = -2c$.

16-3-5) Given that the numbers x, y, z form a geometric progression, show that $\frac{(x+y)^2}{(y+z)^2} = \frac{x}{z}$.

16-3-6) Given that the numbers x, y, z, w form a geometric progression, show that $(y-z)^2 + (z-x)^2 + (w-y)^2 = (x-w)^2$.

16-3-7) Determine the numbers a, b, c given that they form a geometric progression, while the numbers ab, bc, ca and abc form an arithmetic progression, **(Ans: $(1,1,1)$ or $\left(-5, \frac{5}{2}, -\frac{5}{4}\right)$).**

Hint: Set $\frac{b}{a} = \frac{c}{b} = \lambda$, i.e. $b = a\lambda, c = b\lambda = a\lambda^2$, etc.

13-6-8) If A, B, C are the sums of the first $n, 2n, 3n$ terms of a geometric progression, show that $A^2 + B^2 = A(B + C)$.

16-3-9) If $x \neq 1$ is a given number, express in closed form the sum:

$$S = \left(x + \frac{1}{x}\right)^2 + \left(x^2 + \frac{1}{x^2}\right)^2 + \left(x^3 + \frac{1}{x^3}\right)^2 + \cdots + \left(x^n + \frac{1}{x^n}\right)^2$$

(Ans: $S = (x^{2n} - 1)(x^{2n+2} + 1)/(x^{2n}(x^2 - 1)) + 2n$).

16-3-10) Find an infinite decreasing geometric progression, given that the sum of its infinite number of terms is $2/3$, while the sum of its first two terms is $1/2$. Notice that the solution is not unique.

16-3-11) Find three numbers x, y, z which form a geometric progression, given that their sum is 26 and their product is 216, **(Ans: 2,6,18).**

Hint: Set $x = \frac{y}{\omega}, y = y, z = y\omega$ (the symmetric representation of x, y, z).

16-3-12) Given that $x > 3$, find a closed form expression of the sum:

$$S = \frac{1}{x-2} - \frac{1}{(x-2)^2} + \frac{1}{(x-2)^3} - \frac{1}{(x-2)^4} + \cdots$$

Hint: Notice that if $x > 3$ the sum S represents the sum of the infinite number of terms of a degreasing geometric progression, with first term $1/(x-2)$ and ratio $\omega = -1/(x-2)$.

16-3-13) For what values of x is finite the sum of the infinite number of terms

$$x + \frac{x^2}{1+x} + \frac{x^3}{(1+x)^2} + \frac{x^4}{(1+x)^3} + \cdots$$

(Ans: $x > -1/2$).

Hint: The ratio $\omega = x/(x+1)$ must be: $|\omega| < 1$, i.e. $|x/(x+1)| < 1$).

16-3-14) If the real numbers $\{x_1, x_2, \ldots, x_{n-1}, x_n\}$ satisfy the relation

$$(x_1^2 + x_2^2 + \cdots + x_{n-1}^2)(x_2^2 + x_3^2 + \cdots + x_n^2)$$
$$= (x_1 x_2 + x_2 x_3 + \cdots + x_{n-1} x_n)^2$$

show that $\{x_1, x_2, \ldots, x_{n-1}, x_n\}$ form a geometric progression.

Hint: Use the identity of Lagrange.

16-3-15) If the numbers a, b, c form an arithmetic progression, the numbers x, y, z form an harmonic progression and the numbers ax, by, cz form a geometric progression, show that either $ax = by = cz$ or $az = by = cx$.

16-4) Mixed progressions

If we multiply the corresponding terms of an arithmetic and a geometric progression, we obtain a new succession of numbers, called **a mixed progression**. For example, if the arithmetic progression is

$$a, a + \lambda, a + 2\lambda, a + 3\lambda, \ldots$$

and the geometric progression is

$$b, b\omega, b\omega^2, b\omega^3, \ldots$$

then, if we multiply the corresponding terms, we obtain the following succession of numbers,

$$ab, \quad (a+\lambda)b\omega, \quad (a+2\lambda)b\omega^2, \quad (a+3\lambda)b\omega^3, \ldots \qquad (*)$$

which is called **a mixed progression**. Let us call S_n the sum of the first n terms of the mixed progression in (*), i.e.

$$S_n = ab + (a+\lambda)b\omega + (a+2\lambda)b\omega^2 + \cdots + (a+(n-1)\lambda)b\omega^{n-1} \qquad (**)$$

In order to compute the sum S_n, **we multiply both sides of (**) by the ratio ω of the geometric progression**, the result being,

$$\omega S_n = ab\omega + (a+\lambda)b\omega^2 + \cdots + (a+(n-2)\lambda)b\omega^{n-1} \\ + (a+(n-1)\lambda)b\omega^n \qquad (***)$$

Subtracting the equation (***) from the equation (**) results in the following:

$$(1-\omega)S_n = ab + \underbrace{b\lambda\omega + b\lambda\omega^2 + \cdots + b\lambda\omega^{n-1}}_{\text{Geometric Progression}} - (a+(n-1)\lambda)b\omega^n$$

We notice that **all the terms, except the first and the last terms, form a geometric progression**, and therefore the sum of these terms is given by equation (16-3-3), i.e.

$$(1-\omega)S_n = ab + \frac{b\lambda\omega^n - b\lambda\omega}{\omega - 1} - (a+(n-1)\lambda)b\omega^n \Rightarrow$$

$$S_n = \frac{ab - (a+(n-1)\lambda)b\omega^n}{1-\omega} + \frac{b\lambda(\omega - \omega^n)}{(1-\omega)^2} \qquad (16-4-1)$$

Example 16-4-1: Find the sum

$$S = \frac{1}{3} + \frac{2}{3^2} + \frac{3}{3^3} + \cdots + \frac{n-1}{3^{n-1}} + \frac{n}{3^n} \qquad (*)$$

Solution

The terms of the sum result if we multiply the terms of the arithmetic progression $\{1,2,3,\ldots,n\}$ by the corresponding terms of the geometric progression $\left\{\frac{1}{3}, \frac{1}{3^2}, \frac{1}{3^3}, \ldots, \frac{1}{3^n}\right\}$. To find the sum S we apply the method developed previously, i.e. we multiply both terms of S by the ratio of the geometric progression ($\omega = \frac{1}{3}$), the result being:

$$\frac{1}{3}S = \frac{1}{3^2} + \frac{2}{3^3} + \frac{3}{3^4} + \cdots + \frac{n-1}{3^n} + \frac{n}{3^{n+1}} \quad (**)$$

and subtracting (**) from (*) we find,

$$\left(1 - \frac{1}{3}\right)S = \frac{1}{3} + \frac{1}{3^2} + \frac{1}{3^3} + \cdots + \frac{1}{3^n} - \frac{n}{3^{n+1}} \Rightarrow$$

$$\frac{2}{3}S = \frac{\frac{1}{3^n} \cdot \frac{1}{3} - \frac{1}{3}}{\frac{1}{3} - 1} - \frac{n}{3^{n+1}} = -\frac{1 - 3^n}{2 \cdot 3^n} - \frac{n}{3^{n+1}} \Rightarrow$$

$$S = \frac{3}{2}\left(-\frac{1 - 3^n}{2 \cdot 3^n} - \frac{n}{3^{n+1}}\right) = \frac{3}{4} - \frac{1}{4 \cdot 3^{n-1}} - \frac{n}{2 \cdot 3^n}$$

PROBLEMS

16-4-1) Find the sum: $S = 1 + 2x + 3x^2 + 4x^3 \ldots + nx^{n-1}$, $(x \neq 1)$.

(Ans: $S = nx^n/(x-1) - (x^n - 1)/(x-1)^2$).

16-4-2) Find the sum: $S = \frac{1}{2} + \frac{3}{2^3} + \frac{5}{2^5} + \frac{7}{2^7} + \cdots + \frac{2n+1}{2^{2n+1}}$.

16-4-3) Find the sum : $S = 1 - \frac{3}{2} + \frac{5}{4} - \frac{7}{8} + \cdots + (-1)^{n-1}\frac{2n-1}{2^{n-1}}$.

(Ans: $S = (2^n + (-1)^{n+1}(6n+1))/(9 \cdot 2^{n-1})$).

16-4-4) Find the sum: $S = 1 + 3x + 5x^2 + 7x^3 + \cdots + (2n-1)x^{n-1}$. The sum S is a function of x, i.e. $S = S(x)$. Then find the number $S\left(-\frac{1}{2}\right)$ and thus verify the result obtained in the preceding example.

16-5) The arithmetic mean, the geometric mean, the harmonic mean and Cauchy's inequality

Let $\{x_1, x_2, x_3, \ldots, x_n\}$ be n positive numbers. **The arithmetic mean, the geometric mean and the harmonic mean** of these numbers are defined as follows:

$$\begin{cases} \text{Arithmetic Mean } (A.M) \stackrel{\text{def}}{=} \dfrac{x_1 + x_2 + x_3 + \cdots + x_n}{n} \\[2mm] \text{Geometric Mean } (G.M) \stackrel{\text{def}}{=} \sqrt[n]{x_1 x_2 x_3 \cdots x_n} \\[2mm] \text{Harmonic Mean}(H.M) \stackrel{\text{def}}{=} \dfrac{n}{\dfrac{1}{x_1} + \dfrac{1}{x_2} + \dfrac{1}{x_3} + \cdots \dfrac{1}{x_n}} \end{cases} \qquad (16-5-1)$$

Between the A.M, the G.M. and the H.M. the following fundamental inequality (known as the **Cauchy's inequality**) holds true:

$$\frac{x_1 + x_2 + x_3 + \cdots + x_n}{n} \geq \sqrt[n]{x_1 x_2 x_3 \cdots x_n} \geq \frac{n}{\dfrac{1}{x_1} + \dfrac{1}{x_2} + \dfrac{1}{x_3} + \cdots \dfrac{1}{x_n}}$$

the equality obtained only if $x_1 = x_2 = x_3 = \cdots = x_n$.

A proof of Cauchy's inequality is given in my book, "**College Algebra, Vol. 1**", Chapter 9, section 9-3.

Corollary: If $x_1, x_2, x_3, \ldots, x_n$ are n positive numbers, then

$$(x_1 + x_2 + x_3 + \cdots + x_n)\left(\frac{1}{x_1} + \frac{1}{x_2} + \frac{1}{x_3} + \cdots \frac{1}{x_n}\right) \geq n^2 \qquad (16-5-2)$$

Equation (16-5-2) holds as an equality only if $x_1 = x_2 = x_3 = \cdots = x_n$.

Example 16-5-1: Show that $1 + \frac{1}{2} + \frac{1}{3} + \cdots + \frac{1}{n} > n\sqrt[n]{\frac{1}{n!}}$, $(n \geq 2)$. Recall that the symbol $n!$, (read $\boldsymbol{n}$ **factorial**) is defined as: $n! \stackrel{\text{def}}{=} 1 \cdot 2 \cdot 3 \cdots n$.

Solution

Applying Cauchy's inequality for $x_1 = 1, x_2 = 2, x_3 = 3, \ldots, x_n = n$, yields:

$$\frac{1 + \frac{1}{2} + \frac{1}{3} + \cdots + \frac{1}{n}}{n} > \sqrt[n]{1 \cdot \frac{1}{2} \cdot \frac{1}{3} \cdot \cdots \cdot \frac{1}{n}} \Rightarrow$$

$$1 + \frac{1}{2} + \frac{1}{3} + \cdots + \frac{1}{n} > n \sqrt[n]{\frac{1}{1 \cdot 2 \cdot 3 \cdots n}} = n \sqrt[n]{\frac{1}{n!}}$$

PROBLEMS

16-5-1) If x, y, z are positive numbers, show that

$$\frac{1}{x} + \frac{1}{y} + \frac{1}{z} \geq \frac{9}{x+y+z}$$

16-5-2) If $x_1 = 1, x_2 = 2, x_3 = 3, x_4 = 4$ find the A.M., the G.M., the H.M., and verify Cauchy's inequality.

16-5-3) If the positive numbers $x_1, x_2, x_3, \ldots, x_n$ form an arithmetic progression, show that

$$\sqrt{x_1 x_n} \leq \sqrt[n]{x_1 x_2 \cdots x_n} \leq \frac{x_1 + x_n}{2}$$

16-5-4) Making use of Problem 16-5-3, show that $\sqrt{n} < \sqrt[n]{n!} < \frac{n+1}{2}$.

16-5-5) If x, y, z are positive integers, show that

$$\frac{x+y+z}{3} \leq x^{\frac{x}{x+y+z}} \cdot y^{\frac{y}{x+y+z}} \cdot z^{\frac{z}{x+y+z}}$$

CHAPTER 17: LOGARITHMS, LOGARITMIC AND EXPONENTIAL EQUATIONS

17-1) Powers with irrational exponents

As we know, if **a is a positive number and n is a positive integer**, then by the symbol **a^n** we mean the multiplication of a by itself n times, i.e.

$$a^n = \underbrace{a \cdot a \cdot \cdots \cdot a}_{n \text{ times}}$$

If **n is a positive rational number**, i.e. if $n = \frac{\kappa}{\lambda}$, where κ and λ are positive integers, then, as we know,

$$a^{\frac{\kappa}{\lambda}} = \sqrt[\lambda]{a^\kappa}$$

If **n is a negative rational number**, i.e. if $n = -\frac{\kappa}{\lambda}$, where κ and λ are positive integers, then, by definition

$$a^{-\frac{\kappa}{\lambda}} = \frac{1}{a^{\frac{\kappa}{\lambda}}} = \frac{1}{\sqrt[\lambda]{a^\kappa}}$$

However, we have not explained what is the meaning of the symbol a^x when x is **an irrational number**. For example, what is the meaning of the symbol $3^{\sqrt{2}}$? For our purposes, it suffices to think of $3^{\sqrt{2}}$ as the value we approach by taking successively closer approximations to $\sqrt{2}$, such as $3^{1.4}, 3^{1.41}, 3^{1.414}, 3^{1.4142}, \ldots$ In general, **any irrational number x can be approximated (as close as we wish), by a sequence of rational numbers, and in loose terms, the number a^x is approximated by a sequence of powers with rational exponents**.

It can be shown that all fundamental properties of powers with rational exponents **still hold true for powers with irrational exponents**. This means that **if a is a positive number and x and y are arbitrary real numbers**, (rational or irrational) the following properties hold true:

$$\begin{cases} a^x \cdot a^y = a^{x+y} \\ a^x \div a^y = a^{x-y} \\ (a^x)^y = a^{xy} \end{cases} \quad a > 0, \quad x, y \in \mathbb{R} \qquad (17-1-1)$$

There is an irrational number, **denoted by the letter e**, which plays an extremely important role in Mathematics. **The number e, is perhaps the second most important number in Mathematics, after the number π**. The exact definition of e is given in Calculus. For now, it suffices to say that the expression $\left(1+\frac{1}{n}\right)^n$ gets closer and closer to the number e, as the positive integer n gets larger and larger. (In more technical terms, we say that **the quantity $\left(1+\frac{1}{n}\right)^n$ approaches the number e as n approaches infinity**, and we write $e = \lim_{n \to \infty} \left(1+\frac{1}{n}\right)^n$). We may find **the first few digits** of e using a pocket calculator and find the arithmetic value of $\left(1+\frac{1}{n}\right)^n$ for $n = 1,2,3,4, \ldots$. We find that $e \cong 2.718281 \ldots$.

17-2) The logarithm of a positive number

Let a be **a positive number**, other than 1, $(a \neq 1)$, and b be another **positive number**. It can be shown that there exists another number x, **rational or irrational**, such that when a is raised to x yields b, i.e.

$$a^x = b \quad (a \neq 1), a > 0, \quad b > 0 \quad (17-2-1)$$

Given (17-2-1) we say that **x is the logarithm, base a, of b**, and we write $x = \log_a b$, i.e.

$$a^x = b \Leftrightarrow x = \log_a b, \quad a > 0, \quad b > 0 \text{ and } a \neq 1 \quad (17-2-2)$$

We may say that, **the logarithm, base a of b, is the power to which a (the base of the logarithm) must be raised to obtain b**. Based on the definition in (17-2-2) we obtain the following two **identities**:

$$\{a^{\log_a b} \equiv b, \quad \log_a(a^x) \equiv x\} \quad (17-2-3)$$

Logarithms are used extensively for the computation of complicated expressions involving multiplications and divisions.

We call logarithms to the base 10, $(a = 10)$, **common logarithms**, and as a rule, we write $\log b$ instead of $\log_{10} b$, i.e. $\log b = \log_{10} b$. Logarithms to the base e are called **natural logarithms (or Napierian logarithms)**, and we write $\ln b$ in place of $\log_e b$, i.e. $\ln b = \log_e b$.

Notice: Logarithms of negative numbers are not defined. When we take the logarithm of a number, we shall always assume that the number **is positive**.

Let us consider a few examples:

Since $2^4 = 16$, we say that: $4 = \log_2 16$.

Since $3^3 = 27$, we say that: $3 = \log_3 27$.

Since $2^{-5} = \frac{1}{2^5} = \frac{1}{32}$, we say that: $-5 = \log_2\left(\frac{1}{32}\right)$.

Since $10^3 = 1000$, we say that: $3 = \log 1000$.

By virtue of (17-2-3): $5^{\log_5 13} = 13$.

By virtue of (17-2-3): $\log_{15}(15^4) = 4$.

If $2 = e^x$, then $x = \ln 2$.

If $3 = \ln x$, then $x = e^3$.

17-3) Properties of logarithms

Theorem 17-3-1: In every base of logarithms, the logarithm of 1 is zero and the logarithm of the base is 1.

Proof: Since $a^0 = 1$, according to the definition (17-2-2), $0 = \log_a 1$. Also, since $a^1 = a$, $1 = \log_a a$.

Theorem 17-3-2: If the base of the logarithm is greater than 1, then numbers greater than 1 have positive logarithms, while numbers smaller than 1 have negative logarithms.

Proof: Let $b = a^x$. If $a > 1$ and $b > 1$, then x must be positive, i.e. $x = \log_a b > 0$. If $a > 1$ and $0 < b < 1$, then x must be negative, i.e. $x = \log_a b < 0$.

Theorem 17-3-3: Provided that x and y are positive numbers, the following properties hold true:

$$\begin{cases} \log_a(xy) = \log_a x + \log_a y \\ \log_a\left(\dfrac{x}{y}\right) = \log_a x - \log_a y \\ \log_a(x^y) = y \log_a x \end{cases} \qquad (17-3-1)$$

Proof: In general, **the properties of logarithms result from the definition of the logarithm and the corresponding properties of powers**. For example, to show the first property in (17-3-1), we call, $A = \log_a(xy)$, $B = \log_a x$ and $C = \log_a y$. Then, $xy = a^A$, $x = a^B$ and $y = a^C$, i.e.

$$a^A = a^B \cdot a^C \Rightarrow a^A = a^{B+C} \Rightarrow A = B + C \Rightarrow$$

$$\log_a(xy) = \log_a x + \log_a y$$

To show the second property in (17-3-1), we set, $A = \log_a\left(\dfrac{x}{y}\right)$, $B = \log_a x$ and $C = \log_a y$. Then, $\left(\dfrac{x}{y}\right) = a^A$, $x = a^B$ and $y = a^C$, i.e.

$$a^A = \dfrac{a^B}{a^C} = a^{B-C} \Rightarrow A = B - C \Rightarrow$$

$$\log_a\left(\dfrac{x}{y}\right) = \log_a x - \log_a y$$

To show the third property, again we set, $A = \log_a(x^y)$, $B = \log_a x$ and $C = \log_a y$. Then, $x^y = a^A$, $x = a^B$ and $y = a^C$, i.e.

$$a^A = (a^B)^{a^C} = a^{B \cdot (a^C)} \Rightarrow A = (a^C) \cdot B \Rightarrow$$

$$\log_a(x^y) = y \log_a x$$

Corollary 1: Two reciprocal numbers have opposite logarithms, i.e.

$$\log_a\left(\dfrac{1}{x}\right) = -\log_a x \qquad (17-3-2)$$

Indeed, $\log_a\left(\dfrac{1}{x}\right) = \log_a 1 - \log_a x = 0 - \log_a x = -\log_a x$.

Corollary 2: The logarithm of the $n^{\underline{th}}$ root of a positive number, is equal to the logarithm of the number divided by n, i.e.

$$\log_a \sqrt[n]{x} = \frac{1}{n}\log_a x \qquad (17-3-3)$$

Indeed, $\log_a \sqrt[n]{x} = \log_a(x^{1/n}) = \frac{1}{n}\log_a x$.

For example, $\log_3 \sqrt[25]{7} = \frac{1}{25}\log_3 7$, $\log \sqrt[3]{1000} = \frac{1}{3}\log 1000 = \frac{1}{3}\log 10^3 = \frac{1}{3}\cdot 3\cdot \log 10 = 1$, since $\log 10 = \log_{10} 10 = 1$.

Theorem 17-3-4: **Provided that the base a of the logarithm is greater than 1, the bigger of two positive numbers has bigger logarithm, and conversely, i.e.**

$$\text{If } a > 1, \text{then:} \quad y > x \Leftrightarrow \log_a y > \log_a x \qquad (17-3-4)$$

Proof:

$$\log_a y > \log_a x \Leftrightarrow \log_a y - \log_a x > 0 \Leftrightarrow \log_a\left(\frac{y}{x}\right) > 0 \overset{(a>1)}{\Longleftrightarrow} \frac{y}{x} > 1$$

or, equivalently, $y > x$, (since x, y are positive numbers).

Theorem 17-3-5, (Change of the base): **If we know the logarithm of a number b to the base a, we may find the logarithm of b to another base $c \neq 1$, by means of the following formula:**

$$\log_c b = \frac{\log_a b}{\log_a c} \qquad (17-3-5)$$

Proof: Let us call, $A = \log_c b$, $B = \log_a b$, $C = \log_a c$. Then, $b = c^A$, $b = a^B$ and $c = a^C$. Then,

$$c^A = a^B (= b) \Rightarrow \log_a(c^A) = \log_a(a^B) \Rightarrow$$

$$A\log_a c = B\log_a a \xRightarrow{(\log_a a = 1)} \underbrace{\log_c b \cdot \log_a c}_{A} = \underbrace{\log_a b}_{B} \Rightarrow \log_c b = \frac{\log_a b}{\log_a c}$$

Corollary: If $a = e$, **(Napierian logarithms)**, formula (17-3-5) yields, the following, easily memorisable formula:

$$\log_c b = \frac{\ln b}{\ln c} \qquad (17-3-6)$$

Example 17-3-1: Simplify the expression

$$A = \log \frac{3^2 \cdot \sqrt[5]{7} \cdot 2^6}{5^4 \cdot \sqrt[3]{13}}$$

Solution

Applying properties of logarithms, we find:

$$A = \log(3^2 \cdot \sqrt[5]{7} \cdot 2^6) - \log(5^4 \cdot \sqrt[3]{13}) \Rightarrow$$

$$A = 2\log 3 + \frac{1}{5}\log 7 + 6\log 2 - 4\log 5 - \frac{1}{3}\log 13$$

Example 17-3-2: Evaluate (approximately) the number $X = \frac{\sqrt[57]{89} \cdot \sqrt[21]{8}}{\sqrt[3]{17} \cdot 7^{5/2}}$.

Solution

We first find the $\log X$, using the properties of logarithms.

$$\log X = \log\left(\frac{\sqrt[57]{89} \cdot \sqrt[21]{8}}{\sqrt[3]{17} \cdot 7^{5/2}}\right) = \log(\sqrt[57]{89} \cdot \sqrt[21]{8}) - \log(\sqrt[3]{17} \cdot 7^{5/2}) \Rightarrow$$

$$\log X = \frac{1}{57}\log 89 + \frac{1}{21}\log 8 - \frac{1}{3}\log 17 - \frac{5}{2}\log 7$$

Using a pocket calculator, we find easily that $\log X \cong -2.4456$, and $X = 10^{-2.4456} \cong 0.00358$.

Example 17-3-3: Solve for x the equation:

$$2(\log_x 9)^2 + 3\log_x 81 = 5\log_x 9 + 10$$

Solution

If we set $a = \log_x 9$, then $\log_x 81 = \log_x 9^2 = 2\log_x 9 = 2a$, and the equation becomes,

$$2a^2 + 3 \cdot 2a = 5a + 10 \Leftrightarrow 2a^2 + a - 10 = 0 \Rightarrow \begin{Bmatrix} a = 2 \\ or \\ a = -5/2 \end{Bmatrix} \quad (*)$$

1) When $a = 2$, $\log_x 9 = 2$, i.e. $x^2 = 9$ and $x = 3$, (the root $x = -3$ is rejected, since x (the base of the logarithm) must be positive).

2) When $a = -5/2$, $\log_x 9 = -5/2$, i.e. $x^{-5/2} = 9$ and $x = 9^{-2/5} = 1/9^{2/5} = 1/\sqrt[5]{81} = \sqrt[5]{3}/3$.

Example 17-3-4: Find the sum of the first 50 terms of the sum:

$$S = \log 3 + \log 3^2 + \log 3^3 + \log 3^4 + \cdots.$$

Solution

$$S = \log 3 + \log 3^2 + \log 3^3 + \log 3^4 + \cdots + \log 3^{50} \Rightarrow$$

$$S = \log 3 + 2\log 3 + 3\log 3 + 4\log 3 + \cdots + 50\log 3 \Rightarrow$$

$$S = (1 + 2 + 3 + \cdots + 50)\log 3 = \frac{50 \cdot 51}{2}\log 3 = 1275 \log 3$$

Example 17-3-5: Solve the equation: $\log_x 9 = \log_3(3x)$.

Solution

By virtue of formula (17-3-5), $\log_x 9 = \frac{\log_3 9}{\log_3 x} = \frac{\log_3 3^2}{\log_3 x} = \frac{2\log_3 3}{\log_3 x} = \frac{2}{\log_3 x}$, and the given equation becomes:

$$\frac{2}{\log_3 x} = \log_3(3x) = \log_3 3 + \log_3 x = 1 + \log_3 x$$

or, if we set $y = \log_3 x$,

$$\frac{2}{y} = 1 + y \Leftrightarrow y^2 + y - 2 = 0 \Leftrightarrow \left\{\begin{matrix} y = 1 \\ \text{or} \\ y = -2 \end{matrix}\right\} \quad (*)$$

1) When $y = 1$, $\log_3 x = 1$, i.e. $x = 3$.

2) When $y = -2$, $\log_3 x = -2$, i.e. $x = 3^{-2} = \frac{1}{9}$.

Example 17-3-6: Provided that $\frac{\log x}{y-z} = \frac{\log y}{z-x} = \frac{\log z}{x-y}$, show that $x^x y^y z^z = 1$.

Solution

Let us call k **the common value of the equal fractions**, i.e.

$$\frac{\log x}{y-z} = \frac{\log y}{z-x} = \frac{\log z}{x-y} = k \Longrightarrow$$

$$\begin{cases} \log x = k(y-z) \\ \log y = k(z-x) \\ \log z = k(x-y) \end{cases} \Longrightarrow \begin{cases} x\log x = kx(y-z) \\ y\log y = ky(z-x) \\ z\log z = kz(x-y) \end{cases} \Longrightarrow \begin{cases} \log x^x = kx(y-z) \\ \log y^y = ky(z-x) \\ \log z^z = kz(x-y) \end{cases}$$

and adding these equations term wise, results in the following equation,

$$\log x^x + \log y^y + \log z^z = 0 \Longrightarrow \log(x^x y^y z^z) = 0 \Longrightarrow x^x y^y z^z = 1$$

Example 17-3-7: Given that $x > 0$ and n and m are positive integers with $n > m$, show that

$$\frac{\log(1+x^n)}{n} < \frac{\log(1+x^m)}{m}$$

Solution

1) Let us consider the case $0 < x \leq 1$. Then $x^m \geq x^n$, $1 + x^m \geq 1 + x^n$, and by virtue of Theorem 17-3-4,

$$\log(1+x^m) \geq \log(1+x^n) \Longrightarrow \frac{\log(1+x^m)}{m} \geq \frac{\log(1+x^n)}{m} > \frac{\log(1+x^n)}{n}$$

(since, by assumption, $n > m$ and $\log(1+x^n) > 0$, by virtue of Theorem 17-3-2).

2) When $x > 1$, then $0 < \frac{1}{x} < 1$, and as we proved in case 1, we must have:

$$\frac{\log\left(1+\left(\frac{1}{x}\right)^n\right)}{n} < \frac{\log\left(1+\left(\frac{1}{x}\right)^m\right)}{m} \Longrightarrow \frac{\log\left(\frac{1+x^n}{x^n}\right)}{n} < \frac{\log\left(\frac{1+x^m}{x^m}\right)}{m} \Longrightarrow$$

$$\frac{\log(1+x^n) - \log x^n}{n} < \frac{\log(1+x^m) - \log x^m}{m} \Longrightarrow$$

$$\frac{\log(1+x^n) - n\log x}{n} < \frac{\log(1+x^m) - m\log x}{m} \Longrightarrow$$

$$\frac{\log(1+x^n)}{n} - \log x < \frac{\log(1+x^m)}{m} - \log x \Rightarrow \frac{\log(1+x^n)}{n} < \frac{\log(1+x^m)}{m}$$

So, the inequality holds true for **all positive values of x**, and this completes the proof.

Example 17-3-8: For what values of k the following equation has two real and distinct roots?

$$x^2 + 2(1 + \log k)x + 1 - (\log k)^2 = 0$$

Solution

First of all, **k must be positive**, since logarithms of negative numbers are not defined, i.e. $k > 0$.

The roots of the given equation will be real and distinct, when its discriminant is positive, i.e. when

$$4(1 + \log k)^2 - 4\{1 - (\log k)^2\} > 0 \Leftrightarrow$$

$$4(1 + (\log k)^2 + 2\log k - 1 + (\log k)^2) > 0 \Leftrightarrow 8((\log k)^2 + \log k) > 0 \Leftrightarrow$$

$$\log k\,(\log k + 1) > 0 \Leftrightarrow \begin{cases} \log k > 0 \\ or \\ \log k < -1 \end{cases} \Leftrightarrow \begin{cases} k > 1 \\ or \\ 0 < k < 1/10 \end{cases}$$

Example 17-3-9: Provided that x and y are two positive numbers, show that $x^{\log y} = y^{\log x}$.

Solution

It suffices to show that $\log(x^{\log y}) = \log(y^{\log x})$, i.e. $\log y \log x = \log x \log y$, which obviously is true.

PROBLEMS

17-3-1) Find the decimal logarithm of the numbers: $100, 1000, 0.1, 0.001$.

(**Ans:** $2, 3, -1, -3$).

17-3-2) Find the numbers: $\log_5 25$, $\log_4 64$, $\log_3 81$, $\log_2 32$.

17-3-3) Show the following equalities:

$$\log 20 = 2\log 2 + \log 5, \qquad \log_2 504 = 2\log_2 3 + \log_2 7 + 3$$

$$\log 9 - \log 16 + \log 40 - \log 81 = \log\left(\frac{5}{18}\right)$$

17-3-4) Show that $\log_b a \cdot \log_c b \cdot \log_a c = 1$.

Hint: You may use formula (17-3-6).

17-3-5) If $\log(x^3 y^5) = 22$ and $\log x - \log y = 2$, find $\log x$ and $\log y$.

(Ans: $\log x = 4, \log y = 2$).

Hint: $\log(x^3 y^5) = 3\log x + 5\log y$.

17-3-6) If $a > b > 0$ and $a^2 + b^2 = 7ab$, show that

$$\log\left(\frac{a-b}{\sqrt{5}}\right) = \frac{1}{2}(\log a + \log b)$$

17-3-7) Given that $xy^{a-1} = \kappa, xy^{b-1} = \lambda, xy^{c-1} = \mu$, show that:

$$(a-b)\log\mu + (b-c)\log\kappa + (c-a)\log\lambda = 0$$

17-3-8) For what values of k the roots of $x^2 - 4x + \log k = 0$ are real and distinct?

17-3-9) Provided that $y > x > 0$, show that $\log\left(\frac{1+x}{1+y}\right) > \log\left(\frac{x}{y}\right)$.

Hint: It suffices to show that $\frac{1+x}{1+y} > \frac{x}{y}$.

17-3-10) Provided that $x > 3$ and $y > 3$, show that

$$\log 2 + \log x + \log y > \log 3 + \log(x+y)$$

17-3-11) Provided that the positive numbers x, y, z form an harmonic progression, show that $\log(x+z) + \log(x - 2y + z) = 2\log(x-z)$.

17-3-12) Given that the numbers a, b, c are the $k^{\underline{th}}, m^{\underline{th}}, n^{\underline{th}}$ terms respectively, of a geometric progression, show that

$$(n - m)\log a + (m - k)\log c + (k - n)\log b = 0$$

17-3-13) If $\log x$ and $\log y$ are the first and second term respectively of an arithmetic progression, show that the sum S of the first n terms is

$$S = \frac{1}{2}\log\left(y^{n(n-1)}/x^{n(n-3)}\right)$$

Hint: The ratio of the progression is $\omega = \log y - \log x = \log(y/x)$.

17-3-14) Determine k so that the sum of the two roots of the equation $\log k\, x^2 - (2 + \log k)x + 2 = 0$ to be equal to 3, and then find the two roots.

17-4) Logarithmic and exponential equations and systems

If in at least one term in an equation, **the unknown x, or a function of x, appears as an exponent**, the equation is called **exponential equation**. For example, the equations, $3^x = 15, 2^{x^2-5x+3} = 8$, etc, are exponential equations.

In a **logarithmic equation, the unknown x or a function of x, appears inside a logarithm**. For example, the equations, $\log x = 7, \log_3(5x + 7) = 2$, etc, are logarithmic equations.

The solutions of exponential and logarithmic equations are obtained with the aid of properties of powers and logarithms. In very general terms, when working with exponential equations, we try to convert the equation to one of the following forms:

1) $a^x = a^y$, which implies that $x = y$.

2) $a^x = 1$, which has solution $x = 0$.

3) $a^x = b$, the solution of which is $x = \log_a b$.

4) If the equation is of the form $Aa^{2x} + Ba^x + C = 0$, then the substitution $y = a^x$, transforms the given exponential equation into $Ay^2 + By + C = 0$, which is quadratic in y, etc.

When the equation is logarithmic, we try to bring it to the form $\log a = \log b$, which implies that $a = b$, (a and b must be positive numbers).

Let us consider a few illustrative examples.

Example 17-4-1: Solve the equation: $2^{7x+2} = 16$.

Solution

$$2^{7x+2} = 16 = 2^4 \Leftrightarrow 7x + 2 = 4 \Leftrightarrow 7x = 4 - 2 = 2 \Leftrightarrow x = 2/7$$

Example 17-4-2: Solve the equation: $9^x - 2 \cdot 3^x - 15 = 0$.

Solution

$$9^x - 2 \cdot 3^x - 15 = 0 \Leftrightarrow 3^{2x} - 2 \cdot 3^x - 15 = 0$$

or, if we set $y = 3^x$, $y^2 - 2y - 15 = 0 \Leftrightarrow \begin{pmatrix} y_1 = 5 \\ or \\ y_2 = -3 \end{pmatrix}$

When $y = 5, 3^x = 5$, or $x = \log_3 5$.

When $y = -3, 3^x = -3$, and this equation does not have any real solution.

Therefore the original equation has one solution, $x = \log_3 5$.

Example 17-4-3: Solve the equation: $\log(2x - 3) = \log 15 - \log(x + 2)$.

Solution

$$\log(2x - 3) = \log 15 - \log(x + 2) \Leftrightarrow \log(2x - 3) = \log\left(\frac{15}{x + 2}\right) \Leftrightarrow$$

$$2x - 3 = \frac{15}{x + 2} \Leftrightarrow (2x - 3)(x + 2) = 15 \Leftrightarrow 2x^2 + x - 21 = 0$$

Solving this equation we find, $x = 3$ or $x = -\frac{7}{2}$. The first solution ($x = 3$) is a valid solution, while the second solution $x = -7/2$ is rejected, (since at $x = -7/2$ the arguments of the logarithms are negative).

Example 17-4-4: Solve the system: $\{2^x - 3^y = 7, \ 3 \cdot 2^x + 2 \cdot 3^y = 66\}$.

Solution

If we set, $A = 2^x, B = 3^y$, the system becomes: $\begin{cases} A - B = 7 \\ 3A + 2B = 66 \end{cases}$.

This is **a linear system in A and B**, and its solution is easily found to be $A = 16$ and $B = 9$, i.e.

$$\begin{cases} A = 16 \\ B = 9 \end{cases} \Rightarrow \begin{cases} 2^x = 16 = 2^4 \\ 3^y = 9 = 3^2 \end{cases} \Rightarrow x = 4 \text{ and } y = 2$$

Example 17-4-5: Solve the system: $\{x^y = y^x, \; 3^x = 2^y\}$.

Solution

Taking the logarithms of both sides of the equations we find:

$$(S): \begin{cases} y \log x = x \log y \\ y \log 2 = x \log 3 \end{cases} \Rightarrow \frac{y \log x}{y \log 2} = \frac{x \log y}{x \log 3} \Rightarrow \frac{\log x}{\log 2} = \frac{\log y}{\log 3} \quad (*)$$

Also, taking the logarithms of the second equation of the system (S), we obtain,

$$\log(y \log 2) = \log(x \log 3) \Rightarrow \log y + \log(\log 2) = \log x + \log(\log 3) \quad (**)$$

Equations (*) and (**) form a system of two equations in the two unknowns **log x** and **log y**, the solution of which yields,

$$\begin{cases} \log x = \log 2 \cdot \frac{\log\left(\frac{\log 3}{\log 2}\right)}{\log\left(\frac{3}{2}\right)}, & \log y = \log 3 \cdot \frac{\log\left(\frac{\log 3}{\log 2}\right)}{\log\left(\frac{3}{2}\right)} \end{cases} \quad (***)$$

Having found $\log x$ and $\log y$, we obtain x and y.

Example 17-4-6: Solve the inequality: $\log\left(\log \frac{2x+4}{-x+6}\right) > 0$.

Solution

First of all, the quantity $\left(\frac{2x+4}{-x+6}\right)$ must be positive, since logarithms of negative numbers are not defined (in the set of real numbers), i.e.

$$\frac{2x+4}{-x+6} > 0 \Leftrightarrow (2x+4)(-x+6) > 0 \Leftrightarrow 2(x+2)(x-6) < 0 \Leftrightarrow$$

$$-2 < x < 6 \quad (*)$$

So, for every x in the interval $(2,6)$, the quantity $\left(\frac{2x+4}{-x+6}\right)$ is positive, and the number $\left(\log \frac{2x+4}{-x+6}\right)$ does exist, (is a real number). Also, since $\log\left(\log \frac{2x+4}{-x+6}\right) > 0$, the quantity $\log \frac{2x+4}{-x+6}$ must be greater than 1, (by virtue of Theorem 17-3-2), i.e.

$$\log \frac{2x+4}{-x+6} > 1 = \log 10 \Leftrightarrow \frac{2x+4}{-x+6} > 10 \Leftrightarrow \frac{2x+4}{-x+6} - 10 > 0 \Leftrightarrow$$

$$\frac{2x+4+10x-60}{-x+6} > 0 \Leftrightarrow \frac{12x-56}{-x+6} > 0 \Leftrightarrow (12x-56)(-x+6) > 0 \Leftrightarrow$$

$$4(3x-14)(x-6) < 0 \Leftrightarrow (3x-14)(x-6) < 0 \Leftrightarrow$$

$$\frac{14}{3} < x < 6 \quad (**)$$

The values of x that satisfy both inequalities (*) and (**) **simultaneously**, (and therefore satisfy the original inequality), are thus the intersection of the intervals in (*) and (**), i.e.

$$Solution\ set = (-2 < x < 6) \cap \left(\frac{14}{3} < x < 6\right) = \left(\frac{14}{3} < x < 6\right)$$

PROBLEMS

17-4-1) Solve the equations:

1) $12^{x-5} = 144$
2) $3^{x^2-2x} = 27$
3) $5^{(x-1)(x+3)(x-7)} = 1$
4) $x^{x^2-7x+12} = 1$
5) $\log(x+1) + 2\log\sqrt{5x} = 2$
6) $\log(x-1)^2 + \log(x-10)^2 = 2$
7) $x^{3+\log x} = 15$

(Ans: **1)** $x = 7$, **2)** $x = -1\ or\ 3$, **3)** $x = -3\ or\ 1\ or\ 7$, **4)** $x = 1\ or\ 3\ or\ 4$, **5)** $x = 4$, **6)** $x = 11$, **7)** $x = \left(-3 + \sqrt{9 + 4 \cdot \log 15}\right)/2$).

17-4-2) Solve the equations:

1) $3^x = 9^{x+1}$
2) $3^{2^x} = 25$
3) $3^{2x} + 2 \cdot 3^x \cdot 5^x - 3 \cdot 5^{2x} = 0$
4) $10x^{\log x} = \sqrt{x^5}$
5) $6\sqrt[x]{64} - 8 = \left(\sqrt[x]{16}\right)^3$

Hint: In (3) divide through by 5^{2x}, and set $y = 3^x/5^x$; the equation reduces to a quadratic in y, etc.

17-4-3) Solve the system: $\{x^y = y^x, \quad x = y^2\}$.

(Ans: $(x = 1, y = 1)$ or $(x = 4, y = 2)$).

17-4-4) Solve the system: $\{x + y = 2^y, \quad (x+y) \cdot 3^y = 216\}$.

17-4-5) Solve the system: $\{x^2 + y^2 = 25, \quad \log x + \log y = \log 12\}$.

(Ans: $(x = 3, y = 4)$ or $(x = 4, y = 3)$).

17-4-6) Solve the system: $\{3^{\log x} + 2^{\log y} = 7, \quad 9^{\log x} - 4^{\log y} = -7\}$.

Hint: If we set $A = 3^{\log x}, B = 2^{\log y}$, the system reduces to a linear system in A and B, etc.

17-4-7) Solve the system: $\{x^8 = y^{\log y + 2}, \quad xy = 10\sqrt{10}\}$.

(Ans: $\log x = \frac{13}{2} - \sqrt{37}, \log y = -5 + \sqrt{37}$).

17-4-8) Solve the system: $\{y = 4x^2 y^{\log(2x)}, \quad y^{\log(2x)} + (2x)^{\log y} = 8x^2\}$.

Hint: Use the result of Example 17-3-9, i.e. $y^{\log(2x)} = (2x)^{\log y}$, etc. The answer is $(x = 1/2, y = 10)$, or $(x = \sqrt{10}/2, y = 100)$.

17-4-9) Solve the system: $\{y = 3x, \quad x^y = y^x\}$.

(Ans: $x = \sqrt{3}, y = 3\sqrt{3}$).

17-4-10) Solve the inequality: $\log\left(\log \frac{6x-21}{x-2}\right) < 0$.

Hint: See Example 17-4-6.

17-4-11) Solve the system: $\{(2x)^{\log 2} = (3y)^{\log 3},\ \ 3^{\log x} = 2^{\log y}\}$.

(**Ans:** $x = 1/2, y = 1/3$).

CHAPTER 18: CONDITIONAL MAXIMA AND MINIMA

18-1) Introduction

a) Let us, for definiteness, consider a function w of three variables x, y and z. As we know we may write, $w = f(x, y, z)$, where f is the rule which assigns a unique real number w to each ordered triad of numbers (x, y, z). The allowed values of (x, y, z) constitute **the domain D** of the function, while the corresponding values of w constitute **the range R** of the function. Examples of real functions of three real variables are the following:

$$w = x^2 + y^2 + z^2 - xy - yz - zx$$
$$w = xyz + z + x + y + z$$
$$w = 2^x + 3^y + 4^z - 5xyz$$

One of the most important problems in Mathematics is to find the **maximum or the minimum value of the function $w = f(x, y, z)$**, when the three independent variables are allowed to vary within their domain of definition.

If there exists a constant number M such that $f(x, y, z) \leq M$ for all x, y, z in the domain of definition of the function, and if, in addition, there exists a point (x_0, y_0, z_0) at which $f(x_0, y_0, z_0) = M$, then we say that M is a maximum of the function, attained at (x_0, y_0, z_0).

For example, let us consider the function

$$w = f(x, y, z) = 5 - (x - 1)^2 - (y - 3)^2 - (z + 4)^2$$

The maximum value of the function is $M = 5$, attained at $(x_0, y_0, z_0) = (1, 3, -4)$. For all other values of the independent variables, $f(x, y, z) < 5$.

Similarly, **if there exists a constant number m such that $f(x, y, z) \geq m$ for all x, y, z in the domain of definition of the function, and if, in addition, there exists a point (x_0, y_0, z_0) at which $f(x_0, y_0, z_0) = m$, then we say that m is a minimum of the function, attained at (x_0, y_0, z_0)**.

For example, the function

$$w = f(x, y, z) = (x - 2)^2 + (y - 7)^2 + (z + 10)^2 + 15$$

attains its minimum value $m = 15$ at $(x_0, y_0, z_0) = (2,7,-10)$, while for all other values of (x, y, z), $f(x, y, z) > 15$.

b) There are problems where we seek the maximum or the minimum of a function $w = f(x, y, z)$, **provided that the unknowns (variables) satisfy an additional condition.** These problems are referred to as **"conditional maxima or minima"**. For example, given that $x + y + z = 25$, what is the maximum value possible of the function $f(x, y, z) = xyz$, $(x > 0, y > 0, z > 0)$, or, given that $xyz = 125, x > 0, y > 0, z > 0$) what is the minimum value possible of the function $8x + 2y + 5z$?, etc.

In the following section we present some simple Theorems which provide the general guidelines to find the conditional maxima or minima of a given function.

18-2) General theorems on conditional maxima and minima

Theorem 18-2-1: **Provided that the sum of two variables x and y is constant, their product xy takes its maximum value possible, when the absolute value of their difference, becomes as small as possible.**

Proof: Let x and y be two variables, whose sum is equal to the constant number c, i.e. $x + y = c$. Using the elementary identity

$$xy = \left(\frac{x+y}{2}\right)^2 - \left(\frac{x-y}{2}\right)^2 = \left(\frac{c}{2}\right)^2 - \left(\frac{x-y}{2}\right)^2$$

we see that the product xy takes its maximum value possible, when $(x-y)^2/4$ assumes its minimum value possible, i.e. when $|x - y|$ assumes its minimum value possible.

Theorem 18-2-2: **Provided that the sum of two variables x and y is constant, $(x + y = c)$, then the product xy becomes maximum when $x = y = \frac{c}{2}$, i.e. $Max(xy) = c^2/4$.**

Proof: Follows immediately from Theorem 18-2-1.

Theorem 18-2-3: **Provided that the product of two positive variables x and y is constant, their sum $x + y$ takes its minimum value possible, when the absolute value of their difference, becomes as small as possible.**

Proof: Let x and y be two **positive variables**, whose product is constant ($xy = c^2$). Using the elementary identity

$$(x + y)^2 = 4xy + (x - y)^2 = 4c^2 + (x - y)^2$$

we see that the sum $(x + y)^2$ takes its minimum value possible, when the quantity $(x - y)^2$ becomes as small as possible, or, since x and y are positive variables, the sum $(x + y)$ takes its minimum value possible, when the term $|x - y|$ becomes as small as possible.

Theorem 18-2-4: **Given that $x > 0, y > 0$ and that $xy = c^2$ (constant), the sum $(x + y)$ takes its minimum value when $x = y = c$, i.e. $Min(x + y) = 2c$.**

Proof: Follows immediately from Theorem 18-2-3.

Example 18-2-1: From all the equivalent right triangles, (right triangles having the same area c^2), which one has the smallest hypotenuse?

(Two plane figures in geometry, like two triangles or two rectangles, etc., are called **equivalent, if they have the same area**. For example, two triangles that have the same base and the same altitude (corresponding to the base), are equivalent, even though they may have different the other two sides).

Solution

If x and y are the two **perpendicular sides** of the triangle, then the hypotenuse will be $a = \sqrt{x^2 + y^2}$. The area of this triangle is $\frac{1}{2}xy = c^2$, (constant). We may thus formulate our problem as follows:

Provided that $xy = 2c^2$, (constant), find the minimum value of the function $a = \sqrt{x^2 + y^2}$. The quantity $\sqrt{x^2 + y^2}$ becomes minimum, when $x^2 + y^2$ becomes minimum. However, the product $x^2 \cdot y^2 = 4c^4$, (**constant**), and therefore, by virtue of Theorem 18-2-4, the sum $(x^2 + y^2)$ takes its minimum value when $x^2 = y^2$, or since x and y are positive numbers (sides of a triangle), $x = y = \sqrt{2}\,c$, i.e. when **the right triangle is an isosceles right triangle**. The smallest hypotenuse is $a = \sqrt{2c^2 + 2c^2} = 2c$.

Theorem 18-2-5: If the sum of n positive variables $x_1, x_2, \cdots, x_n$ is constant and equal to C, i.e. if $(x_1 + x_2 + \cdots + x_n) = C$, (constant), then their product $P = x_1 x_2 \cdots x_n$ becomes maximum when all the variables become equal to each other, i.e. when $x_1 = x_2 = \cdots = x_n = \frac{C}{n}$. The maximum value of the product is $P_{max} = (C/n)^n$.

Proof: The proof of this Theorem is based on **Cauchy's inequality**, (see section 16-5), i.e.

$$\frac{x_1 + x_2 + x_3 + \cdots + x_n}{n} \geq \sqrt[n]{x_1 x_2 x_3 \cdots x_n} \qquad (*)$$

the equality obtained when $x_1 = x_2 = \cdots = x_n$.

If the sum of the variables is C, (constant), equation (*) implies that

$$\frac{C}{n} \geq \sqrt[n]{x_1 x_2 x_3 \cdots x_n} \qquad (**)$$

and this shows that the maximum value of the product $\sqrt[n]{x_1 x_2 x_3 \cdots x_n}$ is $\left(\frac{C}{n}\right)$, and this maximum value is attained when $x_1 = x_2 = \cdots = x_n = \frac{C}{n}$. The maximum value of the product is

$$P_{max} = \frac{C}{n} \cdot \frac{C}{n} \cdot \ldots \cdot \frac{C}{n} = \left(\frac{C}{n}\right)^n$$

Theorem 18-2-6: If the product of n positive variables $x_1, x_2, \cdots, x_n$ is constant and equal to P, i.e. if $(x_1 x_2 \cdots x_n) = P$, (constant), then their sum $S = x_1 + x_2 + \cdots + x_n$ becomes minimum when all the variables become equal to each other, i.e. when $x_1 = x_2 = \cdots = x_n = \sqrt[n]{P}$. The minimum value of the sum is $S_{min} = n\sqrt[n]{P}$.

Proof: The proof is similar to the proof of the preceding theorem. Let the reader complete the details.

Example 18-2-2: From all the orthogonal parallelepipeds having the same volume C^3, which one has the least area?

Solution

If x, y, z are the three sides of the parallelepiped, we have:

$$\begin{cases} \text{Volume: } xyz = C^3 \text{ (constant)} \\ \text{Area: } 2(xy + yz + zx) \end{cases} \quad (*)$$

Since $(xy) \cdot (yz) \cdot (zx) = x^2 y^2 z^2 = (xyz)^2 = (C^3)^2 = C^6$, **(constant)**, Theorem 18-2-6 implies that the sum $(xy + yz + zx)$ becomes minimum when $xy = yz = zx$, i.e. when $x = y = z$, i.e. **when the parallelepiped becomes a cube**. Then $x = y = z = C$, and the minimum area is $2 \cdot 3C^2 = 6C^2$.

Theorem 18-2-7: **If the sum of n positive variables $x_1, x_2, \cdots, x_n$ is constant and equal to C, i.e. if $(x_1 + x_2 + \cdots + x_n) = C$, (constant), then the product $x_1{}^{m_1} x_2{}^{m_2} \cdots x_n{}^{m_n}$, where $m_1, m_2, \cdots, m_n$ are positive rational numbers, becomes maximum when the numbers $x_1, x_2, \cdots, x_n$ become proportional to $m_1, m_2, \cdots, m_n$, i.e. when**

$$\frac{x_1}{m_1} = \frac{x_2}{m_2} = \cdots = \frac{x_n}{m_n} = \frac{x_1 + x_2 + \cdots + x_n}{m_1 + m_2 + \cdots + m_n} = \frac{C}{m_1 + m_2 + \cdots + m_n}$$

Proof: a) Let us first assume that $m_1, m_2, \cdots, m_n$ are **positive integers**, and consider the product

$$P = \left(\frac{x_1}{m_1}\right)^{m_1} \left(\frac{x_2}{m_2}\right)^{m_2} \cdots \left(\frac{x_n}{m_n}\right)^{m_n} \quad (*)$$

The product P consists of the product of m_1 terms equal to $\left(\frac{x_1}{m_1}\right)$, m_2 terms equal to $\left(\frac{x_2}{m_2}\right)$, ..., m_n terms equal to $\left(\frac{x_n}{m_n}\right)$. The sum of the terms of the product is,

$$m_1 \cdot \frac{x_1}{m_1} + m_2 \cdot \frac{x_2}{m_2} + \cdots + m_n \cdot \frac{x_n}{m_n} = x_1 + x_2 + \cdots + x_n = C, (constant)$$

and according to Theorem 18-2-5, the product

$$\left(\frac{x_1}{m_1}\right)^{m_1} \left(\frac{x_2}{m_2}\right)^{m_2} \cdots \left(\frac{x_n}{m_n}\right)^{m_n} = \frac{x_1{}^{m_1} x_2{}^{m_2} \cdots x_n{}^{m_n}}{m_1{}^{m_1} m_2{}^{m_2} \cdots m_n{}^{m_n}} \quad (**)$$

becomes maximum when all its terms are equal, i.e. when

$$\frac{x_1}{m_1} = \frac{x_2}{m_2} = \cdots = \frac{x_n}{m_n} \qquad (***)$$

and since the term $(m_1{}^{m_1} m_2{}^{m_2} \cdots m_n{}^{m_n})$ is **a constant number**, the product $(x_1{}^{m_1} x_2{}^{m_2} \cdots x_n{}^{m_n})$ assumes its maximum value when the product $\left(\frac{x_1}{m_1}\right)^{m_1} \left(\frac{x_2}{m_2}\right)^{m_2} \cdots \left(\frac{x_n}{m_n}\right)^{m_n}$ assumes its maximum value, i.e. when equation (***) holds true, and this completes the proof.

b) Let us now assume that $m_1, m_2, \cdots, m_n$ are **positive fractions (rational numbers)**. We may make them to have **a common denominator d**, i.e. we may write,

$$m_1 = \frac{k_1}{d}, m_2 = \frac{k_2}{d}, \cdots, m_n = \frac{k_n}{d} \qquad (****)$$

where $k_1, k_2, \cdots, k_n$ and d are **positive integers**. Then the product $x_1{}^{m_1} x_2{}^{m_2} \cdots x_n{}^{m_n}$ may be written as

$$x_1^{\frac{k_1}{d}} x_2^{\frac{k_2}{d}} \cdots x_n^{\frac{k_n}{d}} = \sqrt[d]{x^{k_1} x^{k_2} \cdots x^{k_n}}$$

and this product becomes maximum when the radicand becomes maximum, i.e. (according to part (a)), when

$$\frac{x_1}{k_1} = \frac{x_2}{k_2} = \cdots = \frac{x_n}{k_n} \Rightarrow \frac{x_1}{\left(\frac{k_1}{d}\right)} = \frac{x_2}{\left(\frac{k_2}{d}\right)} = \cdots = \frac{x_n}{\left(\frac{k_n}{d}\right)} \stackrel{(****)}{\Rightarrow}$$

$$\frac{x_1}{m_1} = \frac{x_2}{m_2} = \cdots = \frac{x_n}{m_n}$$

and this completes the proof.

Remark: Application of the **Theorem of equal fractions**, yields:

$$\frac{x_1}{m_1} = \frac{x_2}{m_2} = \cdots = \frac{x_n}{m_n} = \frac{x_1 + x_2 + \cdots + x_n}{m_1 + m_2 + \cdots + m_n} = \frac{C}{m_1 + m_2 + \cdots + m_n}$$

To prove the Theorem, set $\frac{x_1}{m_1} = \frac{x_2}{m_2} = \cdots = \frac{x_n}{m_n} = \lambda$, and then show that $\frac{x_1 + x_2 + \cdots + x_n}{m_1 + m_2 + \cdots + m_n} = \lambda$, etc.

Theorem 18-2-8: If the product $x_1^{m_1} x_2^{m_2} \cdots x_n^{m_n}$ is constant, where $x_1, x_2, \cdots, x_n$ are positive variables and $m_1, m_2, \cdots, m_n$ are positive rational numbers, then the sum $(x_1 + x_2 + \cdots + x_n)$ becomes minimum, when the numbers $x_1, x_2, \cdots, x_n$ become proportional to $m_1, m_2, \cdots, m_n$, i.e. when

$$\frac{x_1}{m_1} = \frac{x_2}{m_2} = \cdots = \frac{x_n}{m_n}$$

Proof: Similar to the proof of Theorem 18-2-7, (see Problem 18-2-1).

Example 18-2-3: Divide the number 20 into two parts such that the product of the square of the first times the cube of the second to as the maximum as possible.

Solution

We want to divide the number 20 into two parts x and y, such that the product $x^2 y^3$ to be as maximum as possible. Since $(x + y) = 20$, Th. 18-2-7 implies that we must have,

$$\frac{x}{2} = \frac{y}{3} = \frac{x+y}{2+3} = \frac{20}{5} = 4 \Rightarrow \begin{Bmatrix} x = 2 \cdot 4 = 8 \\ y = 3 \cdot 4 = 12 \end{Bmatrix}$$

The maximum product is $8^2 \cdot 12^3$.

Example 18-2-4: From all triangles having the same perimeter 2τ, $(a + b + c = 2\tau)$, which one bounds the maximum area. What is this maximum area?

Solution

The area A of a triangle in terms of its sides a, b, c is given by **Heron's formula**,

$$A = \sqrt{\tau(\tau - a)(\tau - b)(\tau - c)} = \sqrt{\tau} \cdot \sqrt{(\tau - a)(\tau - b)(\tau - c)} \qquad (*)$$

The area becomes maximum when the product $(\tau - a)(\tau - b)(\tau - c)$ becomes maximum. However, since

$$(\tau - a) + (\tau - b) + (\tau - c) = 3\tau - (a + b + c) = 3\tau - 2\tau = \tau = const.$$

application of Theorem 18-2-5, implies that the product becomes maximum when $(\tau - a) = (\tau - b) = (\tau - c)$, i.e. when $a = b = c$, (**equilateral triangle**). In this case, the maximum area is $\left(\sqrt{3}\,\tau^2/9\right)$.

Example 18-2-5: In a sphere of radius R inscribe a cone having the maximum volume possible. What is this maximum volume (in terms of R)?

Solution

Let ABC be a cone inscribed in the sphere of radius R, as shown in Fig. 18-1. The volume of the cone is given by the formula

$$V = \frac{1}{3}\pi(BM)^2(AM) \qquad (*)$$

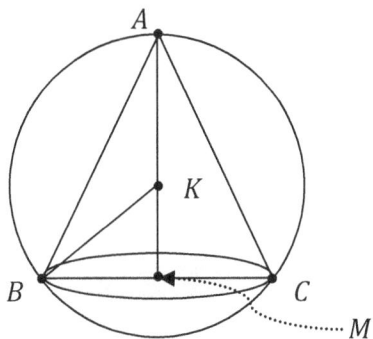

Fig. 18-1: Cone of a maximum volume inscribed in a given sphere.

From Fig. 18-1, we have:

$$(BM)^2 = (BK)^2 - (KM)^2 = R^2 - (KM)^2$$

$$AM = AK + KM = R + KM$$

and equation (*) implies,

$$V = \frac{1}{3}\pi(R^2 - (KM)^2)(R + KM) \qquad (**)$$

and the volume V becomes maximum when the product
$$(R^2 - (KM)^2)(R + KM) = (R - KM)(R + KM)(R + KM)$$
$$= (R - KM)(R + KM)^2$$

becomes maximum. But, since $(R - KM) + (R + KM) = 2R = const.$, Theorem 18-2-7 implies that the product is maximum when

$$\frac{R - KM}{1} = \frac{R + KM}{2} = \frac{2R}{3} \Rightarrow KM = \frac{R}{3}$$

In this case, the maximum volume, from equation (**) is found to be $V_{max} = 32\pi R^3/81$.

Example 18-2-6: If the sum of the positive numbers x, y and z is constant and equal to C, what is the minimum value of $f(x, y, z) = x^2 + y^2 + z^2$?

Solution

If we call $x + y + z = C$, (constant), and square both sides, we find:

$$x^2 + y^2 + z^2 + 2xy + 2yz + 2zx = C^2 \tag{*}$$

and taking into consideration the identities

$$\begin{cases} 2xy = (x^2 + y^2) - (x - y)^2 \\ 2yz = (y^2 + z^2) - (y - z)^2 \\ 2zx = (z^2 + x^2) - (z - x)^2 \end{cases}$$

and substituting in (*), results in the following,

$$3(x^2 + y^2 + z^2) = C^2 + (x - y)^2 + (y - z)^2 + (z - x)^2 \tag{**}$$

Equation (**) shows that the quantity $3(x^2 + y^2 + z^2)$, and hence $(x^2 + y^2 + z^2)$ becomes **minimum** when $(x - y)^2 + (y - z)^2 + (z - x)^2 = 0$, i.e. when $x = y = z = C/3$, (Theorem 4-1-2). The minimum value of $f(x, y, z) = (x^2 + y^2 + z^2)$ is $3x^2 = 3(C/3)^2 = C^2/3$.

Example 18-2-7: If the positive numbers x, y, z satisfy the relation $x^n + y^n = z^n$, where n is a positive integer, show that $\left(\frac{xy}{z^2}\right)^n < \frac{3}{11}$.

Solution

$$x^n + y^n = z^n \Leftrightarrow \frac{x^n + y^n}{z^n} = 1 \Leftrightarrow \left(\frac{x}{z}\right)^n + \left(\frac{y}{z}\right)^n = 1 \tag{*}$$

Since the sum of the numbers $\left(\frac{x}{z}\right)^n$ and $\left(\frac{y}{z}\right)^n$ is 1, (**constant**), their product becomes maximum when the numbers become equal, i.e. when $\left(\frac{x}{z}\right)^n = \left(\frac{y}{z}\right)^n = \frac{1}{2}$, and then $Max\ \left(\frac{x}{z}\right)^n \cdot \left(\frac{y}{z}\right)^n = \frac{1}{4}$, i.e.

$$\left(\frac{xy}{z^2}\right)^n = \left(\frac{x}{z}\right)^n \cdot \left(\frac{y}{z}\right)^n \leq \frac{1}{4} < \frac{3}{11}$$

Example 18-2-8: Assuming that $0 < x < 1$, find the maximum value of $x\sqrt{1-x^4}$.

Solution

Since $x > 0$ and $\sqrt{1-x^4} > 0$, it suffices to find the maximum value of the quantity, $x^4(1-x^4)^2 = \left(x\sqrt{1-x^4}\right)^4$. Since $x^4 + (1-x^4) = 1$, (**constant**), the product $x^4(1-x^4)^2$ becomes maximum when

$$\frac{x^4}{1} = \frac{1-x^4}{2} = \frac{x^4 + (1-x^4)}{1+2} = \frac{1}{3} \Rightarrow x^4 = \frac{1}{3} \Rightarrow x = \frac{1}{\sqrt[4]{3}}$$

The maximum value of $x\sqrt{1-x^4}$ is attained at $x = \frac{1}{\sqrt[4]{3}}$, and is equal to

$$\frac{1}{\sqrt[4]{3}} \cdot \sqrt{1-\frac{1}{3}} = \sqrt[4]{\frac{4}{27}}.$$

Example 18-2-9: If $0 < x < \frac{\pi}{2}$, find the maximum value of the product $(\sin x)^{\frac{2}{3}}(\cos x)^{\frac{4}{5}}$.

Solution

$$(\sin x)^{\frac{2}{3}}(\cos x)^{\frac{4}{5}} = ((\sin x)^2)^{\frac{1}{3}} \cdot ((\cos x)^2)^{\frac{2}{5}} \qquad (*)$$

and since $(\sin x)^2 + (\cos x)^2 = 1$, (**constant**), the product in (*) becomes maximum when

$$\frac{(\sin x)^2}{\frac{1}{3}} = \frac{(\cos x)^2}{\frac{2}{5}} = \frac{(\sin x)^2 + (\cos x)^2}{\frac{1}{3}+\frac{2}{5}} = \frac{1}{\frac{11}{15}} = \frac{15}{11} \Rightarrow$$

$$(\sin x)^2 = \frac{5}{11}, \quad (\cos x)^2 = \frac{6}{11}$$

The maximum value of the product is $\left(\frac{5}{11}\right)^{\frac{1}{3}} \cdot \left(\frac{6}{11}\right)^{\frac{2}{5}}$.

PROBLEMS

18-2-1) Find the maximum value of $4x^2(1 - 2x^2)$, (**Ans: 4/27**).

18-2-2) If $x^2 + y^2 + z^2 = 25$, what is the maximum value of $xy + yz + zx$?

Hint: $x^2 + y^2 \geq 2xy, y^2 + z^2 \geq 2yz, z^2 + x^2 \geq 2zx$, then add term wise, etc. The maximum value is 25, attained when $x = y = z$.

18-2-3) Find the minimum value of $(3^4 + x^4)/x^2$, $(x > 0)$, (**Ans: 18**).

Hint: $\frac{3^4 + x^4}{x^2} = 3^4 x^{-2} + x^2$, notice that $(3^4 x^{-2}) \cdot (x^2) = 3^4$, (**constant**), and apply Theorem 18-2-4.

18-2-4) From all the isosceles triangles inscribed in a circle of radius R, which one bounds the maximum area possible? Show that this maximum area is $3\sqrt{3}R^2/4$.

18-2-5) If the sum of three positive variables x, y, z is 6, find the maximum value of the quantity $x^2 y^3 \sqrt{z}$, (**Ans:** $54\sqrt{2}(12/11)^{11/2}$).

Hint: $x^2 y^3 \sqrt{z} = x^2 y^3 z^{1/2}$, and the product becomes maximum when

$$\frac{x}{2} = \frac{y}{3} = \frac{z}{\frac{1}{2}} = \frac{x + y + z}{2 + 3 + \frac{1}{2}} = \frac{6}{\frac{11}{2}} = \frac{12}{11}$$

18-2-6) If $x^2 + y^2 = C^2$, (constant), show that the product xy becomes maximum when $x = y$, (assume x, y positive).

18-2-7) From all the right circular cylinders with the same total area S, which one bounds the maximum volume? What is this maximum volume?

(**Ans:** $V_{max} = \frac{2}{9}\sqrt{\frac{3}{8\pi}} S^{3/2}$).

18-2-8) Divide the number 40 in two parts such that the product of the fourth power of the first part times the sixth power of the second, to be the maximum possible.

Hint: See Example 18-2-3.

18-2-9) From all the right circular cones with the same lateral area πa^2, (constant), which one bounds the maximum volume?

(**Ans:** Radius of base: $a \sqrt[4]{27}/3$, height: $a\sqrt{2\sqrt{3}/\sqrt{3}}$).

18-2-10) From all the right circular cones with the same volume V, (constant), which one has the least lateral surface?

(**Ans:** Radius of base: $\sqrt[3]{3V/(\pi\sqrt{2})}$, height: $\sqrt[3]{6V/\pi}$).

18-2-11) Provided that $x + y = C$, (constant), show that $x^2 + y^2$ becomes minimum when $x = y$.

18-2-12) From all the right triangles with the same area c^2, (constant), which one has the least perimeter? (**Ans:** The isosceles right triangle).

18-2-13) In a given sphere of radius R inscribe a cylinder bounding the maximum volume possible.

(**Ans:** radius of base: $R\sqrt{6}/3$, height: $2R\sqrt{3}/3$).

18-2-14) Find the maximum of $x - 2y$, provided that $5x^2 + 8y^2 = 10$.

18-2-15) Determine b so that the sum of the squares of the roots of the equation $x^2 + (2 - b)x - b - 3 = 0$ to be minimum, (**Ans:** $b = 1$).

18-2-16) Given $AB = 80\ m$, find a point K such that $AK^2 + 3KB^2$ to be as minimum as possible.

18-2-17) From all the rectangles with the same perimeter 2τ, which one bounds the maximum area?, (**Ans:** A square with side $\tau/2$).

CHAPTER 19: SOME SPECIAL TOPICS ON COMPLEX NUMBERS, POLYNOMIALS AND TRIGONOMETRY

Complex numbers are closely related to trigonometry, as it becomes apparent from the De Moivre's formula

$$\{r(\cos\theta + i\sin\theta)\}^n = r^n(\cos n\theta + i\sin n\theta), \quad n = integer$$

In this chapter we shall use various concepts and Theorems developed in the preceding chapters to derive some useful identities and Theorems, such as **the product of trigonometric functions, Cote's Theorem**, etc. It is worth mentioning, that without the use of complex numbers, most of the Theorems and formulas in this chapter, would be very difficult (if not impossible) to be proved, (see for example Cote's Theorem).

19-1) The Σ and Π notation

Let $x_1, x_2, \cdots, x_n$ be n numbers, **real or complex**. The sum $x_1 + x_2 + \cdots + x_n$ can be expressed briefly, using the Greek symbol Σ, as follows:

$$\sum_{k=1}^{n} x_k = x_1 + x_2 + \cdots + x_n \qquad (19-1-1)$$

The index k is **a dummy variable**, in the sense that

$$\sum_{k=1}^{n} x_k = \sum_{m=1}^{n} x_m = \sum_{i=1}^{n} x_i = x_1 + x_2 + \cdots + x_n$$

The sum in (19-1-1) could also be written as

$$\sum_{k=3}^{n+2} x_{k-2} \quad or \quad \sum_{m=5}^{n+4} x_{m-4}, \quad etc$$

Similarly, the product $x_1 x_2 \cdots x_n$ can be expressed briefly using the Greek symbol Π, as follows:

$$\prod_{k=1}^{n} x_k = x_1 x_2 \cdots x_n \qquad (19-1-2)$$

Obviously, k is **a dummy index** again, since

$$\prod_{k=1}^{n} x_k = \prod_{m=1}^{n} x_m = \prod_{i=1}^{n} x_i = x_1 x_2 \cdots x_n$$

The product in (19-1-2) could also be written as

$$\prod_{k=3}^{n+2} x_{k-2} \quad \text{or} \quad \prod_{m=7}^{n+6} x_{m-6}, \quad \text{etc}$$

It is easily shown that, if c_1 and c_2 are any two constants, then

$$\sum_{k=1}^{n}(c_1 x_k + c_2 y_k) = c_1 \sum_{k=1}^{n} x_k + c_2 \sum_{k=1}^{n} y_k \qquad (19-1-3)$$

Using, for example, the Σ and Π notation, we may express Theorem 8-6-1, (multiplication of complex numbers), in the following compact form:

If $z_1 = r_1(\cos\theta_1 + i\sin\theta_1), z_2 = r_2(\cos\theta_2 + i\sin\theta_2), \cdots, z_n = r_n(\cos\theta_n + i\sin\theta_n)$, then

$$\prod_{k=1}^{n} z_k = \left(\prod_{k=1}^{n} r_k\right)\left\{\cos\left(\sum_{k=1}^{n} \theta_k\right) + i\sin\left(\sum_{k=1}^{n} \theta_k\right)\right\} \qquad (19-1-4)$$

PROBLEMS

19-1-1) Show that $\sum_{k=1}^{n} k = 1 + 2 + \cdots + n = \frac{n(n+1)}{2}$.

19-1-2) Show that $\sum_{k=1}^{n} k^2 = \frac{n(n+1)(2n+1)}{6}$.

19-1-3) Show that $\sum_{m=1}^{n} m^3 = (\sum_{m=1}^{n} m)^2$.

19-1-4) If $c \neq 0$ is a constant number, show that $\prod_{i=1}^{n}(c x_i) = c^n \prod_{i=1}^{n} x_i$.

19-1-5) Show the trigonometric identity:

$$\prod_{n=0}^{k} \cos(2^n x) = \frac{1}{2^{k+1}} \cdot \frac{\sin(2^{k+1} x)}{\sin x}$$

Hint: Use the identity, $\cos x = \dfrac{\sin 2x}{2 \sin x}$.

19-2) Evaluation of trigonometric sums of a special type

Suppose we want to find the following sums:

$$B = a_1 \sin \theta + a_2 \sin 2\theta + \cdots + a_n \sin(n\theta)$$

$$A = a_1 \cos \theta + a_2 \cos 2\theta + \cdots + a_n \cos(n\theta)$$

where $a_1, a_2, \cdots, a_n$ are known **real numbers**, and θ **is a given angle**.

To evaluate these sums, we consider **the complex number $w = A + iB$**, which is

$$w = A + iB = a_1(\cos \theta + i \sin \theta) + a_2(\cos 2\theta + i \sin 2\theta) + \cdots \\ + a_n(\cos(n\theta) + i \sin(n\theta))$$

or, if we set $z = \cos \theta + i \sin \theta$ and make use of De Moivre's formula,

$$w = a_1 z + a_2 z^n + \cdots + a_n z^n \qquad (19-2-1)$$

If the sum in (19-2-1) can be evaluated in **closed form**, say $w = K + iL$, then **equating the real and the imaginary parts** we have, $A = K$ and $B = L$, and we have thus found the sums A and B.

Example 19-2-1: Find the sum $A = \cos\dfrac{2\pi}{9} + \cos\dfrac{4\pi}{9} + \cos\dfrac{6\pi}{9} + \cos\dfrac{8\pi}{9}$.

Solution

Let us call $B = \sin\dfrac{2\pi}{9} + \sin\dfrac{4\pi}{9} + \sin\dfrac{6\pi}{9} + \sin\dfrac{8\pi}{9}$, and consider the sum

$$w = A + iB = \left(\cos\dfrac{2\pi}{9} + i \sin\dfrac{2\pi}{9}\right) + \left(\cos\dfrac{4\pi}{9} + i \sin\dfrac{4\pi}{9}\right) \\ + \left(\cos\dfrac{6\pi}{9} + i \sin\dfrac{6\pi}{9}\right) + \left(\cos\dfrac{8\pi}{9} + i \sin\dfrac{8\pi}{9}\right) \Rightarrow$$

$$w = A + iB = \left(\cos\dfrac{2\pi}{9} + i \sin\dfrac{2\pi}{9}\right) + \left(\cos\dfrac{2\pi}{9} + i \sin\dfrac{2\pi}{9}\right)^2 \\ + \left(\cos\dfrac{2\pi}{9} + i \sin\dfrac{2\pi}{9}\right)^3 + \left(\cos\dfrac{2\pi}{9} + i \sin\dfrac{2\pi}{9}\right)^4$$

or, if we set $z = \cos\frac{2\pi}{9} + i\sin\frac{2\pi}{9}$,

$$w = z + z^2 + z^3 + z^4 = \frac{z \cdot z^4 - z}{z - 1} = \frac{z^5 - z}{z - 1} \quad (*)$$

(since the terms in the sum in (*) form a geometric progression), and finally,

$$w = \frac{\left(\cos\frac{2\pi}{9} + i\sin\frac{2\pi}{9}\right)^5 - \left(\cos\frac{2\pi}{9} + i\sin\frac{2\pi}{9}\right)}{\cos\frac{2\pi}{9} + i\sin\frac{2\pi}{9} - 1} \Rightarrow$$

$$w = \frac{\cos\frac{10\pi}{9} + i\sin\frac{10\pi}{9} - \cos\frac{2\pi}{9} - i\sin\frac{2\pi}{9}}{\left(\cos\frac{2\pi}{9} - 1\right) + i\sin\frac{2\pi}{9}} \Rightarrow$$

$$w = \frac{\left(\cos\frac{10\pi}{9} - \cos\frac{2\pi}{9}\right) + i\left(\sin\frac{10\pi}{9} - \sin\frac{2\pi}{9}\right)}{\left(\cos\frac{2\pi}{9} - 1\right) + i\sin\frac{2\pi}{9}} \quad (**)$$

To find the Cartesian form of w, we multiply numerator and denominator by the complex conjugate of the denominator, and carry out the calculations. We find that $w = -\frac{1}{2} + i\frac{\cot(\pi/18)}{2}$, (let the reader verify the calculations).

We have thus found that

$$w = A + iB = -\frac{1}{2} + i\frac{\cot(\pi/18)}{2} \Rightarrow \begin{cases} A = -\frac{1}{2} \\ B = \frac{\cot(\pi/18)}{2} \end{cases} \Rightarrow$$

$$A = \cos\frac{2\pi}{9} + \cos\frac{4\pi}{9} + \cos\frac{6\pi}{9} + \cos\frac{8\pi}{9} = -\frac{1}{2}$$

$$B = \sin\frac{2\pi}{9} + \sin\frac{4\pi}{9} + \sin\frac{6\pi}{9} + \sin\frac{8\pi}{9} = \frac{\cot(\pi/18)}{2}$$

Remark: Using complex numbers, we evaluate **simultaneously** the sum of the cosines and the sum of sines. This is a typical situation when evaluating sums with complex numbers.

Example 19-2-2: Find the sum of the infinite number of terms:

$$A = 1 + \frac{\cos\theta}{2} + \frac{\cos 2\theta}{2^2} + \frac{\cos 3\theta}{2^3} + \cdots + \frac{\cos(n\theta)}{2^n} + \cdots$$

$$B = \frac{\sin\theta}{2} + \frac{\sin 2\theta}{2^2} + \frac{\sin 3\theta}{2^3} + \cdots + \frac{\sin(n\theta)}{2^n} + \cdots$$

Such infinite sums are termed "**infinite series**" and play an extremely important role in Mathematics. The sum of the infinite number of terms of a decreasing geometric progression, (formula 16-3-7), is an elementary example of "**a infinite converging series**". Note that formula (16-3-7) is still valid, **for complex numbers z, provided that $|z| < 1$**, i.e.

$$\sum_{n=0}^{\infty} z^n = 1 + z + z^2 + z^3 + \cdots + z^n + \cdots = \frac{1}{1-z}, \quad if \ |z| < 1 \quad (*)$$

Solution

Let us form the complex number

$$w = A + iB = 1 + \frac{\cos\theta + i\sin\theta}{2} + \frac{\cos 2\theta + i\sin 2\theta}{2^2} + \frac{\cos 3\theta + i\sin 3\theta}{2^3} + \cdots + \frac{\cos(n\theta) + i\sin(n\theta)}{2^n} + \cdots$$

or, if we set $z = \cos\theta + i\sin\theta$, and make use of De Moivre's Theorem,

$$w = A + iB = 1 + \frac{z}{2} + \frac{z^2}{2^2} + \frac{z^3}{2^3} + \cdots + \frac{z^n}{2^n} + \cdots = \sum_{n=0}^{\infty} \left(\frac{z}{2}\right)^n = \frac{1}{1 - \frac{z}{2}} = \frac{2}{2-z}$$

since $\left|\frac{z}{2}\right| = \frac{|\cos\theta + i\sin\theta|}{2} = \frac{1}{2} < 1$. We now express the number $\left(\frac{2}{2-z}\right)$ in Cartesian form, as follows:

$$\frac{2}{2-z} = \frac{2}{(2-\cos\theta) - i\sin\theta} = \frac{2\{(2-\cos\theta) + i\sin\theta\}}{(2-\cos\theta)^2 + (\sin\theta)^2} \Rightarrow$$

$$\frac{2}{2-z} = \frac{4 - 2\cos\theta}{5 - 4\cos\theta} + i\frac{2\sin\theta}{5 - 4\cos\theta}$$

and finally we obtain,

$$w = A + iB = \frac{4 - 2\cos\theta}{5 - 4\cos\theta} + i\frac{2\sin\theta}{5 - 4\cos\theta} \Rightarrow$$

$$A = 1 + \frac{\cos\theta}{2} + \frac{\cos 2\theta}{2^2} + \frac{\cos 3\theta}{2^3} + \cdots + \frac{\cos(n\theta)}{2^n} + \cdots = \frac{4 - 2\cos\theta}{5 - 4\cos\theta}$$

$$B = \frac{\sin\theta}{2} + \frac{\sin 2\theta}{2^2} + \frac{\sin 3\theta}{2^3} + \cdots + \frac{\sin(n\theta)}{2^n} + \cdots = \frac{2\sin\theta}{5 - 4\cos\theta}$$

PROBLEMS

19-2-1) If n is a positive integer and $x = \frac{2\pi}{n}$, show the formulas:

$$\sum_{k=1}^{n}(2k-1)\cos\{(k-1)x\} = -n, \quad \sum_{k=2}^{n}(2k-1)\sin\{(k-1)x\} = -n\cot\frac{x}{2}$$

19-2-2) Show that $\sum_{k=0}^{n-1}\cos\left(x + \frac{2\pi k}{n}\right) = 0$, where n is a positive integer and x an arbitrary angle.

19-2-3) Find the following finite sums:

$$A = \sum_{k=1}^{n} x^{k-1}\cos((k-1)\theta), \quad B = \sum_{k=1}^{n} x^{k-1}\sin((k-1)\theta)$$

(Ans:
$$A = \frac{1 - x\cos\theta - x^n\cos(n\theta) + x^{n+1}\cos((n-1)\theta)}{1 - 2x\cos\theta + x^2}$$

$$B = \frac{x\sin\theta - x^n\sin(n\theta) + x^{n+1}\sin((n-1)\theta)}{1 - 2x\cos\theta + x^2}$$

19-2-4) Assuming that $|x| < 1$, show that:

$$\sum_{k=1}^{\infty} x^{k-1}\cos((k-1)\theta) = \frac{1 - x\cos\theta}{1 - 2x\cos\theta + x^2}$$

$$\sum_{k=1}^{\infty} x^{k-1}\sin((k-1)\theta) = \frac{x\sin\theta}{1 - 2x\cos\theta + x^2}$$

19-2-5) Show that:

$$\sum_{k=1}^{n}(\cos\theta)^k \sin(k\theta) = \cot\theta\,\{1 - \cos(n\theta)(\cos\theta)^n\}$$

$$\sum_{k=1}^{n}(\cos\theta)^k \cos(k\theta) = \cot\theta\,(\cos\theta)^n \sin(n\theta)$$

Hint: Consider the sum,

$$\sum_{k=1}^{n}(\cos\theta)^k\{\cos(k\theta) + i\sin(k\theta)\} = \sum_{k=1}^{n}(\cos\theta)^k(\cos\theta + i\sin\theta)^k$$

$$= \sum_{k=1}^{n}(z\cos\theta)^k$$

where $z = \cos\theta + i\sin\theta$, then find the sum of the geometric series, etc.

19-3) Factorization of the polynomials $x^n - 1$ and $x^n + 1$

We know that if $r_1, r_2, \cdots, r_n$ are the roots of the polynomial $a_n x^n + a_{n-1} x^{n-1} + \cdots + a_1 x + a_0$, then:

$$a_n x^n + a_{n-1} x^{n-1} + \cdots + a_1 x + a_0 \equiv a_n(x - r_1)(x - r_2)\cdots(x - r_n) \qquad (*)$$

The roots of $x^n - 1$ are **the nth roots of unity**, i.e. (see formula (8-7-2)),

$$r_k = \cos\left(k\frac{2\pi}{n}\right) + i\sin\left(k\frac{2\pi}{n}\right), \quad k = 0,1,2,\ldots,n-1 \qquad (**)$$

Similarly the roots of $x^n + 1$ are **the nth roots of** $-1 = 1\cdot(\cos\pi + i\sin\pi)$ and are given by the formula,

$$r_k = \cos\left(\frac{\pi}{n} + k\frac{2\pi}{n}\right) + i\sin\left(\frac{\pi}{n} + k\frac{2\pi}{n}\right), \quad k = 0,1,2,\ldots,n-1 \qquad (***)$$

Knowing the roots and using (*) we may **factor the polynomials $x^n - 1$ and $x^n + 1$**. We shall consider separately the cases where n is even or odd. The result of the factorization is shown in formulas (19-3-1) and (19-3-2).

$$\begin{cases} x^{2n} - 1 \equiv (x-1)(x+1)\prod_{k=1}^{n-1}\left(x^2 - 2x\cos\dfrac{k\pi}{n} + 1\right) \\ \\ x^{2n+1} - 1 \equiv (x-1)\prod_{k=1}^{n}\left(x^2 - 2x\cos\dfrac{2k\pi}{2n+1} + 1\right) \end{cases} \quad (19-3-1)$$

$$\begin{cases} x^{2n} + 1 \equiv \prod_{k=1}^{n}\left(x^2 - 2x\cos\dfrac{(2k-1)\pi}{2n} + 1\right) \\ \\ x^{2n+1} + 1 \equiv (x+1)\prod_{k=1}^{n}\left(x^2 - 2x\cos\dfrac{(2k-1)\pi}{2n+1} + 1\right) \end{cases} \quad (19-3-2)$$

Let us, for example, show the second identity in (19-3-1). The other equations are proved similarly.

According to equation (*), we have, (the coefficient $a_{2n+1} = 1$):

$$x^{2n+1} - 1 \equiv (x - r_0)(x - r_1)\cdots(x - r_{2n-1})(x - r_{2n})$$

where r_k are the $(2n+1)^{th}$ roots of unity, given by the formulas

$$r_k = \cos\left(k\frac{2\pi}{2n+1}\right) + i\sin\left(k\frac{2\pi}{2n+1}\right), \quad k = 0,1,2,\ldots,2n$$

Obviously, $r_0 = 1$, the **only real root of $x^{2n+1} - 1 = 0$**, and therefore,

$$x^{2n+1} - 1 \equiv (x-1)\underbrace{(x-r_1)(x-r_2)\cdots(x-r_{2n-1})(x-r_{2n})}_{2n\ terms} \quad (****)$$

Let us consider the product of the terms $(x - r_1)$ and $(x - r_{2n})$. We have,

$$(x - r_1)(x - r_{2n}) =$$

$$\left(x - \cos\frac{2\pi}{2n+1} - i\sin\frac{2\pi}{2n+1}\right)\left(x - \cos\frac{(2n)2\pi}{2n+1} - i\sin\frac{(2n)2\pi}{2n+1}\right)$$

We notice that the sum of the angles $\left(\frac{2\pi}{2n+1}\right)$ and $\left(\frac{(2n)2\pi}{2n+1}\right)$ is equal to 2π,

i.e. $\left(\frac{2\pi}{2n+1}\right) + \left(\frac{(2n)2\pi}{2n+1}\right) = \frac{(2n+1)2\pi}{2n+1} = 2\pi$, and therefore $\cos\left(\frac{(2n)2\pi}{2n+1}\right) = \cos\left(\frac{2\pi}{2n+1}\right)$ and $\sin\left(\frac{(2n)2\pi}{2n+1}\right) = -\sin\left(\frac{2\pi}{2n+1}\right)$, and this results in the following:

$$(x - r_1)(x - r_{2n}) =$$

$$\left(x - \cos\frac{2\pi}{2n+1} - i\sin\frac{2\pi}{2n+1}\right)\left(x - \cos\frac{2\pi}{2n+1} + i\sin\frac{2\pi}{2n+1}\right) =$$

$$\left(x - \cos\frac{2\pi}{2n+1}\right)^2 + \left(\sin\frac{2\pi}{2n+1}\right)^2 = x^2 - 2x\cos\frac{2\pi}{2n+1} + 1$$

Similarly, if we consider the products
$(x - r_2)(x - r_{2n-1})$, $(x - r_3)(x - r_{2n-2})$, ... , $(x - r_n)(x - r_{n+1})$

we find:

$$(x - r_2)(x - r_{2n-1}) = x^2 - 2x\cos\frac{2(2\pi)}{2n+1} + 1$$

$$(x - r_3)(x - r_{2n-2}) = x^2 - 2x\cos\frac{3(2\pi)}{2n+1} + 1$$

$$\dots \quad \dots \quad \dots$$

$$(x - r_n)(x - r_{n+1}) = x^2 - 2x\cos\frac{n(2\pi)}{2n+1} + 1$$

and substituting these expressions in (****) we obtain:

$$x^{2n+1} - 1 = (x - 1)\left(x^2 - 2x\cos\frac{2\pi}{2n+1} + 1\right)\left(x^2 - 2x\cos\frac{2(2\pi)}{2n+1} + 1\right)$$
$$+1\right)\left(x^2 - 2x\cos\frac{3(2\pi)}{2n+1} + 1\right)\cdots\left(x^2 - 2x\cos\frac{n(2\pi)}{2n+1} + 1\right)$$

or, using the Π notation,

$$x^{2n+1} - 1 \equiv (x - 1)\prod_{k=1}^{n}\left(x^2 - 2x\cos\frac{k(2\pi)}{2n+1} + 1\right)$$

and this completes the proof.

Identities (19-3-1) lead to the following identities:

$$\begin{cases} x^{2n} + x^{2n-1} + \cdots + x + 1 \equiv \prod_{k=1}^{n} \left(x^2 - 2x \cos \frac{2k\pi}{2n+1} + 1 \right) \\ x^{2n-1} + x^{2n-2} + \cdots + x + 1 \equiv (x+1) \prod_{k=1}^{n-1} \left(x^2 - 2x \cos \frac{k\pi}{n} + 1 \right) \end{cases}$$

$$(19-3-3)$$

The first identity in (19-3-3) results from the second equation in (19-3-1) if we divided both sides by $(x - 1)$, assuming that $x \neq 1$, i.e. it is valid for an infinite number of values of x, i.e. **is an identity for x, and therefore it is valid for $x = 1$ as well**. Similarly, the second identity in (19-3-3) results from the first equation in (19-3-1).

We now consider some interesting applications of the identities just derived.

Example 19-3-1: If n is any positive integer ≥ 2, show that

$$P(n) = \sin\frac{\pi}{n} \sin\frac{2\pi}{n} \sin\frac{3\pi}{n} \cdots \sin\frac{(n-2)\pi}{n} \sin\frac{(n-1)\pi}{n} = \frac{n}{2^{n-1}}$$

Solution

a) We shall first show that $P(2n + 1) = \frac{2n+1}{2^{2n}}$, (i.e. we shall prove the proposition for **odd** values of n, ($n = 3, 5, 7, \ldots$)).

The first identity in (19-3-3), when applied for $x = 1$, yields:

$$2n + 1 = \prod_{k=1}^{n} \left(1^2 - 2 \cdot 1 \cdot \cos \frac{2k\pi}{2n+1} + 1 \right) = \prod_{k=1}^{n} 2 \left(1 - \cos \frac{2k\pi}{2n+1} \right)$$

and if we apply the trigonometric identity $1 - \cos 2x = 2(\sin x)^2$, we obtain:

$$2n + 1 = \prod_{k=1}^{n} 2 \left(1 - \cos \frac{2k\pi}{2n+1} \right) = \prod_{k=1}^{n} 2^2 \left(\sin \frac{k\pi}{2n+1} \right)^2 \Rightarrow$$

$$2n+1 = 2^{2n} \prod_{k=1}^{n} \left(\sin \frac{k\pi}{2n+1} \right)^2 \qquad (*)$$

We notice that since $\frac{\pi}{2n+1} + \frac{(2n)\pi}{2n+1} = \pi$, $\sin\frac{\pi}{2n+1} = \sin\frac{(2n)\pi}{2n+1}$, and therefore, $\left(\sin\frac{\pi}{2n+1}\right)^2 = \sin\frac{\pi}{2n+1}\sin\frac{(2n)\pi}{2n+1}$, and similarly, since $\frac{2\pi}{2n+1} + \frac{(2n-1)\pi}{2n+1} = \pi$, $\sin\frac{2\pi}{2n+1} = \sin\frac{(2n-1)\pi}{2n+1}$, and therefore $\left(\sin\frac{2\pi}{2n+1}\right)^2 = \sin\frac{2\pi}{2n+1}\sin\frac{(2n-1)\pi}{2n+1}$, etc, and equation (*) yields,

$$2n+1 = 2^{2n} \prod_{k=1}^{2n} \sin\frac{k\pi}{2n+1} \Rightarrow$$

$$P(2n+1) = \prod_{k=1}^{2n} \sin\frac{k\pi}{2n+1} = \frac{2n+1}{2^{2n}}$$

and this proves the proposition for n odd.

b) If n is **even**, ($n = 2,4,6, \ldots$), then we apply the second identity in (19-3-3) for $x = 1$, and working similarly, we obtain the desired result.

Example 19-3-2: If $A_1 A_2 \cdots A_n$ is a regular polygon with n sides, inscribed in a circle of radius R, show that the product $(A_1 A_2) \cdot (A_1 A_3) \cdot \cdots \cdot (A_1 A_n) = nR^{n-1}$.

Solution

We set up the axes Ox and Oy, so that the vertex A_1 lies on the Ox axis, as shown in Fig. 19-1. The vertices of the regular polygon with n sides, on the Argand's plane, (the complex plane), correspond to the complex numbers,

$$A_1 \to z_0 = R(\cos 0 + i \sin 0) = R$$

$$A_2 \to z_1 = R\left(\cos\frac{2\pi}{n} + i\sin\frac{2\pi}{n}\right)$$

$$A_3 \to z_2 = R\left(\cos 2\cdot\frac{2\pi}{n} + i\sin 2\cdot\frac{2\pi}{n}\right)$$

... ...

$$A_n \to z_{n-1} = R\left(\cos(n-1)\cdot\frac{2\pi}{n} + i\sin(n-1)\cdot\frac{2\pi}{n}\right)$$

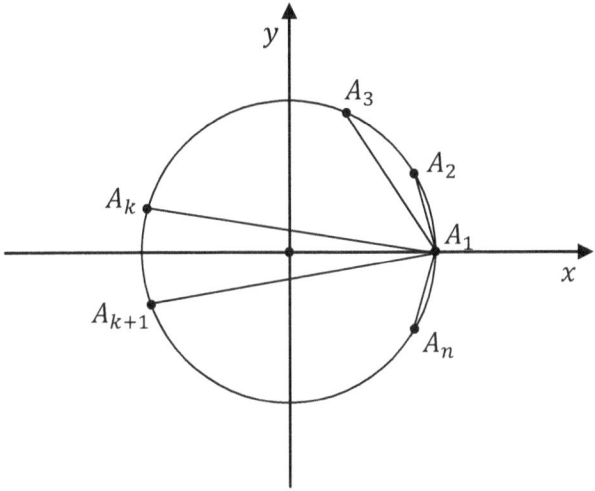

Fig. 19-1: Regular polygon inscribed in a circle of radius R.

We notice that the complex numbers $z_0(A_1), z_1(A_2), \ldots, z_{n-1}(A_n)$ are actually **the roots of the equation $z^n - R^n = 0$**. In terms of its roots, we may write,

$$z^n - R^n \equiv (z - z_0)(z - z_1)\ldots(z - z_{n-1}) \quad (*)$$

The side $A_1 A_2$ is the magnitude of the complex number $(z_1 - z_0)$, i.e. $A_1 A_2 = |z_1 - z_0| = |z_1 - R|$, (since $z_0 = R$), and similarly, $A_1 A_3 = |z_2 - R|$, etc. Since the two complex numbers in (*) are equal, their magnitudes will be equal as well, i.e.

$$|z^n - R^n| = |(z - z_0)(z - z_1)\ldots(z - z_{n-1})| \overset{(z_0 = R)}{\Longrightarrow}$$

$$|z^n - R^n| = |z - R||z - z_1||z - z_2|\cdots|z - z_{n-1}| \Rightarrow$$

$$\left|\frac{z^n - R^n}{z - R}\right| = |z - z_1||z - z_2|\cdots|z - z_{n-1}| \Rightarrow$$

$$|z^{n-1} + z^{n-2}R + z^{n-3}R^2 + \cdots + R^{n-1}| = |z - z_1||z - z_2|\cdots|z - z_{n-1}| \quad (**)$$

The equality in (**) is **an identity**, i.e. is true for all values of z, and if we apply it for $z = R$, we obtain,

$$nR^{n-1} = |R - z_1||R - z_2| \cdots |R - z_{n-1}| = (A_1A_2) \cdot (A_1A_3) \cdot \cdots \cdot (A_1A_n)$$

and this completes the proof.

PROBLEMS

19-3-1) Show that

$$\sin 1° \sin 2° \sin 3° \cdots \sin 89° \sin 90° = \frac{\sqrt{45}}{\left(\sqrt{2}\right)^{177}}$$

Hint: Use Example 19-3-1 with $n = 180$, noting that $\sin 1° = \sin 179°$, $\sin 2° = \sin 178°$, etc.

19-3-2) Using the factorization of $x^{2n} + 1$, as shown in (19-3-2), show that

$$\prod_{k=1}^{n} \cos \frac{(2k-1)\pi}{4n} = \frac{\sqrt{2}}{2^n}$$

19-3-3) Show that

$$\prod_{k=1}^{n} \sin \frac{k\pi}{2n+1} = \frac{\sqrt{2n+1}}{2^n}$$

Hint: Apply the first identity in (19-3-3), for $x = 1$.

19-3-4) Show that the roots of the equation $(x+1)^n = (x-1)^n$, (n is a positive integer), are: $x_k = i \cot(k\pi/n)$, $k = 1,2,3,\ldots,(n-1)$.

Hint: Since neither $x = -1$ nor $x = 1$ are roots, the given equation is equivalent to $\left(\frac{x-1}{x+1}\right)^n = 1$, or if we set $y = \frac{x-1}{x+1}$, $y^n = 1$, i.e. y is the nth root of unity, find y and then find x.

19-3-5) Show the formulas:

$$\prod_{k=1}^{n-1} \cos\frac{k\pi}{2n} = \frac{\sqrt{n}}{2^{n-1}}, \quad \prod_{k=1}^{n-1} \sin\frac{k\pi}{2n} = \frac{\sqrt{n}}{2^{n-1}}$$

19-4) Factorization of the polynomial $x^{2n} - 2a^n \cos(n\theta) x^n + a^{2n}$, where $a > 0$ and $n = 1, 2, 3, ...$

a) The polynomial $x^{2n} - 2a^n \cos(n\theta) x^n + a^{2n}$ is a polynomial in x of degree $2n$. If we set $y = x^n$, it becomes $y^2 - 2a^n \cos(n\theta) y + a^{2n}$, which is a **quadratic trinomial in y**. The roots of the trinomial are:

$$y_{1,2} = \frac{2a^n \cos(n\theta) \pm \sqrt{4a^{2n}(\cos(n\theta))^2 - 4a^{2n}}}{2} \Rightarrow$$

$$y_{1,2} = \frac{2a^n \cos(n\theta) \pm i2a^n \sin(n\theta)}{2} \Rightarrow \begin{cases} y_1 = a^n(\cos(n\theta) + i\sin(n\theta)) \\ y_2 = a^n(\cos(n\theta) - i\sin(n\theta)) \end{cases}$$

The values of x corresponding to y_1 result from the equation $x^n = y_1$, i.e. x is the nth root of y_1, i.e.

$$x_{1k} = \sqrt[n]{y_1} = a\left\{\cos\left(\theta + k\frac{2\pi}{n}\right) + i\sin\left(\theta + k\frac{2\pi}{n}\right)\right\}, k = 0, 1, ..., (n-1) \quad (*)$$

and similarly, from $x^n = y_2$, we find,

$$x_{2\lambda} = \sqrt[n]{y_2} = a\left\{\cos\left(\theta + \lambda\frac{2\pi}{n}\right) - i\sin\left(\theta + \lambda\frac{2\pi}{n}\right)\right\}, \lambda = 0, 1, ..., (n-1) \quad (**)$$

Expressing the polynomial **in terms of its roots**, we obtain the following expression,

$$x^{2n} - 2a^n \cos(n\theta) x^n + a^{2n}$$
$$\equiv (x - x_{10})(x - x_{20}) \underbrace{(x - x_{11})(x - x_{21})} \cdots \underbrace{(x - x_{1(n-1)})(x - x_{2(n-1)})}$$

We note that

$$(x - x_{10})(x - x_{20}) = (x - a(\cos\theta + i\sin\theta))(x - a(\cos\theta - i\sin\theta)) \Rightarrow$$

$$(x - x_{10})(x - x_{20}) = (x - a\cos\theta)^2 + (a\sin\theta)^2 = x^2 - 2ax\cos\theta + a^2$$

and similarly we find,

$$(x - x_{11})(x - x_{21}) = x^2 - 2ax \cos\left(\theta + \frac{2\pi}{n}\right) + a^2$$

$$\cdots \quad \cdots \quad \cdots$$

$$(x - x_{1(n-1)})(x - x_{2(n-1)}) = x^2 - 2ax \cos\left(\theta + (n-1)\frac{2\pi}{n}\right) + a^2$$

and finally we obtain the following expression for the polynomial $x^{2n} - 2a^n \cos(n\theta) x^n + a^{2n}$,

$$x^{2n} - 2a^n \cos(n\theta) x^n + a^{2n} \equiv \prod_{k=0}^{n-1} \left\{ x^2 - 2ax \cos\left(\theta + k\frac{2\pi}{n}\right) + a^2 \right\}$$

$$(19 - 4 - 1)$$

In particular, if we apply (19-4-1) for $a = 1$, we obtain the identity:

$$x^{2n} - 2\cos(n\theta) x^n + 1 \equiv \prod_{k=0}^{n-1} \left\{ x^2 - 2x \cos\left(\theta + k\frac{2\pi}{n}\right) + 1 \right\} \quad (19 - 4 - 2)$$

b) The distance between two points on the complex plane:

Theorem 19-4-1: **Provided that $x > 0$, the quantity $x^2 - 2x \cos \theta + 1$ expresses (in the complex plane), the square of the distance between the complex numbers $z_0 = x(\cos 0 + i \sin 0)$ and $z = \cos \theta + i \sin \theta$.**

Proof: Indeed, the square of the distance d^2 between the complex numbers z_0 and z, is given by the formula,

$$d^2 = |z - z_0|^2 = |(\cos \theta - x) + i \sin \theta|^2 = (\cos \theta - x)^2 + (\sin \theta)^2 \Rightarrow$$

$d^2 = x^2 - 2x \cos \theta + 1$, and this completes the proof.

Example 19-4-1: Let $A_0 A_1 \ldots A_{n-1}$ be a regular polygon with n sides inscribed in a unit circle ($R = 1$), and let $M(z)$ be a point in the complex plane corresponding to the complex number $z = x(\cos 0 + i \sin 0), x > 0$. Evaluate the product $(MA_0)^2 (MA_1)^2 \cdots (MA_{n-1})^2$.

Solution

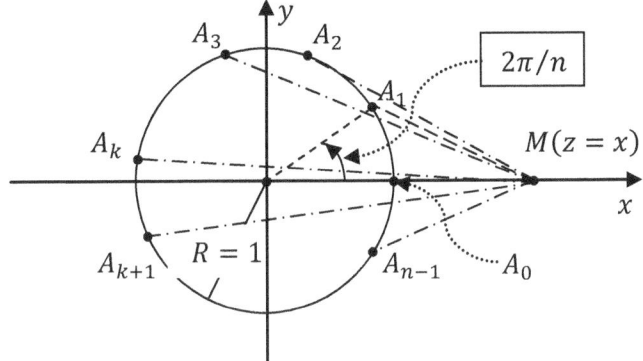

Fig. 19-2: Regular polygon inscribed in a unit circle.

The vertex A_0 corresponds to the complex number $z_0 = 1(\cos 0 + i \sin 0)$, the vertex A_1 corresponds to the complex number $z_1 = 1 \cdot \left(\cos \frac{2\pi}{n} + i \sin \frac{2\pi}{n}\right)$, the vertex A_2 corresponds to $z_2 = 1 \cdot \left(\cos \frac{2 \cdot 2\pi}{n} + i \sin \frac{2 \cdot 2\pi}{n}\right)$,..., and the vertex A_{n-1} corresponds to $z_{n-1} = 1 \cdot \left(\cos \frac{(n-1)2\pi}{n} + i \sin \frac{(n-1)\pi}{n}\right)$.

By virtue of Theorem 19-4-1,

$$(MA_k)^2 = x^2 - 2x \cos\left(k \cdot \frac{2\pi}{n}\right) + 1, \qquad k = 0,1,2,\ldots,n-1 \qquad (*)$$

and the sought for product is:

$$(MA_0)^2 (MA_1)^2 \cdots (MA_{n-1})^2 = \prod_{k=0}^{n-1}\left\{x^2 - 2x \cos\left(k \cdot \frac{2\pi}{n}\right) + 1\right\}$$
$$= x^{2n} - 2x^n + 1 = (x^n - 1)^2$$

(by virtue of formula (19-4-2), with $\theta = 0$).

PROBLEMS

19-4-1) If n is a natural number, show the formula:

$$\sin n\theta = 2^{n-1} \prod_{k=0}^{n-1} \sin\left(\theta + \frac{k\pi}{n}\right)$$

Hint: Apply identity (19-4-2) for $x = 1$.

19-4-2) Show the formula, (n is a natural number):

$$\cos(n\phi) - \cos(n\theta) = 2^{n-1} \prod_{k=0}^{n-1} \left\{ \cos\phi - \cos\left(\theta + k \cdot \frac{2\pi}{n}\right) \right\}$$

Hint: Show that (19-4-2) can be written, equivalently, as

$$x^n - 2\cos(n\theta) + x^{-n} \equiv \prod_{k=0}^{n-1} \left\{ x - 2x\cos\left(\theta + k\frac{2\pi}{n}\right) + x^{-1} \right\}$$

and then apply this identity for $x = \cos\phi + i\sin\phi$. Note that $x + x^{-1} = 2\cos\phi$ and $x^n + x^{-n} = 2\cos(n\phi)$, etc.

19-4-3) Let M be a point on the circumference of a circle of radius R and center O. We call (d) the diameter determined by the points O and M. We consider the chords passing through M and forming angles $\frac{\pi}{2n}, \frac{2\pi}{2n}, \ldots, \frac{(n-1)\pi}{2n}$ with the diameter (d). Show that the product of these chords is equal to $R^{n-1}\sqrt{n}$.

Hint: The lengths of the cords are: $2R\cos\frac{\pi}{2n}, 2R\cos\frac{2\pi}{2n}, \ldots, 2R\cos\frac{(n-1)\pi}{2n}$. The product of the cosines has been evaluated in Problem 19-3-5.

19-4-4) Show that the quantity

$$Q = \prod_{k=0}^{n-1} \left(\sin\left(\theta + k\frac{\pi}{n}\right) \right)^2 + \prod_{k=0}^{n-1} \left(\cos\left(\theta + k\frac{\pi}{n}\right) \right)^2$$

is equal to $2^{3-2n}(\sin n\theta)^2$ if n is even or 2^{2-2n} if n is odd.

19-4-5) For the readers familiar with the theory of limits, the formula obtained in Problem 19-4-1, can be used to derive the expression for the product of sines, obtained in Example 19-3-1. Divide both sides by $\sin\theta$ and then pass to the limit as $\theta \to 0$, (recall that $\lim_{\theta \to 0}\left(\frac{\sin n\theta}{\theta}\right) = \lim_{\theta \to 0}\left(\frac{\sin n\theta}{n\theta}n\right) = 1 \cdot n = n$), etc.

19-5) Cote's Theorem

Cote's theorem is actually a slight modification of the problem studied in Example 19-4-1.

Theorem 19-5-1, (Cote's Theorem): Let $A_0A_1 \ldots A_{n-1}$ be a regular polygon with n sides inscribed in a circle of radius R, and let $M(z)$ be a point in the complex plane corresponding to the complex number
$z = x(\cos 0 + i \sin 0), x > 0$. Then, the product $(MA_0)(MA_1) \cdots (MA_{n-1})$ is given by the formula:

$$(MA_0)(MA_1) \cdots (MA_{n-1}) = \begin{cases} x^n - R^n & \text{if } M \text{ lies out of the circle} \\ \text{or} \\ R^n - x^n & \text{if } M \text{ lies inside the circle} \end{cases}$$

Proof: The proof is identical to the proof in Example 19-4-1. We easily find that

$$(MA_0)^2(MA_1)^2 \cdots (MA_{n-1})^2 = (R^n - x^n)^2 \Rightarrow$$

$$(MA_0)(MA_1) \cdots (MA_{n-1}) = |R^n - x^n| \Rightarrow$$

$$(MA_0)(MA_1) \cdots (MA_{n-1}) = \begin{cases} x^n - R^n & \text{if } x > R \ (M \text{ outside}) \\ \text{or} \\ R^n - x^n & \text{if } x < R \ (M \text{ inside}) \end{cases}$$

Remark: For small values of n, say $n = 4$, the inscribed regular polygon is a square, and a geometric proof of the Theorem, could, perhaps, be given. But can you imagine a geometric proof for large values of n, say $n = 1695$ or $n = 7638$? I think that it would be extremely difficult, if not impossible. However, using complex numbers, the proof follows quite easily.

PROBLEMS

19-5-1) Let $A_0A_1A_2 \cdots A_{n-1}$ be a regular polygon with n sides inscribed in a circle of radius $R = 1$, and let $M(z)$ be a point in the complex plane corresponding to the complex number $z = x(\cos 0 + i \sin 0), x > 0$. If the vertex A_0 corresponds to the complex number $1 \cdot (\cos \theta + i \sin \theta)$, show that

$$(MA_0)^2(MA_1)^2 \cdots (MA_{n-1})^2 = x^{2n} - 2x^n \cos(n\theta) + 1$$

Note that for $\theta = 0$, we obtain the formula derived in Example 19-4-1.

19-5-2) Assuming that $x = 2/3$ and $\theta = 20°$ and $n = 195$, find the product $(MA_0)(MA_1)\cdots(MA_{n-1})$. Repeat, assuming now that $x = 3, \theta = 10°$ and $n = 80$, (in both cases the regular polygon is inscribed in a unit circle).

Hint: Apply the result of Problem 19-5-1.

19-5-3) Show that the distance AB between the two complex numbers $A(z_1)$ and $B(z_2)$, where $z_1 = a(\cos\phi + i\sin\phi), a > 0$ and $z_2 = b(\cos\phi + i\sin\phi), b > 0$, is given by the formula,
$(AB)^2 = a^2 + b^2 - 2ab\cos(\phi - \theta)$.

Hint: $(AB)^2 = |z_1 - z_2|^2$.

19-5-4) A regular polygon $A_0A_1\ldots A_{n-1}$ with n sides, is inscribed in a circle, center at O and radius R. If M is an arbitrary point inside the circle, and call $OM = d$, show that the sum of the squares of the distances of the point M from the vertices of the polygon is:
$$(MA_0)^2 + (MA_1)^2 + (MA_2)^2 + \cdots + (MA_{n-1})^2 = n(d^2 + R^2)$$

Hint: Use the formulas derived in Problem 19-5-3 and Problem 19-2-2.

19-5-5) A regular polygon $A_0A_1\ldots A_{n-1}$ with n sides is inscribed in a circle of radius $R = 1$. Show that the product of the diagonals $(A_0A_2), (A_0A_3), (A_0A_4), \cdots, (A_0A_{n-2})$ is $n/\{4(\sin(\pi/n))^2\}$.

SUPPLEMENTARY PROBLEMS

1) Simplify the expression: $A = \dfrac{|x|+1}{x^2-1} - \dfrac{x^2-|x|}{x^2-2|x|+1}$, (**Ans:** $A = -1$).

2) If $x < y < z < w$ show that $|z - y| < |w - x|$.

3) If n is an odd positive integer, show the equivalency: $x^n < y^n \Leftrightarrow x < y$.

4) The product of n positive numbers $x_1, x_2, \ldots, x_n$ is equal to 1. Show that $(1 + x_1)(1 + x_2)\cdots(1 + x_n) \geq 2^n$.

Hint: $1 + x_1 \geq 2\sqrt{x_1}$, etc.

5) Find the integers x and y, given that $2^{2x-y-3} \cdot 7^{x+y-6} = 28^{2y-x+1}$.

(**Ans:** $x = 5, y = 3$).

6) Given that $x - \dfrac{1}{x} = 5$, find the value of the expressions:

$$x^2 + \frac{1}{x^2}, \quad x^3 - \frac{1}{x^3}, \quad x^4 + \frac{1}{x^4}$$

7) Determine the parameter k, so that the following homogeneous system has a solution other than the trivial none ($x = 0, y = 0, z = 0$), (**Ans:** $k = 4$).

$$\{x + 2y + 3z = 0, \quad kx + 5y + 6z = 0, \quad 5x + 7y + 9z = 0\}$$

Hint: The determinant of the coefficients must be zero.

8) Show the inequalities: $\sqrt{11} + \sqrt{5} < \sqrt{10} + \sqrt{8}$, $\sqrt[4]{5} < (\sqrt{2} + \sqrt{3})/2$.

9) Given that the real numbers x and y satisfy the relation $y(1 + |x|)(1 - |x| + x^2) = x^3$, show that $x^3(|y| - 1) + y = 0$.

10) Provided that the real numbers a and b satisfy the inequality $||ab| - 3a| < |a - b| - 4$, show that $(1 - |a|)(4 - |b|) < 0$.

Hint: $|ab| - 3|a| \leq ||ab| - 3a| < |a - b| - 4 \leq |a| + |b| - 4$, etc.

11) If $a_1, a_2, \ldots, a_n$ are real numbers, show that

$$\frac{|a_1| + |a_2| + \cdots + |a_n|}{n} \leq \sqrt{\frac{a_1^2 + a_2^2 + \cdots + a_n^2}{n}}$$

Hint: Apply Schwarz's inequality (see my book, College Algebra, Vol. 1).

12) For what values of the parameter k, the quadratic equation $x^2 - 6x + 3k = 0$, has complex roots?

13) For what values of k the equation $x^2 - (k+2)x + k + 5 = 0$ has a double real root?   (**Ans:** $k = \pm 4$).

14) Show that for all real values of a, b, c, the roots of the equation $(a - b + c)x^2 + 4(a - b)x - (b + c - a) = 0$ are real.

15) Show that the equation $\frac{x^2}{x^2 - k^2} + \frac{x^2}{x^2 - \lambda^2} = 4$ has real roots for all values of the real numbers k and λ.

16) Solve the equation: $\frac{(x^2-1)^2}{x^4+1} + \frac{2k^2\lambda^2}{k^4+\lambda^4} - 1 = 0$,   (**Ans:** $\pm\frac{k}{\lambda}, \pm\frac{\lambda}{k}$).

17) Form a biquadratic equation having roots: **a)** ± 5 and $\pm 3i$, **b)** $\pm 2, 3, -5$, **c)** $\pm\sqrt{2} \pm \sqrt{3}$, **d)** $\pm 4 \pm 7i$.

18) Solve the system: $\{x^2 + y^2 = 34, \quad x - y = -2\}$,   (**Ans:** 3,5 or $-3, -5$).

19) Show that $\sqrt{a + b + \sqrt{b^2 + 2ab}} = \sqrt{(2a+b)/2} + \sqrt{b/2}$.

20) Show that the expression $\sqrt{x + 3\sqrt{2x - 9}} - \sqrt{x - 3\sqrt{2x - 9}}$ is independent of x, provided that $x \geq 9$.

21) Simplify the expressions:

$$\frac{1}{\sqrt{6 + 2\sqrt{5}}} - \frac{1}{\sqrt{6 - 2\sqrt{5}}}, \quad \frac{2 + \sqrt{3}}{\sqrt{2} + \sqrt{2 + \sqrt{3}}} + \frac{2 - \sqrt{3}}{\sqrt{2} + \sqrt{2 - \sqrt{3}}}$$

$$\sqrt{xz^2 + yw^2 - 2zw\sqrt{xy}}, \quad \sqrt[4]{28 - 16\sqrt{3}}$$

22) Solve the system: $\{a + b = 5, \quad a^4 + b^4 = 97\}$.

(Ans: $(a, b) =$
(2,3) or (3,2) or $\left(\frac{5+i\sqrt{151}}{2}, \frac{5-i\sqrt{151}}{2}\right)$ or $\left(\frac{5-i\sqrt{151}}{2}, \frac{5+i\sqrt{151}}{2}\right)$).

23) Show that $x^8 - x^5 + x^2 - x + 1 > 0, \quad \forall x \in \mathbb{R}$.

Hint: Consider two cases, a) $x \in (-\infty, 0] \cup [1, \infty)$ and b) $x \in (0,1)$, and show that in both cases the inequality holds true.

24) Solve the following systems:

$$\begin{cases} x - y = 1 \\ x^3 - y^3 = 7 \end{cases} \quad \begin{cases} \frac{1}{x} - \frac{1}{y} = 2 \\ \frac{1}{x^2} + \frac{1}{y^2} = 34 \end{cases} \quad \begin{cases} x^3 + y^3 = 56 \\ x + y = 2 \end{cases}$$

25) Solve the inequality: $x^3 + \frac{1}{x^3} \geq -3\left(x + \frac{1}{x}\right)$, (**Ans:** $x > 0$).

26) At what values of k does the equation $2(k + 1)x^2 - (k^3 + 8k + 7)x + k^2 - 4k = 0$, possess roots of opposite signs?

27) For what values of k is the inequality $kx^2 - 6kx + 5k + 4 > 0$ satisfied for any $x \in \mathbb{R}$?, (**Ans:** $0 < k \leq 1$).

Hint: It suffices, $k > 0$ and $\Delta < 0$.

28) Solve the inequality: $\frac{1}{x+3} + \frac{1}{x-5} + \frac{1}{x-3} + \frac{1}{x+5} > 0$.

29) Find all real values of x and y satisfying the equation:
$x^2 + 4x \sin(xy) + 4 = 0$.

(**Ans:** $x = 2, y = -\frac{\pi}{4} + k\pi, k$ integer, or, $x = -2, y = \frac{\pi}{4} + \lambda\pi, \lambda$ integer).

30) Solve the system: $\left\{\frac{xyz}{x+y} = 6, \quad \frac{xyz}{y+z} = \frac{15}{4}, \quad \frac{xyz}{z+x} = \frac{30}{7}\right\}$.

Hint: Take the reciprocals of the given equations, which form a linear system for $u = \frac{1}{xy}, v = \frac{1}{yz}, w = \frac{1}{zx}$, find u, v, w, etc.

31) Solve the system: $\left\{\frac{xyz}{w} = 1, \frac{yzw}{x} = 2, \frac{zwx}{y} = 5, \frac{wxy}{z} = 10\right\}$, (real solutions).

Hint: Multiply together to compute the product $xyzw$, etc.

32) Find the integers x and y that satisfy simultaneously the inequalities:
$\left\{y - |x^2 - 2x| + \frac{1}{2} > 0, \; y + |x - 1| < 2\right\}$, (**Ans:** (x, y): $(0,0), (1,1), (2,0)$).

33) Given that $2x + 4y = 1$, show that $x^2 + y^2 \geq \frac{1}{20}$.

34) If $a + b = 1$, show that $a^4 + b^4 \geq \frac{1}{8}$.

35) Solve the inequality: $(x - 5)^7 (x + 1)^5 (x - 3)^9 (x + 2)^{11} \leq 0$.

(**Ans:** $(-2 \leq x \leq -1)$ or $(3 \leq x \leq 5)$).

Hint: The given inequality is equivalent to $(x - 5)(x + 1)(x - 3)(x + 2) \leq 0$, (why?).

36) Solve the inequality: $\frac{x^3 - 2x^2 + 5x + 2}{x^2 + 3x + 2} \geq 1$.

37) Find the domain of definition of the functions:

$y = \sqrt{9 - x^2} + \sqrt{x^2 - x}, \quad y = \sqrt{x^2 - |x|} + \frac{1}{\sqrt{16 - x^2}}, \quad y = 2x + \sqrt{1 - 3x^2}$

(**Ans:** $-3 \leq x \leq 0$ or $1 \leq x \leq 3, -4 \leq x \leq -1$ or $1 \leq x \leq 4$ or $x = 0$, $-\frac{1}{\sqrt{3}} \leq x \leq \frac{1}{\sqrt{3}}$).

38) If r_1, r_2, r_3 are the roots of $x^3 + 3x - 2 = 0$, find the numbers: $r_1^2 + r_2^2 + r_3^2$ and $r_1^3 + r_2^3 + r_3^3$, (use Vieta's formulas).

39) If n is a natural number, show that: $\frac{1}{n+1} + \frac{1}{n+2} + \cdots + \frac{1}{n+n} > \frac{1}{2}$.

Hint: $\frac{1}{n+k} > \frac{1}{n+n}$ for all $k = 1, 2, \ldots, n - 1$, and add term wise the resulting inequalities.

40) Show that $\frac{1}{2^2} + \frac{1}{3^2} + \cdots + \frac{1}{n^2} < \frac{n-1}{2}$.

41) Solve the equation: $\sqrt[3]{x-1} + \sqrt[3]{x+1} = x\sqrt[3]{2}$.

(Ans: $0, \pm 1, \pm\sqrt{1+\frac{\sqrt{27}}{2}}, \pm i\sqrt{\frac{\sqrt{27}}{2}-1}$).

Hint: See Example 7-6-1.

42) Find the positive solution of the system: $\{3^{2x} + 3^{2y} = 20, \ 3^{x+y} = 9\}$.

43) Solve the system: $\{\log(x+y) - \log(x-y) = 2, \ x^2 - y^2 = \frac{1}{4}\}$.

(Ans: $x = 5.05/2, y = 4.95/2$).

44) Solve the equations:

$$\{8x^3 = 16, \quad x^3 + x^2 + x + 1 = 0, \quad (x+3)(x^2 + 2x - 3) = 0\}$$

45) Solve the system: $\{|x-2| + |y-4| = 2, \ y = 4 + |x-2|\}$.

(Ans: (x, y): $(3,5)$ or $(1,5)$).

46) Solve the systems:

$$\begin{cases}(x+y)(x^2-y^2) = 147 \\ (x-y)(x^2+y^2) = 87\end{cases} \quad \begin{cases}x+y = 1 \\ x^5 + y^5 = 31\end{cases}$$

47) Solve the system: $\{\frac{x^2}{y} + \frac{y^2}{x} = \frac{217}{6}, \ \frac{1}{x} + \frac{1}{y} = \frac{7}{6}\}$.

(Ans: (x, y): $(1,6)$ or $(6,1)$ or $(0.735, -5.163)$ or $(-5.163, 0.735)$).

48) Show the identity: $\frac{\log_a x}{\log_{ab} x} = 1 + \log_a b$.

49) Let S_n be the sum of the first n terms of a geometric progression with ratio ω, $(S_n \neq 0)$. Show that $S_n/(S_{2n} - S_n) = (S_{2n} - S_n)/(S_{3n} - S_{2n})$.

50) Using De Moivre's formula compute:

$$(1+i)^6, \quad (2 - i\sqrt{12})^5, \quad \left(1 + \cos\frac{\pi}{3} - i\sin\frac{\pi}{3}\right)^6$$

51) If the numbers $\log_k x, \log_m x, \log_n x$ $(x \neq 1)$, form an arithmetic progression, show that $n^2 = (kn)^{\log_k m}$.

52) Starting with $(\cos\theta + i\sin\theta)^5 = \cos 5\theta + i\sin 5\theta$, find expressions of $\cos 5\theta$ and $\sin 5\theta$ in terms of $\cos\theta$ and $\sin\theta$.

53) Compute the sum $S = 1 + 2x + 3x^2 + \cdots + (n+1)x^n$ in closed form.

(**Ans:** $S = \{1 - (n+2)x^{n+1} + (n+1)x^{n+2}\}/(x-1)^2$).

Hint: It is a mixed progression.

54) Solve the system : $\{(x^2+1)(y^2+1) = 30, \ (x+y)(1-xy) = 3\}$, using the substitution, $x + y = a$ and $xy = b$, (find a and b and then find x and y).

55) Solve the equations:

$$\begin{cases} x^4 + 11x^2 + 10 + 7x(x^2+1) = 0 \\ 6x^4 - x^3 - 5x^2 + x - 1 = 0 \end{cases}$$

(**Ans:** First eq. $(\pm i, -2, -5)$, Second eq. $(\pm 1, (1 \pm \sqrt{23})/2)$).

56) The numbers $x_1, x_2, \ldots, x_n$ form an arithmetic progression. Provided that $\sum_{k=1}^{n} x_k = X$, (given) and $\sum_{k=1}^{n} x_k^2 = Y^2$, (given), find the progression, i.e. find the first term and the ratio of the progression.

57) Construct the domain of points z, (complex numbers), according to the conditions: $1 < |z| < 5$ and $\pi/3 < \theta < 2\pi/3$.

58) Provided that the real numbers x, y, z satisfy the inequality $|x+y| + |y+z| + |z+x| \geq xyz(|x| + |y| + |z|)$, show that $xyz \leq 2$.

59) Provided that the real numbers x, y, z satisfy the inequality $|x+y-z| + |y+z-x| + |z+x-y| = |x+y+z|$, show that they will satisfy the inequality $x^2 + y^2 + z^2 \leq 2(xy + yz + zx)$, as well.

Hint: If $|a| + |b| + |c| = |a+b+c|$, $(a, b, c \in \mathbb{R})$, then the three numbers a, b, c are either, all positive or all negative, i.e. $ab > 0, ac > 0, bc > 0$.

60) Determine the constant k so that $x^3 + y^3 + z^3 + kxyz$ is divided exactly by $x + y + z$.

61) Provided that the successive terms a, b, c of a geometric progression satisfy $a^x = b^y = c^z$, show that x, y, z are successive terms of an harmonic progression.

62) Find the roots of the polynomial $x^3 + (k+1)x^2 + 8x + k + 2$, given that this polynomial is divided exactly by $(x+1)$.

63) Find the non negative integers x, y, z, w which satisfy the inequalities $0 \leq x < 4, 0 \leq y < 6, 0 \leq z < 5$ and the equation $x + 4y + 24z + 120w = 782$, (**Ans:** $x = 2, y = 3, z = 2, w = 6$).

64) Assuming that a, b, c, d form an arithmetic progression, and that the numbers $a^2 + b^2, b^2 + c^2, c^2 + d^2$ form a geometric progression, show that $ad + 3bc = 0$.

Hint: Set, $a = x - 3\omega, b = x - \omega, c = x + \omega, d = x + 3\omega$.

65) Assuming that the three sides of a triangle form a geometric progression with ratio ω, show that: $(\sqrt{5} - 1)/2 < \omega < (\sqrt{5} + 1)/2$.

66) The numbers a, b, c form an arithmetic progression, while the numbers ab, bc, ca form a geometric progression. Given that the sums of the three terms of the arithmetic progression is equal to the sum of the three terms of the geometric progression, find a, b, c.

67) Given that S_1, S_2, S_3 are the sums of the first $n, 2n, 3n$ terms respectively, of an arithmetic progression, show that $S_3 = 2(S_2 - S_1)$.

68) Given that x, y, z form a geometric progression, show that:
$$x^3 y^3 + y^3 z^3 + z^3 x^3 = xyz(x^3 + y^3 + z^3)$$
$$(xy + yz + zx)^3 = xyz(x + y + z)^3$$

Hint: $\frac{y}{x} = \frac{z}{y} = \lambda$, i.e. $y = \lambda x, z = \lambda^2 x$.

69) For what values of k, the equation $x^4 - 3kx^2 + 5 - k = 0$ has four real roots? (**Ans:** $0 < k \leq 5$).

70) Determine c so that the sum of the squares of the roots of the trinomial $x^2 - 7x + 2c$ to be equal to 25. Then find the two roots.

71) Given that one root of the equation $x^2 + ax + b = 0$ is equal to the square of the other root, show that $b^3 + a^2 = b(3a - 1)$.

72) If the numbers a, b, c form an arithmetic progression and the numbers a, kb, c form a geometric progression, show that the numbers a, k^2b, c form an harmonic progression.

73) Assuming that $x \in (-\infty, \infty)$, show that the corresponding values of the function $f(x) = \frac{(x+\lambda)^2 - 4k\lambda}{2(x-k)}$ lie outside of the interval $(2k \ldots 2\lambda)$.

74) Provided that $ab \neq 0$, show the trinomial $x^2 - (a+b)x + ab$ is not divided by $(x - a - b)$.

75) Given the trinomials $(k-1)x^2 + 3x + \frac{1}{k}$ and $k^2x^2 + 2(k-1)x - \frac{3}{k-1}$, find k, so that the product of the roots of the first lies between the roots of the second, (Ans: $-\sqrt{2} < k < 0$, or $\sqrt{2} < k < \infty$).

76) If n is a natural number, show that $\frac{1}{\sqrt{n}} > 2\sqrt{n+1} - 2\sqrt{n}$, and then show that

$$\frac{1}{\sqrt{1}} + \frac{1}{\sqrt{2}} + \frac{1}{\sqrt{3}} + \cdots + \frac{1}{\sqrt{n}} > 2(\sqrt{n+1} - 1)$$

77) Consider the fractions $\frac{x_1}{y_1}, \frac{x_2}{y_2}, \frac{x_3}{y_3}, \ldots, \frac{x_n}{y_n}$ where $y_1, y_2, y_3, \ldots, y_n$ are all positive. Show that the fraction $\frac{x_1+x_2+x_3+\cdots+x_n}{y_1+y_2+y_3+\cdots+y_n}$ is contained between the least and the greatest of these functions, (This proposition is known as "**the Theorem of the unequal fractions**").

78) Based on Problem 77, show that if $0 < x_1 < x_2 < \cdots < x_n < 90°$, then $\tan x_1 < \frac{\sin x_1 + \sin x_2 + \cdots + \sin x_n}{\cos x_1 + \cos x_2 + \cdots + \cos x_n} < \tan x_n$.

79) If $\frac{y+z}{x} = \sqrt{3} + 1$ and $\frac{z+x}{y} = \sqrt{3} - 1$, find the value of $\frac{y+x}{z}$.

(Ans: $2(\sqrt{3} + 1)$).

80) If x, y, z are positive numbers, show that:

$$(x+y)(y+z)(z+x) \geq 8xyz$$

$$\frac{x}{y+z} + \frac{y}{z+x} + \frac{z}{x+y} \geq \frac{3}{2}$$

Hint: $x + y \geq 2\sqrt{xy}$, etc.

81) Given that $\sum_{k=1}^{n} x_k^2 = 1$ and $\sum_{k=1}^{n} y_k^2 = 1$, show that $|\sum_{k=1}^{n} x_k y_k| \leq 1$

Hint: Apply Lagrange's identity.

82) At what values of $x_1, x_2, \ldots, x_n$ the sum $S = x_1^2 + x_2^2 + \cdots + x_n^2$ takes on the least value, provided that $a_1 x_1 + a_2 x_2 + \cdots + a_n x_n = C$, $(a_1, a_2, \ldots, a_n, C$ are constant numbers).

(Ans: $x_k = a_k C / (a_1^2 + a_2^2 + \cdots + a_n^2), k = 1, 2, \ldots, n)$.

Hint: Use Lagrange's identity to the numbers $(x_1, x_2, \ldots, x_n)$ and $(a_1, a_2, \ldots, a_n)$.

83) Find the sum of the kth powers of the roots of $x^n - 1 = 0$, $(k, n$ are natural numbers).

(Ans: 0 if k is not divisible by n, n if k is divisible by n).

84) For what values of a and b the trinomial $x^2 + ax + b$ divides exactly the polynomial $x^4 + 1$?

85) Show that $x^2 + 1$ divides exactly the polynomial
$P(x) = (\cos\theta + x\sin\theta)^n - \cos n\theta - x\sin n\theta$.

Hint: It suffices to show that $P(i) = P(-i) = 0$.

86) Show that $(x - b)^2$ divides exactly the polynomial
$P(x) = x(x^{n-1} - nb^{n-1}) + b^n(n - 1)$.

87) Given that $\frac{b-c}{1+bc} + \frac{c-b}{1+cb} + \frac{b-a}{1+ba} = 0$, show that two of the numbers a, b, c are equal to each other.

88) Given that $\frac{x}{y} = \frac{y}{z} = \frac{z}{w}$ where x, y, z, w are positive numbers, show that $|x - w| \geq 3|y - z|$.

89) Eliminate κ and λ between the equations:

$$\{x = \kappa(\lambda^2 + 1), \quad y = \lambda(\kappa^2 + 1), \quad z = (\kappa + \lambda)(1 - \kappa\lambda)\}$$

(**Ans:** $(x + y + z)^3(-x + y + z)(x - y + z) = 16z^2(x + y - z)$).

90) Determine a and b, so that $(x - 1)^2$ divides exactly the polynomial $P(x) = ax^{n+1} + bx^n + 1$.

91) If $f(x) = \sqrt{x^2 + 7}$ and $a \neq b$ are any two real numbers, show that $|f(b) - f(a)| < |b - a|$.

92) Considering the non negative expression
$(a_1 x + b_1)^2 + (a_2 x + b_2)^2 + \cdots + (a_n x + b_n)^2$ as a trinomial in x, deduce Schwarz's inequality:

$$(a_1 b_1 + a_2 b_2 + \cdots a_n b_n)^2 \leq (a_1^2 + a_2^2 + \cdots + a_n^2)(b_1^2 + b_2^2 + \cdots + b_n^2)$$

93) Consider the equation $(x^2 + ax + 1)^2 = b^2$, where $b > 0$. **a)** What is the relation between a and b for the equation to have four, real and distinct roots? **b)** Solve the equation, assuming that $a = 3$ and $b = 1$.

(**Ans: (a)** $a^2 > 4(1 + b)$, **(b)** $-3, -2, -1, 0$).

94) Show the inequality:

$$\sqrt{4x + 1} + \sqrt{4y + 1} + \sqrt{4z + 1} + \sqrt{4w + 1} + \sqrt{4t + 1}$$
$$< 5 + 2(x + y + z + w + t)$$

Assume that the radicands are positive, and that $xyzwt \neq 0$.

Hint: $\sqrt{4x + 1} < \sqrt{4x^2 + 4x + 1} < \sqrt{(2x+1)^2} = 2x + 1$, etc.

95) Solve graphically the system of inequalities:

$$\{2x + 3y - 7 > 0, \quad 3x - 4y + 5 < 0\}$$

96) Solve graphically the inequality: $xy(3x - 5y + 15) < 0$.

Hint: The inequality is equivalent to $\{xy > 0 \text{ and } (3x - 5y + 15) < 0\}$, or, $\{xy < 0 \text{ and } (3x - 5y + 15) > 0\}$.

97) Provided that a, b, c are rational numbers, show that the two equations $ax^2 + bx + c = 0$ $(a \neq 0)$ and $x^3 - 2 = 0$ do not have a common real root. As a consequence of this, show that the equality $a + b\sqrt[3]{2} + c\sqrt[3]{4} = 0$ holds true, if and only if $a = b = c = 0$.

98) Show that the number $a = \sqrt{3} + \sqrt[3]{3}$ is irrational.

99) Show that the number $a = \sqrt{3} + \sqrt[3]{3}$ is an algebraic number.

Hint: It suffices to show that a is a root of a polynomial equation, with integer coefficient. From $a - \sqrt{3} = \sqrt[3]{3}$, if we raise both sides to the third power we get: $(a - \sqrt{3})^3 = 3$, i.e. $a^3 - 3a^2\sqrt{3} + 3a(\sqrt{3})^2 - (\sqrt{3})^3 = 3$, i.e. $a^3 - 3\sqrt{3}a^2 + 9a - 3\sqrt{3} = 3$, i.e. $a^3 + 9a - 3 = (3a^2 + 3)\sqrt{3}$, and then squaring both sides we find a polynomial equation satisfied by a, which shows that a is an algebraic number.

100) Show that the following numbers are algebraic:

$$x = 3 + 5i, \quad y = \sqrt{2} + i\sqrt{3}, \quad z = \cos\frac{\pi}{4} + i\sin\frac{\pi}{4}$$

101) Given that $a > b > c, k + c > 0$, show the inequality:
$(b - c)\sqrt{a + k} + (c - a)\sqrt{b + k} + (a - b)\sqrt{c + k} < 0$.

Hint: Set: $(a + k) = x^2, (b + k) = y^2, (c + k) = z^2$.

102) Solve the systems:

$$\begin{cases} \sqrt[3]{x} + \sqrt[3]{y} = 6 \\ x + y = 72 \end{cases}, \quad \begin{cases} \sqrt{x} + \sqrt{y} = 8 \\ x + y - \sqrt{xy} = 19 \end{cases}, \quad \begin{cases} x^5 - y^5 = 992 \\ x - y = 2 \end{cases}$$

103) Solve the inequality: $\sqrt{2x + 1} + \sqrt{x + 1} < 1$.

(**Ans:** $-\frac{1}{2} < x < 3 - 2\sqrt{3}$).

104) If r is a root of the equation $ax^2 + bx + c = 0$, show that $|r| < \frac{|a|+|b|+|c|}{|a|}$.

105) Solve the equation: $\sqrt{x + 4} + \sqrt{x - 3} = \sqrt{4x + 1}$, (**Ans:** $x = 12$).

106) If r is a root of $x^2 + ax + b = 0$ and $|a| < 1$ and $|b| < 1$, show that $|r| < 2$.

107) Find the real numbers x and y satisfying the system:
$\{|x| + |y| = 1, \quad x^2 + y^2 \geq 1\}$, **(Ans:** $(x, y): (1,0), (-1,0), (0,1), (0,-1)$**).**

108) Given that $a + b > 0$ and that the equations $x^2 + ax + b = 0$ and $x^2 + ax + |b| = 0$ do not have a common root, show that $|a - b| = |a| + |b|$.

109) Determine k given that the equations $x^2 + kx + 1 = 0$ and $x^2 + kx + 1 = 0$ have: (a) one common root and (b) two common roots.

(Ans: (a) $k = -2$, **(b)** $k = 1$**).**

110) Solve the equations:

$$x + \frac{1}{x} = 2(\sqrt{5} + 1), \qquad \frac{3x - 4}{x + 1} = x^2 + 2x - \frac{7}{x + 1}$$

$$\frac{x - 1}{x - 3} + \frac{x - 7}{x - 9} = \frac{x - 3}{x - 5} + \frac{x - 5}{x - 7}, \qquad \frac{x + i}{x - i} + \frac{x + 2i}{x - 2i} = \frac{2x + 3i}{x - 3i}$$

111) If $\left(1 - a + \frac{b^2}{2}\right)x^2 + b(1 + a)x + a(a - 1) + \frac{b^2}{2} = 0$ has a double root, show that the equation $x^2 + bx + a = 0$ has also a double root, and conversely.

112) Show that the roots of the equation
$f(x) = x^2 - (a + b + c)x + ab + bc + ca - \mu^2 - \kappa^2 = 0$ are real and distinct, provided that $abc - \mu^2 b - \kappa^2 c = 0$. Assume $abc(a - b)(b - c)(c - a) \neq 0$.

Hint: Find $f(a), f(b), f(c)$. If $bc < 0$, then $f(a) < 0$. If $bc > 0$, find the product $f(b)f(c)$, taking into account the given condition.

113) Determine k given that the equations $x^2 + (2k + 1)x + 1 = 0$ and $x^2 + 2kx + 5 = 0$ have only one common root. What is this common root?

(Ans: $k = -21/8, r = 4$**).**

114) Show that the equation $(a^2 + b^2)x^2 - 2(ac + bd)x + c^2 + d^2 = 0$ has complex roots, unless if $a/b = c/d$.

115) Determine $\phi(x) = ax^2 + bx + c$, such that $\phi(x)$ divides exactly $\phi(x^2)$, (**Ans:** $\phi(x) = x^2 + x + 1$, or $\phi(x) = x^2 - 2x + 1$).

116) Find the domain of definition of the function
$y = \sqrt{x^2 - 11x + 30} + 3\sqrt{-x^2 + 10x - 21}$.

117) Find k given that one root of $x^2 - (2 + \log k)x + \log k + 1 = 0$ is three times the other root, (**Ans:** $k = 2$ or $1/\sqrt[3]{100}$).

118) Provided that x, y satisfy the equations $ax + x^2 = 1, by + y^2 = 1$ and $x^2 + y^2 = 1$, show that $\left(\sqrt[3]{a}\right)^2 \left(\sqrt[3]{b}\right)^2 \left(\left(\sqrt[3]{a}\right)^2 + \left(\sqrt[3]{b}\right)^2\right) = 1$.

119) Solve the equation $\sqrt{x^2 - 4x + 3} + \sqrt{x^2 - 9} + \sqrt{x^2 - 7x + 12} = 0$.

(**Ans:** $x = 3$).

Hint: The sum of three non negative numbers is zero.

120) The real numbers a, b, c, x, y, z satisfy the equation
$(a + b + c)^3 = 3(ab + bc + ca) - 3(x^2 + y^2 + z^2)$. Show that $a = b = c$ and $x = y = z = 0$.

121) The area A of a triangle with sides $a = 2k - 1, b = 2k, c = 2k + 1$ is $A = k(k^2 - 1)/4$. Find k, (**Ans:** $k = 7$).

Hint: Apply Heron's formula.

122) Given that $a, b, c \in \mathbb{R}$, $b^2 - 4ac > 0$ and $|a| > |a - b + c|$, show that at least one root of $ax^2 + bx + c = 0$ is a negative number.

Hint: $1 > |(a - b + c)/a|$, i.e. $1 > |1 + r_1 + r_2 + r_1 r_2|$, i.e. $|(1 + r_1)(1 + r_2)| < 1$, etc...

123) Find the positive numbers x, y which satisfy the equation,
$x^2 y + 2y^2 + 4x - 6xy = 0$, (**Ans:** $0 < x < 6, 0 < y < 2/3$).

Hint: $yx^2 + (4 - 6y)x + 2y^2 = 0$ is treated as a quadratic equation in x, and since x must be positive, the roots must both be positive, and this occurs when the sum and the product of the roots are positive numbers, i.e.

$-((4-6y)/y) > 0$ and $(2y^2/y) > 0$, etc. Then, treat the same equation as a quadratic in y, etc.

124) Show that if one of the equations $x^2 + 2x + k = 0$ and $(1+k)(x^2 + 2x + k) = 2(k-1)(x^2 + 1)$ has real and distinct roots, then the other equation has complex roots.

125) Assuming that a, b, c are rational numbers, show that the equation $(a - b + c)x^2 + 2cx + b + c - a = 0$ has rational roots.

Hint: It suffices to show that the discriminant is a perfect square.

126) Solve the system: $\left\{\frac{x}{y} + \frac{y}{x} = \frac{17}{4}, \quad x^2 + y^2 = \sqrt{273 + x^2 y^2}\right\}$.

Hint: Set $\omega = x + y, \phi = xy$.

127) Solve the equation $x^3 - 3x^2 - 10x + 24 = 0$, given that one of its roots is twice that of another, **(Ans: $2, 4, -3$)**.

128) Find the positive and integer x and y which satisfy the equation, $|x + \sqrt{50}| + \left|\sqrt{y^2 + 1} - x\right| = 10$, **(Ans: $x = 1, y = 7$)**.

129) Solve the inequality: $\sqrt{x^2 - 5x + 6} + \sqrt{x^2 - 7x + 10} > \sqrt{x^2 - 6x + 8}$

(Ans: $(-\infty < x < 2)$ or $(5 \le x < \infty)$).

130) Solve the equation: $2x^2 + 9 + \sqrt{x^2 + 9} = x^4$, **(Ans: $\pm \frac{\sqrt{6 + 2\sqrt{41}}}{2}$)**.

131) If $z = \left(1 + i\tan\frac{\pi}{12}\right)^n + \left(1 - i\tan\frac{\pi}{12}\right)^n$, n positive integer, show that $z = 2(\sqrt{6} - \sqrt{2})^n \cos\frac{n\pi}{12}$.

132) Solve the equation $z^5 + (z+1)^5 = 0$.

(Ans: $z_k = -\frac{1}{2}\left(1 + i\cot\left(\frac{2k+1}{10}\pi\right)\right), k = 0,1,2,3,4$).

133) Show that $(1+i)^n + (1-i)^n = 2\sqrt{2^n}\cos\frac{n\pi}{4}$, n is a natural number.

134) If all the roots of $x^3 + px + q = 0$ are real ($q \neq 0$), then p must be a negative number.

135) Solve the equation: $x^2 - |7x - 4|x - 4 = 0$, (**Ans:** $x = -1/2$).

136) Solve the equation: $2z^2 + 2iz + |z| + 3 = 0$.

(**Ans:** $z = i$ or $z = -(3 + \sqrt{33})i/4$).

137) If z_1, z_2, z_3 are complex numbers such that $|z_1| = |z_2| = |z_3|$ and $z_1 + z_2 + z_3 = 0$, show that $z_1^2 + z_2^2 + z_3^2 = 0$.

138) Solve the equation: $z^2 - 8iz - |z| - 20 = 0$.

139) Show that the polynomial $\phi(x) = x^3 + x^2 - 2$ has three distinct roots r_1, r_2, r_3. Then find a polynomial $P(x) = ax^2 + bx + c$ such that $P(1) = 1$, $P(r_1) = r_2, P(r_2) = r_1$, and show that $\phi(x)$ divides exactly the polynomial $P(P(x)) - x$, (**Ans:** $P(x) = \frac{1}{5}(4x^2 + 3x - 2)$).

140) Using De Moivre's Theorem, show that
$(1 + i\sqrt{3})^5 + (1 - i\sqrt{3})^5 = 32$.

141) Find the two square roots of $z = -40 - 42i$.

(**Ans:** $3 - 7i, -3 + 7i$).

142) If $a \in \mathbb{C} - \{0\}$, show that every root r (real or complex) of the equation $ax^n + x^{n-1} + x^{n-2} + \cdots + x + 1 = 0$ satisfies the inequality $|r| < 1 + \frac{1}{|a|}$.

143) Show that for at least one root r of $z^3 + 5z + 7 = 0$, $|r| > 1$.

144) If r_1, r_2, r_3 are the roots of $z^3 + 1 = 0$, show that

$$\prod_{k=1}^{3}(1 + r_k x + r_k^2 x^2) = (1 + x^3)^2$$

145) If $z \in \left\{\frac{-1+i\sqrt{3}}{2}, \frac{-1-i\sqrt{3}}{2}\right\}$ and $n = 1,2,3,\ldots$, show that $z^{2n} + z^n + 1 \in \{0,3\}$.

146) Simplify the fraction: $(x^2 - 6x - 7)/(3x^2 - 6x - 9)$.

147) Form an equation having roots the reciprocals of the roots of the equation $ax^2 + bx + c = 0, (ac) \neq 0$,   (**Ans:** $cx^2 + bx + a = 0$).

148) If $a_1, a_2, \ldots, a_n$ are positive numbers and $\frac{1}{a_1} + \frac{1}{a_2} + \cdots + \frac{1}{a_n} = 1$, show that $|z_1 + z_2 + \cdots + z_n|^2 \leq a_1|z_1^2| + a_2|z_2^2| + \cdots + a_n|z_n^2|$, $(z_1, z_2, \ldots, z_n$ are complex numbers).

149) For what values of k the equation $|x| + k(x+1)(3-x) = \frac{x+3}{2}$ has four real and distinct roots,   (**Ans:** $k \in \left(0, \frac{1}{2}\right) - \left\{\frac{1}{8}, \frac{3}{8}\right\}$).

150) Show that the function $y = \frac{4x-3}{x^2-1}$ can take on any real value, $(-\infty < y < \infty)$ when x runs from $-\infty$ to ∞.

151) For what values of k the equations $(x-k)^2 = x(k-1)^2$ and $(x-1)^2 = k^2(x+k)$ have a common root?

(**Ans:** $0, -1, \frac{-1 \pm \sqrt{5}}{2}$).

152) Solve the inequality: $(x^4 + 2x^2 - 3)^2 - (x^2 - 5)^2 > 0$.

153) Outline the method of solution of $ax^4 + bx^3 + cx^2 - bx + a = 0$.

Hint: Divide through by x^2 and set $y = x - \frac{1}{x}$.

154) Solve the system: $\{x(3x - 2y) = -3, \quad y(7x - 4y) = 5\}$.

155) Find the real roots of the equation: $\sqrt[4]{x + 27} + \sqrt[4]{55 - x} = 4$.

(**Ans:** $x = -26, or, x = 54$).

Hint: Set $a = \sqrt[4]{x + 27}$ and $b = \sqrt[4]{55 - x}$. Then show that a and b satisfy the system, $\{a + b = 4, \quad a^4 + b^4 = 82\}$, etc.

156) If a, b, c, d are rational numbers and $a\sqrt{6} + b\sqrt{3} + z\sqrt{2} + d = 0$, show that $a = b = c = d = 0$.

157) Solve the system: $\{xz + y = 7z, \quad yz + x = 8z, \quad x + y + z = 12\}$.

(**Ans:** $(x, y, z) = (4, 6, 2)$ or $\left(\frac{60}{7}, \frac{66}{7}, -6\right)$).

158) Solve the system: $\begin{cases} a^3 + b^3 - c^3 - d^3 = 133 \\ a^2 + b^2 - c^2 - d^2 = 15 \\ a + b - c - d = 1 \\ ab = cd \end{cases}$.

(**Ans:** (a, b, c, d): $(6,2,4,3)$ or $(6,2,3,4)$ or $(2,6,4,3)$ or $(2,6,3,4)$).

159) What condition should be satisfied by a, b, c, d if the polynomial $x^4 + ax^3 + bx^2 + cx + d$ can be written in the form $(x^2 + kx)^2 + A(x^2 + kx) + B$? Then, if we set $w = x^2 + kx$, the original equation becomes a quadratic in w, w is found and then x is determined. As an application, solve the equation $x^4 + 6x^3 + 4x^2 - 15x + 4 = 0$.

(**Ans:** $b = \frac{a^2}{4} + \frac{2c}{a}$, $x = -4, 1, \frac{-3 \pm \sqrt{13}}{2}$).

160) If r is a root of $x^2 + ax + b = 0$, where a, b are real or complex numbers $\neq 0$, show that the real number $|r|$ cannot lie between the two real roots of the equation $x^2 + |a|x - |b| = 0$. (Justify why the last equation has two real and distinct roots).

161) If r_1, r_2, r_3 are the roots of $x^3 + 3x^2 - 24x + 1 = 0$, show that $\sqrt[3]{r_1} + \sqrt[3]{r_2} + \sqrt[3]{r_3} = 0$.

162) If the terms of order k, l, m, n of an arithmetic progression form a geometric progression, show that the numbers $k - l, l - m, m - n$ form a geometric progression.

163) Can the numbers $\sqrt{2}, \sqrt{5}, \sqrt{7}$ be terms (not necessarily successive) of an infinite geometric progression? (**Ans:** Not possible).

164) Show that the number $a = \sqrt{2} + \sqrt[3]{3}$ is irrational.

165) If the real coefficients of $ax^3 + bx^2 + cx + d = 0$ satisfy the relations $ad = bc$ and $bd > 0$, show that the equation has one real and two pure imaginary roots.

166) Given that the three heights of a triangle form an increasing geometric progression with ratio ω, show that $1 < \omega < (1+\sqrt{5})/2$.

167) Consider three quadratic trinomials $P(x), Q(x)$ and $R(x)$ with real coefficients. Assuming that $P(x)$ and $Q(x)$ do not have a common root and that $P^2(x) + Q^2(x) = R^2(x)$, show that: **(a)** Each one of $P(x)$ and $Q(x)$ has real and distinct roots while $R(x)$ has complex roots, and **(b)** The polynomials $R(x) - Q(x)$ and $R(x) + Q(x)$ have double roots.

168) Determine a, b, c, d so that the polynomial $P(x) = ax^4 + bx^3 + cx^2 + d$ to satisfy the identity $P(x) - P(x-1) \equiv (5x+1)^3$, and then evaluate the sum, $S_n = 6^3 + 11^3 + 16^3 + \cdots + (5n+1)^3$.

(**Ans:** $S_n = \frac{1}{4}n(125n^3 + 350n^2 + 305n + 84)$).

169) If (-2) is the remainder of the division $P(x) \div (x-1)$ and $-2x^2 + x + 5$ is the remainder of the division $P(x) \div (x^3 - 3)$, find the remainder of the division $P(x) \div (x-1)(x^3 - 3)$.

(**Ans:** $3x^3 - 2x^2 + x - 4$).

170) If the equations $x^3 + kx - \lambda = 0$ and $\lambda x^3 - 2k^2 x^2 - 5k\lambda x - 2k^3 - \lambda^2 = 0$ ($k\lambda \neq 0$), have a common root, show that the first equation has a double root, and then find the roots of the second equation.

(**Ans:** First equation, Roots: r (double), $-2r$, Second equation, Roots: r, $-5r$ (double).

171) Determine a, b, c given that: **1)** The trinomials $\phi(x) = ax^2 + bx + 3c$ and $f(x) = x^2 + 3x + c$ have only one common root, **2)** $\phi(x)$ and $f(x)$ when divided by $x - 2$ leave the same remainder and **3)** $\phi(x)$ and $f(x)$ when divided by $x + 1$ leave the same remainder.

(**Ans:** $a = -1, b = 5, c = 2$, or $a = 11, b = -7, c = -10$).

172) Show that: $\sqrt[3]{45 + 29\sqrt{2}} + \sqrt[3]{45 - 29\sqrt{2}} = 6$.

173) Given that the system (for x and y), $ax + by + c = 0$ and $\dfrac{a}{x} + \dfrac{b}{y} + c = 0$, where a, b, c are positive numbers, has complex roots, show that $|a - c| < b < a + c$.

174) Given that $ax + by + cz = 0$, show that
$ab(x - y)^2 + bc(y - z)^2 + ca(z - x)^2 = (a + b + c)(ax^2 + by^2 + cz^2)$.

175) Show that $x^6 + y^6 \geq xy^5 + x^5y$, $\forall x, y \in \mathbb{R}$.

176) Given that a, b, c are positive numbers and $a + b + c = 1$, show that $(1 - a)(1 - b)(1 - c) > 8abc$.

177) For what values of a, the biquadratic equation
$(a - 1)x^4 - 2(a + 1)x^2 + a - 2 = 0$ has: **1)** Four real roots, **2)** Two real and two complex roots, and **3)** Four complex roots?

(**Ans:** Four real roots when $(2 \leq a < \infty)$, Two real and two complex roots when $(1 < a < 2)$, Four complex roots when $(-\infty < a < 1)$.

Hint: If we set $y = x^2$, the equation reduces to a quadratic equation in y, which has two roots y_1 and y_2. The biquadratic equation will have four real roots, provided that $y_1 > 0$ and $y_2 > 0$, and this occurs when $\Delta > 0$, $(y_1 y_2) > 0$ and $(y_1 + y_2) > 0$. The biquadratic equation will have two real roots and two complex roots when either $y_1 > 0$ and $y_2 < 0$, or $y_1 < 0$ and $y_2 > 0$, and this occurs when $(y_1 y_2) < 0$. Finally, the four roots will be complex when $y_1 < 0$ and $y_2 < 0$, i.e. when $\Delta > 0$, $(y_1 y_2) > 0$, $(y_1 + y_2) < 0$, or when y_1, y_2 are complex, i.e. when $\Delta < 0$.

178) Solve the equation: $\dfrac{1}{\sqrt{1+x} - \sqrt{x}} + \sqrt{1 + x} + \sqrt{x} = 4$.

179) If n is a natural number and a is a real number with $|a| < 1$, show that $\left| n \dfrac{a - n}{a^2 + n} \right| < \dfrac{n}{|a|}$.

180) Show that $\sqrt[3]{\sqrt{5} + 2} - \sqrt[3]{\sqrt{5} - 2} = 1$.

181) If the sides of a triangle are $a = 2x + 1, b = x^2 + x + 1, c = x^2 - 1$, show that one angle of the triangle is $120°$.

Hint: Apply the law of cosines.

182) Assuming that a, b, c are real numbers, show the inequality:
$(1 + ab + bc + ca)^2 - 3abc(a + b + c) \geq 1 + 2ab + 2bc + 2ca$.

183) Solve the system: $\{\sqrt{x} + \sqrt[7]{y} = 8, \quad \sqrt{x^3} + \sqrt[7]{y^3} = 152\}$.

(**Ans:** $x = 25, y = 3^7$, or, $x = 9, y = 5^7$).

184) Show that $r = \sqrt[3]{\sqrt{2} - 1} - \dfrac{1}{\sqrt[3]{\sqrt{2}-1}}$ is a root of $x^3 + 3x + 2 = 0$.

185) Find a closed form expression of the infinite sum:
$S = x - y + \dfrac{y^2}{x} - \dfrac{y^3}{x^2} + \dfrac{y^4}{x^3} - \dfrac{y^5}{x^4} + \cdots$, assuming that $\left|\dfrac{y}{x}\right| < 1$.

(**Ans:** $S = \dfrac{x^2}{x+y}$).

186) Assuming that $abc \neq 0$ and $\dfrac{a+b}{c} = \dfrac{b+c}{a} = \dfrac{c+a}{b}$ find the value of the expression $A = \left(\dfrac{a+b}{c} - 3\right)^3 + \left(\dfrac{b+c}{a} - 4\right)^3 + \left(\dfrac{c+a}{b} - 5\right)^3$.

Hint: Show first that $\dfrac{a+b}{c} = \dfrac{b+c}{a} = \dfrac{c+a}{b} = 2$.

187) Solve the inequality: $\dfrac{x^3-1}{x^2} + \dfrac{3\sqrt[3]{2}}{2} < 0$, (**Ans:** $x < \dfrac{\sqrt[3]{2}}{2}, x \neq -\sqrt[3]{2}$).

Hint: Set $x = \sqrt[3]{2}\, y$.

188) Solve the system: $\{x^{\frac{1}{4}} + y^{\frac{1}{4}} = 5, \quad x^{\frac{3}{4}} + y^{\frac{3}{4}} = 35\}$.

189) For what values of k, $f(x) = 4x^2 + 8(k^2 - 3)x + 12 - 4k^2$ is a perfect square? (**Ans:** $k = \pm\sqrt{3}$, or, $k = \pm\sqrt{2}$).

190) Find the domain of definition of the functions $f(x) = \dfrac{x^2+3x-8}{x^2+3x-12}$ and $g(x) = \dfrac{x^2-2x-6}{x^2-2x-9}$, and then solve the equation $f(x) = g(x)$.

191) If $x + y + z = 1$, show that $xy + yz + zx \leq \dfrac{1}{3}$. When the equality occurs? (**Ans:** Equality when $x = y = z$).

192) Solve the equation:
$$x(x+1) + (x+1)(x+2) + (x+2)(x+3) + \cdots + (x+n-1)(x+n) = n.$$

193) a) If a polynomial $P(x)$ of degree n, takes the same value for $(n+1)$ different values of x, show that $P(x)$ is the constant polynomial, i.e. $P(x) = c$, (constant), $\forall x$.

b) Assuming that $a_1, a_2, \ldots, a_n$ are n different integers, show that the polynomial $P(x) = (x-a_1)^2(x-a_2)^2 \cdots (x-a_n)^2 + 1$ does not have real roots, and cannot be expressed as a product of two polynomials with integer coefficients, unless one of them is ± 1.

194) Given that $r = 1+i$ is a root of the polynomial
$P(x) = 6x^4 + (6\lambda - 5)x^3 + (6k - 5\lambda + 1)x^2 + (\lambda - 5k)x + k$, where k, λ are real numbers, find k and λ, and then find the other three roots of the polynomial.

195) a) Solve the equation $z^n + 1 = 0$, $n = 1, 2, 3, \cdots$.

b) If $x = z + \frac{1}{z}$, show that $z^n + \frac{1}{z^n}$ is a polynomial $P_n(x)$ of degree n.

c) Show that $P_n(x) = 0$ has all its roots real and distinct, and find them.

(Ans: $x_k = 2\cos\left(\frac{\pi}{2n} + k\frac{2\pi}{2n}\right)$, $k = 0, 1, 2, \cdots, n-1$).

196) Consider the function $f: \mathbb{Q} \to \mathbb{R}$, which satisfies the functional equation: $f(xy) = f(x) + f(y)$, $\forall x, y \in \mathbb{Q}$, ($\mathbb{Q}$ is the set of rational numbers). Show that $f(1) = 0$ and that $f\left(\frac{1}{r}\right) = -f(r)$, $\forall r \in \mathbb{Q}$.

197) a) Show that $y = 1$ is a root of the equation $y^3 - 2y^2 - 5y + 6 = 0$, and find the other two roots.

b) Solve the equation $10^{-3x} - 2 \cdot 10^{-2x} - 5 \cdot 10^{-x} + 6 = 0$.

(Ans: $y = 1, 3, -2$, $x = 0, -\log 3/3$).

198) If $|a| + |b| = 1$, show that $\left(|a| + \frac{1}{|a|}\right)^2 + \left(|b| + \frac{1}{|b|}\right)^2 \geq \frac{25}{2}$.

199) a) A polynomial $P(x)$ is such that $P(x) \equiv P(1-x)$. Show that $\phi(x) \equiv P(x) - P(0)$ is divided by $x(x-1)$.

b) A polynomial $P(x)$ is such that $P(x) \equiv P(x-1)$. Show that $P(x)$ is a constant polynomial.

Hint: For part (a), it suffices to show that $\phi(0) = \phi(1) = 0$.

200) Determine all complex numbers z satisfying: $3\bar{z} = z^3$.

201) Solve the equation: $|x^2 - x + 1| + |x^2 - 9x - 10| = 19$.

(**Ans:** $x = -2, or, 1$).

Hint: $x^2 - x + 1 > 0, \forall x \in \mathbb{R}$.

202) Solve and investigate the system (for the various values of the real parameter k), $\{kx + y = 3k - 2, \quad 4x + ky = 4k\}$.

203) Given that $a, b, c \in \mathbb{R}$, and $a^2 + b^2 + c^2 = 1$ and $a + b + c = \sqrt{3}$, show that $a = b = c$.

Hint: Apply Lagrange's identity to the triads (a, b, c) and $(1,1,1)$.

204) If a, b, c are successive terms of a geometric progression, show that
$a^2 b^2 c^2 \left(\frac{1}{a^3} + \frac{1}{b^3} + \frac{1}{c^3}\right) = a^3 + b^3 + c^3$.

Hint: Set $a = \frac{b}{\omega}, b = b, c = b\omega$.

205) If $S = \sum_{k=1}^{89} \log(\tan k°)$ and $P = \log(\prod_{k=1}^{89}(\tan k°))$, show $S = P$.

206) Let $P(x) = a_n x^n + a_{n-1}x^{n-1} + \cdots + a_1 x + a_0$ be a polynomial with real coefficients, and $a_n \neq 0$. **a)** If $max\{|a_n|, |a_{n-1}|, ..., |a_1|, |a_0|\} = |a_n|$, then for every root r of $P(x)$, real or complex, show that $|r| < 2$, and **b)** If all the real roots of the polynomial are $r_1, r_2, ..., r_k$ and for real numbers a and b it is true that $P(a)P(b) > 0$, show that either none or an even number of roots lies in the interval (a, b).

207) Find all complex numbers $z = x + iy$, $(x, y \in \mathbb{R})$, satisfying the equation, $z^2 - 3|z| + a^2 = 0$, where a is a real number $> \frac{3}{2}$.

(**Ans:** $x = 0, |y| = (\sqrt{9 + 4a^2} - 3)/2$).

208) Solve the system: $\left\{\frac{x^5+y^5}{x^3+y^3} = \frac{31}{7}, \quad x^2 + xy + y^2 = 3\right\}$.

209) Given that the equation $f(x) \equiv x^2 + ax + b = 0$ has two real and distinct roots, show that the equation $k(2x + a) + f(x) = 0$ has two real and distinct roots, $\forall k \in \mathbb{R}$.

210) If the set $A = \{|x + y - 2z| = |y + z - 2x| = |z + x - 2y| \neq 0\}$, where x, y, z are real numbers, show that $A = \emptyset$.

211) In the set of complex numbers, find the common roots of the equations: $(z^2 + 1)^2 + z^3 + z = 0$ and $z^{16} + 2z^{14} + 1 = 0$.

(**Ans:** $\pm i$).

212) Find the sum:
$S_n = i + (2 + 3i) + (4 + 5i) + (6 + 7i) + \cdots + \{(2n - 2) + (2n - 1)i\}$.

213) Consider the equation $ax^2 + bxy - cy^2 = 0$, $a, b, c \in \mathbb{R}$. We assume that the roots $x = r_1$ and $x = r_2$ are real, and satisfy the equalities:

$$\left\{ac = -6, \quad 1 - |a| = \frac{2}{|b|}, \quad \frac{|r_1 + r_2|}{|r_1 + r_2| + |r_1 r_2|} = |a|\right\}$$

Find y.

(**Ans:** $|y| = 1/3$).

214) Given $f(x) = \sqrt{x + \sqrt{1 + x^2}} + \sqrt{\sqrt{1 + x^2} - x}$, show that $\{f(x)\}^2 - 2\sqrt{1 + x^2}$ is a constant number.

215) Solve the system: $\{x^{x+y} = y^k, \quad x + y - 2k = 0\}$, $x, y, k > 0$.

(**Ans:** $x = (-1 + \sqrt{1 + 8k})/2$, $y = (1 + 4k - \sqrt{1 + 8k})/2$).

216) In the Argand plane show that the set of points z satisfying the equation $|z - 1|^2 + |z - 3 - 2i|^2 = 6$ represents a circle. Find the center and the radius of this circle.

Hint: For any complex number z, $|z|^2 = z\bar{z}$.

217) Consider the polynomials $P(z) = z^2 - 2z + 2$ and $Q(z) = z^3 + az^2 + bz - 2$, where a, b are real numbers. **a)** Find the roots z_1, z_2 of $P(z)$ and show that $z_1^{12} + z_2^{12} = -2^7$, **b)** If one root of $P(z)$ is a root of $Q(z)$, find a and b.

218) Find the remainder of the division of a polynomial $P(x)$ by $(x - a)(x - b)(x - c)$, where a, b, c are three different numbers.

Hint: The general form of the remainder is $R(x) = Ax^2 + Bx + C$, and according to the identity of the algorithmic division, we have the identity: $P(x) \equiv Q(x)(x - a)(x - b)(x - c) + Ax^2 + Bx + C$. Apply this identity for $x = a, x = b, x = c$, and determine A, B, C in terms of $P(a), P(b), P(c)$.

219) Find the integer n which satisfies the equation:

$$3^{2(n+1)} + 9 \cdot \left(-\frac{1}{2} + i\frac{3\sqrt{3}}{2}\right)^{3n} - 810 \cdot 3^n = 0$$

(Ans: $n = 2$).

220) Solve and investigate the equation: $\dfrac{1}{x-4} = \dfrac{4}{4k-x} - \dfrac{5}{k-x}$, $k \in \mathbb{R}$.

221) Solve in the set of complex numbers the equation:

$$(ax^2 + bx + c)^2 = (bx^2 + cx + a)(cx^2 + ax + b)$$

(Ans: The given equation is equivalent to the following: $(x^3 - 1)\{(a^2 - bc)x + (ab - c^2)\} = 0$.

222) Consider the function $f: \mathbb{R} \to \mathbb{R}$ which satisfies the equality, $f(x + y) = f(x) + f(y), \forall x, y \in \mathbb{R}$. **a)** Show that $f(0) = 0$, and that $f(-x) = -f(x), \forall x \in \mathbb{R}$, and **b)** If r is rational, show that $f(rx) = rf(x), \forall x \in \mathbb{R}$.

223) Solve the system: $\left\{\sqrt[3]{x} - \sqrt[3]{y} = 3, \quad 2(x - y) = 7\left(\sqrt[3]{x^2 y} - \sqrt[3]{xy^2}\right)\right\}$.

224) Consider the equation:

$$(a+1)x^3 - (a^2+5a-5)x^2 + (a^2+5a-5)x - (a+1) = 0$$

where $a \in \mathbb{R} - \{-1\}$. **a)** Show that $\forall a$ the roots of the equation form successive terms of a geometric progression, **b)** If r_2 is the root not depending on a, determine a so that r_1, r_2, r_3 form successive terms of an arithmetic progression, **c)** For the value of a found in (b), show that the equation has three equal roots.

225) If $P(x) = x^2 + 2|z_1 - z_2|x + (1+|z_1|^2)(1+|z_2|^2)$, where z_1 and z_2 are arbitrary complex numbers, show that $P(x) \geq 0, \forall x \in \mathbb{R}$.

226) Determine a, b such that $r = 1$ is a double root of the equation $x^4 + (a-b)x^3 + 2ax^2 - 5x + 4 = 0$.

227) Show that the equation $(1+iz)^n = \frac{1+i\sqrt{3}}{\sqrt{3}+i}$, where n is a natural number and $z \in \mathbb{C}$, does not have real solutions.

228) Show that $\begin{vmatrix} a & b+1 & 1 \\ b & a+1 & 1 \\ a+b & 1 & 1 \end{vmatrix} = 0.$

229) Solve the equation: $\begin{vmatrix} x+3 & 2x & 3x-1 \\ -3 & 2x-6 & -x-1 \\ 1 & 5 & 1 \end{vmatrix} = 0$, **(Ans: $-3, 2$)**.

230) If w is a cubic root of unity, other than 1, find the value of the expression: $(1-w)(1-w^2)(1-w^4)(1-w^5)$.

231) Solve the equation: $\log_x 256 = (\log_x 4)^2 + 3$, **(Ans: $4, \sqrt[3]{4}$)**.

232) Consider the function: $f(z) = \frac{(z-1)(\bar{z}+1)}{z+\bar{z}}$, $z \in \mathbb{C}$. **a)** Show that $f\left(-\frac{1}{\bar{z}}\right) = f(z), \forall z \in \mathbb{C}$, and **b)** Determine the set of points $K(x, y)$ for which the complex numbers $z = ax + iby$, $a, b, x, y \in \mathbb{R}$, $abx \neq 0$, satisfy $\text{Re}(f(z)) = 0$.

233) Let x and y be two real variables satisfying the equation $x + y = 1$. We set $z = xy$ and $S_n = x^n + y^n$, $n = 0,1,2,3, \ldots$ **a)** Show that $S_n = S_{n-1} - zS_{n-2}$, for $n = 2,3, \ldots$ **b)** Determine the numbers a and b, so that the quantity $Q \stackrel{\text{def}}{=} 6S_5 + aS_4 + bS_3$ to be independent from x and y, and **c)**

Based on result obtained in part (b), show that the quantity
$R = 6((\cos\phi)^{10} + (\sin\phi)^{10}) - 15((\cos\phi)^8 + (\sin\phi)^8) + 10((\cos\phi)^6 + (\sin\phi)^6)$, is independent of the value of ϕ. What is the value of $R, \forall \phi$?

(**Ans:** $a = -15, b = 10, Q = 1 \forall \phi$).

234) Assuming that n is a natural number and
$a_n = \left(1 + \frac{1}{\sqrt{2}}\right)^n + \left(1 - \frac{1}{\sqrt{2}}\right)^n$, $b_n = \left(1 + \frac{1}{\sqrt{2}}\right)^n - \left(1 - \frac{1}{\sqrt{2}}\right)^n$, show that
$a_{k+n} = a_k a_n - \frac{a_{k-n}}{2^n}$, $b_{k+n} = b_k a_n - \frac{b_{k-n}}{2^n}$.

235) If $z = x + iy$, $x, y \in \mathbb{R}$, solve and investigate the equation
$z^2 - 3|z| + a^2 = 0, (a > 0)$.

(**Ans:** $z = 0 \pm i\left(-3 + \sqrt{9 + 4a^2}\right)/2$, $\forall a \in \mathbb{R}$, or $z = \pm\left(3 \pm \sqrt{9 - 4a^2}\right)/2$, where $|a| < 3/2$).

236) Solve the system: $\left\{xy = a^2, (\log x)^2 + (\log y)^2 = \frac{5}{2}(\log a)^2\right\}, a > 0$.

237) If $a \in \mathbb{R} - \left\{-\frac{3}{2}\right\}$, show that each one of the following inequalities implies the other two:

$$\left|\frac{2+3a}{3+2a}\right| < 1, \quad |a| < 1, \quad \left|\frac{2a+3a^2}{3+2a}\right| < 1$$

238) Assuming that a is a positive real number, solve the equation $|z|^2 - 2iz + 2a(1+i) = 0$.

239) Find the positive and integer value of b, given that $\frac{3x^2+2x+2}{x^2+x+1} > b$ is satisfied for all real values of x, (**Ans:** $b = 1$).

240) In the set of real numbers solve the inequality: $\frac{(x+a)^2}{(x+b)^2} < \frac{x^2+a^2}{x^2+b^2}$, where $a > b > 0$.

241) a) Find the general form a_n of positive integers, satisfying the relation, $1 + r + \frac{r}{a_n(1+r)} = 0$, where r is a rational number, and **b)** Find the sum $S = \sum_{k=1}^n \frac{1}{a_k}$, (**Ans:** $a_n = n(n+1), n = 1,2,3, \ldots, S = 1 - \frac{1}{n+1}$).

242) For $a \geq 0$, show $(2a+2)^3 < (a+2)^4 - a^4 < (2a+3)^3$. Is there an integer n, such that the quantity $\sqrt[3]{(n+2)^4 - n^4}$ to be a natural number?

243) Solve the equation $z^2 - \sqrt{2}(1-i)z - 2i = 0$ and show that one of the roots of $z^4 + 1 = 0$ lies on the straight line determined by the roots of the first equation (in the Argand plane), **(Ans: $\sqrt{2}, -\sqrt{2}\,i$).**

244) If $x + y + z = a, xyz = b, xy + yz + zx = c$, express $x^3 + y^3 + z^3$ in terms of a, b, c.

245) Show that the polynomial $\phi(x) \equiv x^3 + x^2 + x + 1$ divides exactly the polynomial $f(x) \equiv x^{4a+3} + x^{4b+2} + x^{4c+1} + x^{4d}$ for any values of the positive integers a, b, c, d.

Hint: The roots of $\phi(x)$ are the roots of $x^4 - 1 = 0$, except $x = 1$.

246) If $y = f(x) = (\sqrt[3]{1+x} - \sqrt[3]{1-x})/(\sqrt[3]{1+x} + \sqrt[3]{1-x})$ and a, b are two distinct real numbers, $(a \neq b)$, show $f(a) \neq f(b)$. This shows that $y = f(x)$ is a "**one to one function**" and hence has an inverse function $g(x)$. Find the expression of $g(x)$.

(Ans: $g(x) = ((1+x)^3 - (1-x)^3)/((1+x)^3 + (1-x)^3)$).

247) If r is a root of $z^2 + az + b = 0$, where a, b are real or complex numbers $(ab \neq 0)$, and p is the positive root of $x^2 - |a|x - |b| = 0$, show that $|r| - \frac{|b|}{|r|} \leq p - \frac{|b|}{p}$ and then conclude that, necessarily, $|r| \leq p$.

248) Find all the numbers $z = x + iy$, $x, y \in \mathbb{R}$, satisfying the equation $z^3 = -2 - 7i$.

249) If a polynomial $P(x)$ is divided exactly by $(x-a)^2$ and by $(x-b)^2$, $(a \neq b)$, show that $P(x)$ is divided exactly by $(x-a)^2(x-b)^2$.

250) Provided that $k_1, k_2, \ldots, k_n$ are positive constant numbers and that $x_1 + x_2 + \cdots + x_n = c > 0$ (constant), find the minimum value of the quantity

$$Q = \sqrt{x_1^2 + k_1^2} + \sqrt{x_2^2 + k_2^2} + \cdots + \sqrt{x_n^2 + k_n^2}.$$

(**Ans:** $Q_{min} = \sqrt{(k_1 + k_2 + \cdots + k_n)^2 + c^2}$, attained when $\frac{x_1}{k_1} = \frac{x_2}{k_2} = \cdots = \frac{x_n}{k_n}$).

251) Provided that n is odd integer, show that

$$\cos\frac{\pi}{n} \cos\frac{2\pi}{n} \cos\frac{3\pi}{n} \cdots \cos\frac{(n-1)\pi}{n} = \frac{1}{(2i)^n}$$

Hint: Use the factorization of $x^n - 1$, (formula (19-3-1)).

252) For what values of the parameter k, the equation $kx^4 - (k-3)x^2 + 3k = 0$ has one root smaller that (-2) and three roots greater than (-1)?

(**Ans:** $-\frac{4}{5} < k < 0$).

253) Show that any biquadratic trinomial $ax^4 + bx^2 + c$ with real coefficients a, b, c, can always be expressed as the product of two quadratic trinomials with real coefficients.

254) Factor the polynomial $P(x) \equiv (x^n + x^{n-1} + \cdots + x + 1)^2 - x^n$.

Hint: $x^n + x^{n-1} + \cdots + x + 1 = (x^{n+1} - 1)/(x - 1)$.

255) Provided that $ax + by + cz = k$ and $x + y + z = \lambda$, where a, b, c, k, λ are known numbers, determine x, y, z so that the sum $x^2 + y^2 + z^2$ takes the minimum value possible.

(**Ans:** $\min(x^2 + y^2 + z^2) = k^2/(a^2 + b^2 + c^2)$, attained when $x = (a\lambda)/(a+b+c), y = (b\lambda)/(a+b+c), z = (c\lambda)/(a+b+c)$.

Hint: Apply Lagrange's identity to the two triads (x, y, z) and (a, b, c).

256) Provided that $a, b, k \in \mathbb{R}$ and $k > 0$, show that

$$\left|a + b + \frac{k - ab}{a + b}\right| \geq |\sqrt{3k}|$$

www.ingramcontent.com/pod-product-compliance
Lightning Source LLC
Chambersburg PA
CBHW080335240526
45466CB00029B/2951